Advanced Process Monitoring for Industry 4.0

Advanced Process Monitoring for Industry 4.0

Editors

Marco S. Reis
Furong Gao

MDPI • Basel • Beijing • Wuhan • Barcelona • Belgrade • Manchester • Tokyo • Cluj • Tianjin

Editors

Marco S. Reis
University of Coimbra
Portugal

Furong Gao
The Hong Kong University of Science and Technology
Hong Kong

Editorial Office
MDPI
St. Alban-Anlage 66
4052 Basel, Switzerland

This is a reprint of articles from the Special Issue published online in the open access journal *Processes* (ISSN 2227-9717) (available at: https://www.mdpi.com/journal/processes/special_issues/Monitoring_Industry).

For citation purposes, cite each article independently as indicated on the article page online and as indicated below:

LastName, A.A.; LastName, B.B.; LastName, C.C. Article Title. *Journal Name* **Year**, *Volume Number*, Page Range.

ISBN 978-3-0365-2073-5 (Hbk)
ISBN 978-3-0365-2074-2 (PDF)

Contents

About the Editors

Marco S. Reis is an Associate Professor with Habilitation of Chemical Engineering at the University of Coimbra, Portugal. His research interests are centered on the fields of process systems engineering, industrial data science, sensor fusion, feature engineering, hybrid modeling, fault detection/diagnosis/prognosis, predictive analytics, structured process improvement, AutoML, design of experiments, and chemometrics. He was President of the European Network for Business and Industrial Statistics (ENBIS), and is currently an Honorary Member of this society. He has published 130+ articles in international journals or book series, 5 book chapters, 2 books, and authored or co-authored 160+ presentations in international congresses. He was awarded with a Fulbright scholar fellowship (2020) and with the Professor Almiro e Castro award that distinguishes the scientific merit of a Portuguese researcher or faculty under 45 years old (2018).

Furong Gao is a Chair Professor of Chemical & Biological Engineering and the funding director for the Center for Polymer Processing and Systems (CPPS) at the Hong Kong University of Science and Technology (HKUST). He received his BEng in automation from the East China Institute of Petroleum in 1981 and MEng and PhD in chemical engineering from McGill University in 1989 and 1993. His research interest includes batch process modeling, monitoring, optimization and control, and automation systems applications to polymer processing and energy management. To date, he has published over 500 journal and conference papers and 6 books. He has received numerous awards including a dozen best journal and conference paper awards and the 1st and 2nd prizes in Natural Science from the Chinese Ministry of Education.

Preface to "Advanced Process Monitoring for Industry 4.0"

Since the seminal work of W.A. Shewhart in the mid 1920s, initiated at Western Electric Company and continued afterwards at Bell Labs, process monitoring (PM) has become a cornerstone of modern industry, embodying a coherent and systematic methodology to continuously assess process stability and diagnose the origin of faults and process upsets. Having emerged during the 2nd industrial revolution, PM greatly expanded during the 3rd technological wave, evolving and adapting its body of knowledge to fulfill the same fundamental goals in increasingly diverse and challenging scenarios. Its wide success has significantly contributed to pushing forward the performance and effectiveness of plant operations, positively impacting process efficiency, safety, the environment, and the companies' bottom line results. Today, we are witnessing the dawn of a new industrial revolution, already coined as Industry 4.0. Again, as before, new challenges arise that need to be handled in order to perform PM in more complex multiscale processes, through the analysis of "extreme data" (in terms of volume, velocity, variety, quality, etc.), in real time. In this Special Issue, a collection of 12 contributions provide an updated overview of the challenges that PM is currently facing and how they can be effectively addressed. Real case studies are cited and new solutions proposed and discussed. We sincerely hope these works catalyze the efforts of the research community and inspire more teams to join in and embrace the many PM challenges lying ahead.

Marco S. Reis, Furong Gao

Editors

 processes

MDPI

Editorial

Special Issue "Advanced Process Monitoring for Industry 4.0"

Marco S. Reis [1,*] and Furong Gao [2]

[1] Department of Chemical Engineering, University of Coimbra, CIEPQPF, Rua Sílvio Lima, Pólo II—Pinhal de Marrocos, 3030-790 Coimbra, Portugal
[2] Department of Chemical Engineering, The Hong Kong University of Science and Technology, Hong Kong, China; kefgao@ust.hk
* Correspondence: marco@eq.uc.pt

Industry 4.0 is continually and progressively changing the landscape of manufacturing throughout the world and across different industrial sectors. This movement is catalyzed by the concurrence of three key drivers that, synergistically combined, create the conditions to push forward the performance and effectiveness of plant operations, positively impacting process efficiency, safety, the environmental fingerprint, and the economic outcome through faster and better decision-making processes. These drivers are: the facilitated access to unprecedented amounts of data (both structured and unstructured), new technological developments (smart sensors, IoT, cloud storage, and high-performance computing), and a new wave of advanced analytical solutions (machine learning, artificial intelligence, free programming platforms, and commercial software). As happens in other activities, the key drivers are also impacting Process Monitoring, creating the capability to handle complex processes that generate "extreme data", i.e., data collected at high sampling rates, possibly asynchronously, in large amounts with a variety of structures and variable quality, arising from different places across the value chain.

This Special Issue aims to bring together recent advances in the broad field of Advanced Process Monitoring for Industry 4.0, including all the activities related to fault detection, diagnosis, and prognosis.

All process monitoring activities are critically dependent upon the capability to collect informative data about the state of plant operations and equipment condition. Therefore, new sensors are developed and deployed, transforming quality monitoring from an offline activity conducted in the plants laboratories to a real-time activity made online, in the process, enabling fast product release and decision making, with all the consequent benefits on productivity, quality, inner logistics, and plant economy. Reyes et al., used spectral data in the visible–near infrared (VIS–NIR) range to monitor a combustion process [1], while Hotait et al., reported the use of piezoelectric sensors together with an advanced feature extraction methodology, called AOC-OPTICS [2], for fast and automatic condition monitoring.

Batch processes are always challenging scenarios for process monitoring given their intrinsic non-stationarity and natural tensorial arrangement of data (batch $\times$ variables $\times$ time). These processes become even more difficult to handle when batch operations take place in multiple phases, as covered by Palací-Lopez et al. [3], and show multiple normal operation modes, as addressed by Zhao et al. [4]. Both studies make use of Latent Variable Models as the analytical backbone to address batch modeling. The extreme case of a multistep process (semiconductors) is also covered by Espadinha-Cruz et al. [5], where quality control, monitoring, and diagnosis and other critical tasks are revised under the general umbrella of data mining.

On the other hand, machine learning (ML) and artificial intelligence (AI) methodologies have also been increasingly brought to the process monitoring arena. The papers by Xing Wu et al. [6], Xin Wu et al. [7], and Yumin Liu et al. [8] report applications of convolutional neural networks (CNN) and recurrent neural networks (RNN) for process

Citation: Reis, M.S.; Gao, F. Special Issue "Advanced Process Monitoring for Industry 4.0". *Processes* **2021**, *9*, 1432. https://doi.org/10.3390/pr9081432

Received: 6 August 2021
Accepted: 12 August 2021
Published: 19 August 2021

monitoring and diagnosis, exploring their ability to learn new data representations that efficiently represent the normal operation conditions of the process.

The integration of existing knowledge about the processes for process monitoring and diagnosis through digital twins, is also a current trend in Process Monitoring. Rato et al., present a framework where accurate models for the process common cause variation are used to mitigate the scarcity of data for high-dimensional process monitoring, especially during early monitoring periods, also enhancing the diagnosis activity once the fault is detected [9]. On the other hand, de Menezes et al., use a steady-state model to perform data reconciliation, an operation that is instrumental for the estimation of unmeasured variables in the proposed online monitoring scheme, playing the role of a soft sensor, and paving the way for the future adoption of an accurate digital twin [10].

The adaptation of Quality Engineering [11] and Six-Sigma [3] to the new types of measurements, data structures, processes, and the growing analytical body of knowledge, are opportunely covered in the contributions by Ramezani et al., and Palací-López et al., respectively. Similarly, Sader et al. [12] explored the use of modern methods of machine learning to assist in the implementation of Failure Mode and Effect Analysis (FMEA), and exemplify the proposed methodology in a dataset that includes a one-year historic of over 1500 failures with their respective description.

For all these valuable and insightful contributions, the Guest Editors are deeply grateful to the authors and their teams. We hope this rich and diverse collection of contributions fuel and inspire new developments on Statistical Process Monitoring and related fields, keeping up with the accelerating pace of the technological progress, data resources, and complexity of modern processes.

Finally, we would like to express our deepest gratitude and appreciation to the Section Managing Editor, Ms. Shirley Wang, for all the continuous and diligent support throughout all the phases of preparation of this Special Issue of *Processes*. We are also thankful to all the reviewers for their unconditional generosity in the time and effort dedicated to improve all the contributions with their knowledge, insights, and critical reasoning. The Special Issue "Advanced Process Monitoring for Industry 4.0" is available at https://www.mdpi.com/journal/processes/special_issues/Monitoring_Industry (accessed on 3 August 2021).

Funding: Marco Reis acknowledges support from the Chemical Process Engineering and Forest Products Research Centre (CIEPQPF), which is financed by national funds from FCT/MCTES (reference UID/EQU/00102/2019).

Conflicts of Interest: The authors declare no conflict of interest.

References

1. Reyes, G.; Diaz, W.; Toro, C.; Balladares, E.; Torres, S.; Parra, R.; Vásquez, A. Copper oxide spectral emission detection in chalcopyrite and copper concentrate combustion. *Processes* **2021**, *9*, 188. [CrossRef]
2. Hotait, H.; Chiementin, X.; Rasolofondraibe, L. Aoc-optics: Automatic online classification for condition monitoring of rolling bearing. *Processes* **2020**, *8*, 606. [CrossRef]
3. Palací-López, D.; Borràs-Ferrís, J.; da Silva de Oliveria, L.T.; Ferrer, A. Multivariate six sigma: A case study in industry 4.0. *Processes* **2020**, *8*, 1119. [CrossRef]
4. Zhao, L.; Huang, X.; Yu, H. Quality-analysis-based process monitoring for multi-phase multi-mode batch processes. *Processes* **2021**, *9*, 1321. [CrossRef]
5. Espadinha-Cruz, P.; Godina, R.; Rodrigues, E.M.G. A review of data mining applications in semiconductor manufacturing. *Processes* **2021**, *9*, 305. [CrossRef]
6. Wu, X.; Jin, H.; Ye, X.; Wang, J.; Lei, Z.; Liu, Y.; Wang, J.; Guo, Y. Multiscale convolutional and recurrent neural network for quality prediction of continuous casting slabs. *Processes* **2021**, *9*, 33. [CrossRef]
7. Wu, X.; Jiao, D.; Du, Y. Automatic implementation of a self-adaption non-intrusive load monitoring method based on the convolutional neural network. *Processes* **2020**, *8*, 704. [CrossRef]
8. Liu, Y.; Zhao, Z.; Zhang, S.; Jung, U. Identification of abnormal processes with spatial-temporal data using convolutional neural networks. *Processes* **2020**, *8*, 73. [CrossRef]
9. Rato, T.J.; Delgado, P.; Martins, C.; Reis, M.S. First principles statistical process monitoring of high-dimensional industrial microelectronics assembly processes. *Processes* **2020**, *8*, 1520. [CrossRef]

10. de Menezes, D.Q.F.; de Sá, M.C.C.; Fontoura, T.B.; Anzai, T.K.; Diehl, F.C.; Thompson, P.H.; Pinto, J.C. Modeling of spiral wound membranes for gas separations—Part ii: Data reconciliation for online monitoring. *Processes* **2020**, *8*, 1035. [CrossRef]
11. Ramezani, J.; Jassbi, J. Quality 4.0 in action: Smart hybrid fault diagnosis system in plaster production. *Processes* **2020**, *8*, 634. [CrossRef]
12. Sader, S.; Husti, I.; Daróczi, M. Enhancing failure mode and effects analysis using auto machine learning: A case study of the agricultural machinery industry. *Processes* **2020**, *8*, 224. [CrossRef]

Article

First Principles Statistical Process Monitoring of High-Dimensional Industrial Microelectronics Assembly Processes

Tiago J. Rato [1], Pedro Delgado [2], Cristina Martins [2] and Marco S. Reis [1,*]

[1] Department of Chemical Engineering, CIEPQPF, University of Coimbra, Rua Sílvio Lima, Pólo II-Pinhal de Marrocos, 3030-790 Coimbra, Portugal; trato@eq.uc.pt

[2] Bosch Car Multimedia, SA, Rua Max Grundig 35, Lomar, 4705-820 Braga, Portugal; external.pedro.delgado@pt.bosch.com (P.D.); Cristina.Martins@pt.bosch.com (C.M.)

* Correspondence: marco@eq.uc.pt; Tel.: +351-239-798-700; Fax: +351-239-798-703

Received: 18 October 2020; Accepted: 19 November 2020; Published: 23 November 2020

Abstract: Modern industrial units collect large amounts of process data based on which advanced process monitoring algorithms continuously assess the status of operations. As an integral part of the development of such algorithms, a reference dataset representative of normal operating conditions is required to evaluate the stability of the process and, after confirming that it is stable, to calibrate a monitoring procedure, i.e., estimate the reference model and set the control limits for the monitoring statistics. The basic assumption is that all relevant "common causes" of variation appear well represented in this reference dataset (using the terminology adopted by the founding father of process monitoring, Walter A. Shewhart). Otherwise, false alarms will inevitably occur during the implementation of the monitoring scheme. However, we argue and demonstrate in this article, that this assumption is often not met in modern industrial systems. Therefore, we introduce a new approach based on the rigorous mechanistic modeling of the dominant modes of common cause variation and the use of stochastic computational simulations to enrich the historical dataset with augmented data representing a comprehensive coverage of the actual operational space. We show how to compute the monitoring statistics and set their control limits, as well as to conduct fault diagnosis when an abnormal event is declared. The proposed method, called AGV (Artificial Generation of common cause Variability) is applied to a Surface Mount Technology (SMT) production line of Bosch Car Multimedia, where more than 17 thousand product variables are simultaneously monitored.

Keywords: high-dimensional data; statistical process monitoring; artificial generation of variability; data augmentation; Industry 4.0

1. Introduction

Industry 4.0 is taking its course and continuously raising new challenges to classical process operation and management functions. One pillar of any production system is related to guaranteeing the stability, consistency and predictability ofF industrial processes. This was and it will always be a major concern for industry. Juran considered it as a fundamental function for implementing any quality management systems (the Juran trilogy of Planning, Control and Improvement), and it is an intrinsic part of every Quality standard, deeply embedded in the celebrated ISO 9000 series. Traditionally, this function is operationalized in the shop floor using variation management and reduction methodologies such as Statistical Process Monitoring (SPM), Engineering Process Control (EPC), or combined versions of them [1–4], as well as other approaches like error-proof systems (Poka Yoke). However, the way SPM is implemented in industry is changing, as a result of, among other possible drivers: (i) increasing complexity of the processes/products under monitoring and (ii) the new challenges imposed by data collected from them.

Regarding the first category of drivers, current processes under monitoring can present characteristics (almost) absent from the scenarios where, during over 80 years, most SPM technology was developed, such as: they can present stationary dynamics (autocorrelation) [5–10] or non-stationary dynamics (such as discontinuous or batch processes) [11–14]; multiple set-points and operations modes [15–17]; processes have a complex network structure and are composed of many sub-process in series or parallel [18–22]; processes are intrinsically multiscale in space and time [23–27].

The second category of drivers are connected to the new data-intensive environments industry is currently immersed in. SPM is a data-driven methodology, as its implementation strongly relies on the analysis of historical data (Phase 1 analysis) and data collected during operation (Phase 2 analysis) [28]. The vast majority of SPM methods were developed to handle scalar (0th order tensor) sensor like data, either univariately [29–31] or multivariately [32–35]. More recently, profile monitoring emerged as a new SPM branch dedicated to functional relationships [36–38] or higher-order tensorial data structures such as: near-infrared (NIR) spectra [39–41], surface profilometry [37], grey-level images [42], colour and hyperspectral images [43–47], hyphenated instruments [48] among others, expanding the SPM domain to new processes/products.

The one aspect shared by all methods proposed in the past and the fundamental premise established since the seminal work of Walter A. Shewhart [31], is that all common cause variation must be present in the reference historical Normal Operating Conditions (NOC) dataset collected to conduct Phase 1 analysis, i.e., in order to assess process stability and establish the control limits for the monitoring statistics. However, this fundamental premise does not hold in many processes, such as the assembling processes in the microelectronics industry using Surface Mount Technology (SMT), which is the focus of our paper.

To make our presentation more objective and clearer, in the following sub-sections we will describe our problem and provide and introductory sketch of the proposed solution.

1.1. Problem Statement: Process Monitoring of Surface Mount Technology (SMT) Production Lines

The assembly process of complex electronic devices involves placing, fixing and functionalizing electronic components on Printed Circuit Boards (PCB). This is done through the preliminary deposition of solder past deposits (SPD) in specific positions, with a well-defined target shape and volume. This part of the process is critical for quality, as any defect or misplacement may result in the loss of function of the module and eventually the entire device. Therefore, SPDs should be monitored immediately after being placed, to avoid the well-known consequence of the accumulation of costs as any fault is detected later on in the process (usually the cost rises by roughly a factor of 10 as one moves from one stage to the next without detecting a potential problem, leading to the progression of $1:$10:$100: ..., in the costs due to poor quality of the assembly process). In the present case study, data arises from a modern production line equipped with Surface Mount Technology (SMT) that performs 100% inspection of all paste deposits for each PCB produced (more details about the process are provided in Section 3), which implies the simultaneous analysis of several thousands of SPDs. Handling such a large number of variables for process monitoring raises several important challenges to traditional statistical process monitoring approaches, as addressed elsewhere [49], but the fundamental issue that was not considered so far is indeed the poor coverage of common cause variation in the reference dataset (even when composed of what is usually considered a sufficiently high number of samples).

For a better understanding of the problem under analysis, Figure 1a,b represent the scores for the first three principal components (PC1, PC2 and PC3) for several datasets collected for the production of the same product, all of them regarding NOC conditions. Given the large number of variables involved, we opt to present just these three scores, that represent a significant portion of the overall variability (from the properties of Principal Component Analysis, they are the three linear combinations with maximum explanation power of the variability presented by the original variables; see [50,51]) and are enough to establish the picture we want to convey at this point. Each point concerns a PCB, and these points, as stated above, arise from the reference dataset (CS1) and other two datasets collected afterwards at different periods, also regarding normal operating conditions. All observations should therefore be considered "normal", as confirmed by process experts. Conducting a Phase 1 analysis of the reference NOC dataset (CS1) it is possible to conclude that the process is stable. This can be confirmed by analyzing Figure 1c, where the two multivariate monitoring statistics fall in general within the region limited by the Phase 1 control limits (more details on these monitoring statistics are presented in Section 4). This historic reference dataset can therefore be used to setting up the multivariate control limits from conducting the monitoring activity of future incoming PCBs. However, from the plot in Figure 1a,b, it is clearly visible that many alarms are expected to be issued when applying these control limits to data from future PCBs. This can actually be confirmed by analyzing Figure 1d, where the SPM charts developed using CS1 were applied to dataset CS3, resulting in sustained alarms being issued by the two control charts. Therefore, implementing any SPM methodology based on the information strictly inferred from the historical dataset will inevitably result in frequently signaling as faulty PCBs that are perfectly good, making this activity, as it is currently conducted, of very limited value.

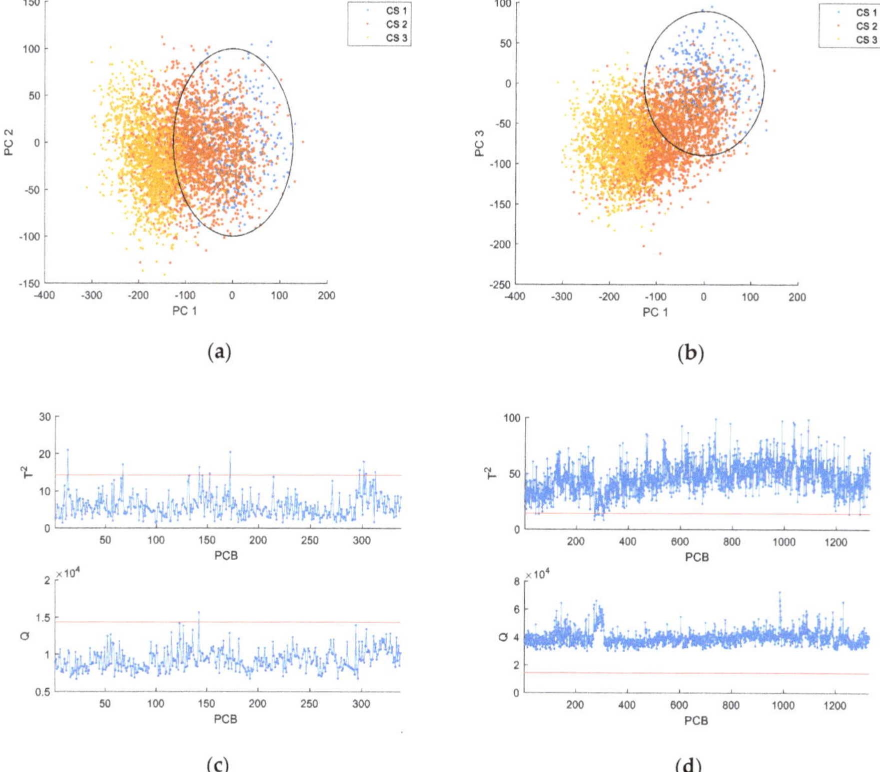

Figure 1. Projection of data from several production periods (designated as CS1, CS2 and CS3) on the principal components space estimated using CS1 as the reference dataset: (**a**) Scores plot of PC1 vs. PC2; (**b**) Scores plot of PC1 vs. PC3. The 99% confidence ellipses for the scores of CS1 are also represented; (**c**) MSPM-PCA monitoring statistics (Hotelling's T^2 of the scores and the Q or SPE statistic of the residuals; see Section 4) for CS1, using CS1 as the reference dataset; (**d**) MSPM-PCA monitoring statistics for CS3, using CS1 as the reference dataset.

The fundamental reason causing the situation described above is the limited information about process variability that can be extracted from a dataset representing a stable period of NOC operation—the reference dataset does not reflect the whole of common cause variation sources, but just a small part of it.

Fortunately, the reference NOC dataset is not the only source of information about common cause variation available.

In fact, the engineering team has been accumulating knowledge over time about the "common" causes of variability that lead to such false alarms, by analyzing case by case what is causing them. The root causes regard aspects that are very common and perfectly normal and expected in process operations, such as slight changes in the settings from different lots, automatic corrections and adjustments introduced by the assembly units, "normal" (or acceptable) tridimensional deformations in the boards fed to the process, rigid body movements (rotations, translations) of the boards as they move in the line, and other known specificities of the paste deposition tools. These aspects of variability are known and expected to happen over a long time frame, but not all of them will be, for sure, present in the initial stages of the process, or in any other single isolated period in the future.

Furthermore, the extensive process knowledge accumulated over time enables not only the identification of which phenomena may cause the variation patterns presented in Figure 1, but also to know their magnitudes. The detailed analysis of past production runs where they took place, does provide information on the amount of variation associated with them. Therefore, besides the reference dataset, there is also information available about the *long term structural components of common cause variability* (at least for the dominant modes of common cause variability), and their stochastic behavior (of which the reference dataset is a particular realization).

Bringing this extra engineering knowledge of common structural cause variation is fundamental to address the present problem rooted in the unavoidable underrepresentation of common cause variation in the reference dataset.

1.2. Proposed Methodology: Artificial Generation of (Common Cause) Variability (AGV)

In this work we introduce a Data Augmentation approach for enriching the reference dataset with structural common cause variation sources. Data is generated by conducting stochastic computational simulation of process behavior using rigorous mechanistic models for the dominant structural modes of common cause variation, whose conditions and parameters are described probabilistically based on data collected from industrial runs for other related assembly processes (dispersion) whereas the targets and settings are those of the current process (central tendency). The simulations will generate patterns of variation in the measurements respecting the physical constrains of the systems/products, leading to the same natural long term and short term correlations patterns found in real process data.

Ultimately, the proposed methodology is even able to create a frame of reference for starting the monitoring right from the onset of the production, using the variation patterns extracted from previous runs with other products (as shown in Section 5). We call it *conditionally expected common cause variation*, i.e., the expected variability under normal operating conditions, *conditioned* on all the past data and knowledge extracted over the years from data regarding the production of related products.

Together with the solution for the unwelcomed false alarms, the proposed methodology also brings new diagnostic tools: the variation from the different simulated phenomena is usually well captured by specific principal components. Therefore, once a potential fault is detected the analysis of the scores my reveal a possible mechanism underlying its origin (the one connected to the principal component where the deviation is more noticeable). Other causes can be explored by other diagnostic tools, such as residual analysis and contribution plots.

The aforementioned concept of *conditionally expected common cause variation* differs from Shewhart's original perspective, but is still inspired in it. It requires and builds up on extensive knowledge about the process physics and accumulated data. As Industry 4.0 takes place, the evolution of Digital Twining technology is likely to create similar opportunities to use accurate models of the process, namely for process monitoring as proposed in this article.

The present article is organized as follows. In Section 2, a brief description of the process and datasets to be analyzed is provided. Then, in Section 3, we introduce in detail the proposed data augmentation methodology (Artificial Generation of common cause Variability, AGV). The monitoring scheme based on the augmented NOC dataset is described in Section 4. The results from the application of the proposed data augmentation methodology for process monitoring to several real world industrial dataset are presented and discussed in Section 5. Section 6 further extends the discussion of results and their consequences. Finally, in Section 7 we provide a brief summary of the main advantages and limitations of the proposed methodology and refer to future work to address these limitations and make the approach more sensitive to localized faults affecting a relatively small number of elements.

2. Process Description

The process considered in this work regards a surface mount industrial unit (Bosch Car Multimedia Portugal), composed by several lines where electronic components are attached to PCBs through reflow soldering. In this process, SPDs are accurately placed in designated points of copper PCBs that will be used to fix and functionalize electronic components further ahead in the assembly line, through reflow soldering under a predefined temperature profile. Several thousands of SPDs are set in place, with a cycle time that can achieve approximately 20s. The resulting PCBs with printed SPDs are subject to 100% 3D Solder Past Inspection (3DSPI). This technology has been a pillar for process monitoring and improvement in this plant, and will be further explored in this work. 3DSPI generates, for each single SPD printed in the pad, a full 3D profile, retaining 5 features for analysis: area (a), height (h), volume (v), offset in the X-direction (x) and offset in the Y-direction (y).

In this work, we consider the production of one specific product from which 3507 SPDs (we will also use the designation of "pads" to refer to such SPDs) are monitored, totaling $3507 \times 5 = 17{,}535$ variables to be handled. Three datasets are available and will be used in this study to test the proposed approach: CS1 contains information for 337 NOC PCBs and is used to validate the AGV methodology and confirm/tune some of its parameters; CS2 and CS3, contain 2080 and 1330 PCBs, respectively, mostly regarding good products, but also containing some PCBs with a few faulty pads. CS2 and CS3 will be used to further test the proposed methodology.

3. Artificial Generation of (Common Cause) Variability (AGV)

The proposed methodology for Artificial Generation of (common cause) Variability (AGV) aims at completing the reference dataset with the long term dominant sources of common cause variability that, from engineering knowledge, are known to exist. This is done by simulating the printing of solder paste deposits in Printed Circuit Boards (PCB) under situations where the dominant sources of common cause variation span their normal range (as extracted from the analysis of historic data from production runs of related products). The simulations are based on accurate mechanistic/first principle models and reproduce the effects of underlying common causes on the 3DSPI measurements of solder paste deposits: area (a), height (h), volume (v), offset-X (x) and offset-Y (y). As part of the common cause variation is introduced when different lots of the same product are produced while others are active during the production of a given lot, the AGV module also accounts for both inter-lot, intra-lot and pad specific variability.

During the process simulation, the AGV module generates data for L lots with B PCBs each, where each PCB is composed by P pads (a pad is a small part of the PCB where SPD should be placed). A more comprehensive description of the simulation settings for each of the variability sources are provided in the following subsections. A schematic representation of the AGV module is presented in Figure 2.

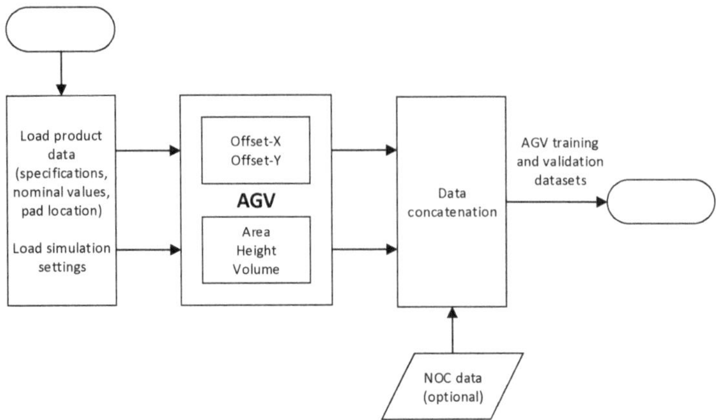

Figure 2. Schematic representation of the AGV module.

The basic inputs of the AGV module are the nominal values and tolerance limits for the measured parameters (a, h, v, x, y) and the geometrical coordinates of each pad (Table 1). Based on this information, the allowed variability given the existing tolerance limits is estimated by,

$$\sigma_{tol}\{parameter, pad\} = \frac{\mathbf{UTL}\{parameter, pad\} - \mathbf{LTL}\{parameter, pad\}}{6}. \tag{1}$$

This parameter is analogous to the standard deviation and is based on the fact that, for normally distributed data, 99.73% of the data falls within the ±3 sigma interval.

Table 1. Tolerance specifications of the PCB (inputs to the AGV module).

Parameter	Description
$\mathbf{N}\{parameter, pad\}$	Nominal value for $\{a, h, v, x, y\}$ of each pad.
$\mathbf{UTL}\{parameter, pad\}$	Upper tolerance limit for $\{a, h, v, x, y\}$ of each pad.
$\mathbf{LTL}\{parameter, pad\}$	Lower tolerance limit for $\{a, h, v, x, y\}$ of each pad.
$\mathbf{c}_x$	x-coordinate of each pad.
$\mathbf{c}_y$	y-coordinate of each pad.

Along with the PCB's tolerance specifications, the user can also define a set of tuning parameters (Table 2) that allow for the full configuration of the simulated conditions of the printing process. The relationship between the tuning parameters and the process phenomena is discussed in the following subsections. It is expected that, in future applications for the same lines and for related products, the settings for most of these parameters will not change. However, the need for some tuning can be readily assessed after collecting the 3DSPI measurements from the first few products.

Table 2. Tuning parameters in the AGV module and the values used in this work.

Parameter	Value (5)	Description
$\alpha_{trans}^{(inter)}$ (1)	0.1000	Scaling factor for the inter-lot component of the translation effect
$\alpha_{trans}^{(intra)}$ (1)	0.0775	Scaling factor for the intra-lot component of the translation effect
$\alpha_{trans}^{(pad)}$ (1)	0.9920	Scaling factor for the pad specific component of the translation effect
$\alpha_{rot}^{(inter)}$ (2)	0.9487	Scaling factor for the inter-lot component of the rotation effect
$\alpha_{rot}^{(intra)}$ (2)	0.3162	Scaling factor for the intra-lot component of the rotation effect
θ	1.57×10^{-4} rad	Allowance angle of the rotation effect

Table 2. *Cont.*

Parameter	Value (5)	Description
Δ_y	5 µm	Allowance deviation in Offset-Y due to squeegee effect
$\alpha_h^{(inter)\,(3)}$	0.9695	Scaling factor for the inter-lot component of the solder mask effect on height
$\alpha_h^{(intra)\,(3)}$	0.2449	Scaling factor for the intra-lot component of the solder mask effect on height
$\Delta_{h,sold}$	6 µm	Allowance deviation in height due to solder mask effect
$\Delta_{h,squee}$	7.5 µm	Allowance deviation in height due to squeegee effect
$\alpha_a^{(inter)\,(4)}$	0	Scaling factor for the inter-lot component of area
$\alpha_a^{(intra)\,(4)}$	0	Scaling factor for the intra-lot component of area
$\alpha_a^{(pad)\,(4)}$	1	Scaling factor for the pad specific component of area
φ_x	0.80	Scaling factor for total variance of offset-X
φ_y	0.80	Scaling factor for total variance of offset-Y
φ_h	0.80	Scaling factor for total variance of height
φ_a	0.80	Scaling factor for total variance of area

(1), (2), (3), (4) These parameters are between 0 and 1 and the sum of their squares is equal to 1. (5) These values were selected based on engineering knowledge and to achieve a realistic representation of the printing process discussed in Section 5.

3.1. Common Cause Variation Affecting Offset-X and Offset-Y

The relative location of the pads can be affected by rigid body translations and rotations of the PCB. Furthermore, offsets in the y-coordinate (offset-Y) can also be attributed to the printing direction of the squeegee.

The translation effects are identical for both offset-X and offset-Y. For the sake of brevity, only the case of offset-X is described. To simulate the rigid body translation effect (Figure 3), we considered that it can be split into three components. These are relative to:

(i). An overall deviation that is common to all PCBs within a lot—inter-lot variation: $\alpha_{trans}^{(inter)} z_{x,l,:,:}^{(inter)} \sigma_{tol}\{x,p\}\varphi_x$;

(ii). A local deviation that affects all pads in a PCB—intra-lot variation: $\alpha_{trans}^{(intra)} z_{x,l,b,:}^{(intra)} \sigma_{tol}\{x,p\}\varphi_x$;

(iii). Unstructured deviations for each pad—pad specific variation: $\alpha_{trans}^{(pad)} z_{x,l,b,p}^{(pad)} \sigma_{tol}\{x,p\}\varphi_x$.

In these simulations, $z_{x,l,:,:}^{(inter)}$, $z_{x,l,b,:}^{(intra)}$ and $z_{x,l,b,p}^{(pad)}$ are generated by taking random draws from the standard normal distribution. The fraction of total variation attributed to each component is determined by the scaling factors $\alpha_{trans}^{(inter)}$, $\alpha_{trans}^{(intra)}$ and $\alpha_{trans}^{(pad)}$. These scaling factors must fall between 0 and 1 and are further restricted to comply with:

$$\left(\alpha_{trans}^{(inter)}\right)^2 + \left(\alpha_{trans}^{(intra)}\right)^2 + \left(\alpha_{trans}^{(pad)}\right)^2 = 1. \quad (2)$$

Thus, the sum of all components amounts to one unit of allowed variance. Finally, the components are scaled by the allowed variance $\sigma_{tol}^2\{x,p\}\varphi_x^2$, where $\sigma_{tol}\{x,p\}$ is the allowed standard deviation for the offset-X of the p-th pad and φ_x is a scaling factor.

Following the above definitions, the translation effect on the offset in the x-coordinate for the p-th pad of the b-th PCB of the l-th lot ($x_{l,b,p}^{(trans)}$) is given by (Table 3):

$$x_{l,b,p}^{(trans)} = \left(\alpha_{trans}^{(inter)} z_{x,l,:,:}^{(inter)} + \alpha_{trans}^{(intra)} z_{x,l,b,:}^{(intra)} + \alpha_{trans}^{(pad)} z_{x,l,b,p}^{(pad)}\right)\sigma_{tol}\{x,p\}\varphi_x. \quad (3)$$

Likewise, the translation effect on the offsets in the y-coordinate ($y^{(trans)}_{l,b,p}$) are simulated by (Table 3):

$$y^{(trans)}_{l,b,p} = \left(\alpha^{(inter)}_{trans} z^{(inter)}_{y,l,:,:} + \alpha^{(intra)}_{trans} z^{(intra)}_{y,l,b,:} + \alpha^{(pad)}_{trans} z^{(pad)}_{y,l,b,p}\right)\sigma_{tol}\{y,p\}\varphi_y. \quad (4)$$

Table 3. Pseudo-code for generating the translation effect on offset-X and offset-Y.

1. For each lot, $l = 1, \ldots, L$:
   - 1.1. Generate inter-lot variability: $z^{(inter)}_{x,l,:,:} \sim N(0,1)$, $z^{(inter)}_{y,l,:,:} \sim N(0,1)$;
   - 1.2. For each PCB, $b = 1, \ldots, B$:
     - 1.2.1. Generate intra-lot variability: $z^{(intra)}_{x,l,b,:} \sim N(0,1)$, $z^{(intra)}_{y,l,b,:} \sim N(0,1)$;
     - 1.2.2. For each pad, $p = 1, \ldots, P$:
       - 1.2.2.1. Generate random variability: $z^{(pad)}_{x,l,b,p} \sim N(0,1)$, $z^{(pad)}_{y,l,b,p} \sim N(0,1)$;
       - 1.2.2.2. Compute translation effect on offset-X: $x^{(trans)}_{l,b,p} = \left(\alpha^{(inter)}_{trans} z^{(inter)}_{x,l,:,:} + \alpha^{(intra)}_{trans} z^{(intra)}_{x,l,b,:} + \alpha^{(pad)}_{trans} z^{(pad)}_{x,l,b,p}\right)\sigma_{tol}\{x,p\}\varphi_x.$
       - 1.2.2.3. Compute translation effect on offset-Y: $y^{(trans)}_{l,b,p} = \left(\alpha^{(inter)}_{trans} z^{(inter)}_{y,l,:,:} + \alpha^{(intra)}_{trans} z^{(intra)}_{y,l,b,:} + \alpha^{(pad)}_{trans} z^{(pad)}_{y,l,b,p}\right)\sigma_{tol}\{y,p\}\varphi_y.$

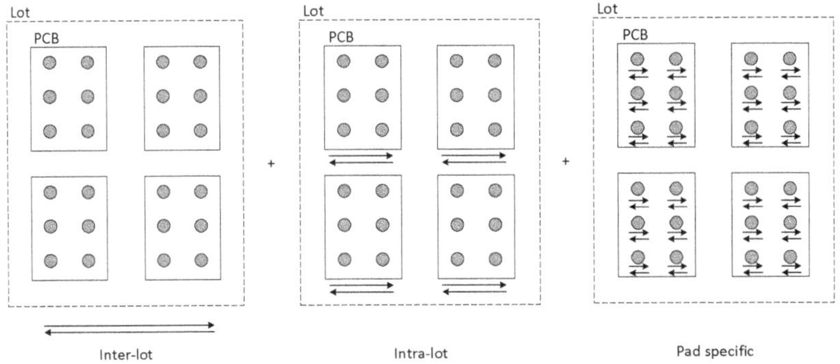

Figure 3. Schematic representation of the inter-lot, intra-lot and pad specific components used to generate the translation effect on offset-X.

The rigid body rotation effects (Figure 4), are taking into account by simulating rotations in the PCB with different centers and magnitudes. This rotation can be further decomposed into:

(i) An inter-lot rotation angle that affects all PCBs within a lot - inter-lot variation: $\alpha^{(inter)}_{rot} z^{(inter)}_{l,:,:}\theta/3$;

(ii) An intra-lot rotation angle that adds specific variation to each PCB—intra-lot variation: $\alpha^{(intra)}_{rot} z^{(intra)}_{l,b,:}\theta/3$.

The $z^{(inter)}_{l,:,:}$ and $z^{(intra)}_{l,b,:}$ variables are generated from the standard normal distribution and scaled by the allowance rotation angle θ, divided by 3 to ensure that the rotation angle falls, with high probability, between $\begin{bmatrix} -\theta & \theta \end{bmatrix}$. Furthermore, the amount of variability attributed to the inter-lot and intra-lot components is defined by the scale factors $\alpha^{(inter)}_{rot}$ and $\alpha^{(intra)}_{rot}$, which are between 0 and 1 and comply to the condition:

$$\left(\alpha^{(inter)}_{rot}\right)^2 + \left(\alpha^{(intra)}_{rot}\right)^2 = 1. \quad (5)$$

Based on this, the rotation angle of the b-th PCB of the l-th lot ($t_{l,b,:}$) is computed as,

$$t_{l,b,:} = \frac{\alpha_{rot}^{(inter)} z_{l,:,:}^{(inter)} \theta + \alpha_{rot}^{(intra)} z_{l,b,:}^{(intra)} \theta}{3}. \quad (6)$$

The rotation matrix is subsequently obtained by,

$$R\left(t_{l,b,:}\right) = \begin{bmatrix} \cos\left(t_{l,b,:}\right) & -\sin\left(t_{l,b,:}\right) \\ \sin\left(t_{l,b,:}\right) & \cos\left(t_{l,b,:}\right) \end{bmatrix}. \quad (7)$$

The center of rotation is randomly generated for each PCB using the uniform distribution. More specifically, the x-coordinate for the center of rotation (r_x) is taken from the uniform distribution on the interval $\begin{bmatrix} \min(\mathbf{c}_x) & \max(\mathbf{c}_x) \end{bmatrix}$, where $\mathbf{c}_x$ is a vector with the x-coordinates of each pad. Similarly, the y-coordinate of the center of rotation (r_y) is generated from the uniform distribution in the interval $\begin{bmatrix} \min(\mathbf{c}_y) & \max(\mathbf{c}_y) \end{bmatrix}$, where $\mathbf{c}_y$ is a vector with the y-coordinates of each pad.

Using the above information, the rotation effects on the offset-X, $x_{l,b,p}^{(rot)}$, and offset-Y, $y_{l,b,p}^{(rot)}$, for the p-th pad of the b-th PCB of the l-th lot, are determined by (see Table 4.):

$$\begin{bmatrix} x_{l,b,p}^{(rot)} & y_{l,b,p}^{(rot)} \end{bmatrix} = \left(\begin{bmatrix} \mathbf{c}_x\{p\} & \mathbf{c}_y\{p\} \end{bmatrix} - \begin{bmatrix} r_x & r_y \end{bmatrix}\right) R^{\mathrm{T}} + \begin{bmatrix} r_x & r_y \end{bmatrix} - \begin{bmatrix} \mathbf{c}_x\{p\} & \mathbf{c}_y\{p\} \end{bmatrix}. \quad (8)$$

Table 4. Pseudo-code for generating the rotation effect on offset-X and offset-Y.

1. For each lot, $l = 1, \ldots, L$:
   - 1.1. Generate inter-lot rotation angle: $z_{l,:,:}^{(inter)} \sim N(0,1)$;
   - 1.2. For each PCB, $b = 1, \ldots, B$:
     - 1.2.1. Generate intra-lot rotation angle: $z_{l,b,:}^{(intra)} \sim N(0,1)$;
     - 1.2.2. Compute overall rotation angle: $t_{l,b,:} = \left(\alpha_{rot}^{(inter)} z_{l,:,:}^{(inter)} \theta + \alpha_{rot}^{(intra)} z_{l,b,:}^{(intra)} \theta\right)/3$;
     - 1.2.3. Compute the rotation matrix: $R\left(t_{l,b,:}\right)$
     - 1.2.4. Generate random center of rotation:
       $r_x \sim U(\min(\mathbf{c}_x), \max(\mathbf{c}_x))$,
       $r_y \sim U\left(\min(\mathbf{c}_y), \max(\mathbf{c}_y)\right)$;

       1.2.4.1 For each pad, $p = 1, \ldots, P$:
       1.2.4.2.Compute rotation effect on offset-X and offset-Y:
       $$\begin{bmatrix} x_{l,b,p}^{(rot)} & y_{l,b,p}^{(rot)} \end{bmatrix} = \left(\begin{bmatrix} \mathbf{c}_x\{p\} & \mathbf{c}_y\{p\} \end{bmatrix} - \begin{bmatrix} r_x & r_y \end{bmatrix}\right) R^{\mathrm{T}} + \ldots + \begin{bmatrix} r_x & r_y \end{bmatrix} - \begin{bmatrix} \mathbf{c}_x\{p\} & \mathbf{c}_y\{p\} \end{bmatrix}$$

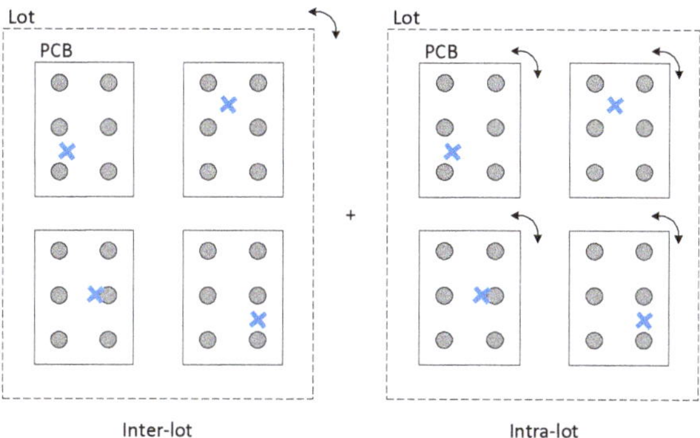

Figure 4. Schematic representation of the inter-lot and intra-lot components used to generate the rotation effect on offset-X and offset-Y. The cross represents the specific center of rotation of each PCB.

Along with the translation and rotation effects, the offset-Y is also affected by an additional systematic bias caused by the printing direction of the squeegee (the squeegee is the device that pushes the solder past deposits placed on the top side of the stencil towards the existing apertures in the stencil). This phenomenon leads to a positive deviation in the offset-Y when the printing is performed in one direction and a negative deviation when the printing direction is reversed. Without loss of generality it is assumed that PCBs with an odd index b are printed in the direction of increasing y-coordinates. Conversely, PCBs with an even index b are assumed to be printed in the opposite direction of decreasing magnitude y-coordinates. To simulate the squeegee effect on offset-Y ($y_{l,b,p}^{(sqee)}$; Table 5), the allowance deviation that can be caused by the squeegee is defined by Δ_y and its impact is alternated with the PCB index. Furthermore, to account for local random variation, for each PCB, the allowance deviation is multiplied by a random variable with uniform distribution in the interval [0,1].

Table 5. Pseudo-code for generating the squeegee effect on offset-Y.

1. For each lot: $l = 1, \ldots, L$
   - 1.1. For each PCB: $b = 1, \ldots, B$:
     - 1.1.1. Generate squeegee variability: $u_{l,b,:} \sim U(0,1)$;
     - 1.1.2. For each pad: $p = 1, \ldots, P$:
     - 1.1.3. Compute squeegee effect on offset-Y: $y_{l,b,p}^{(sqee)} = (-1)^{b+1}\Delta_y u_{l,b,:}$.

Finally, the offset-X ($x_{l,b,p}^{(agv)}$) and offset-Y ($y_{l,b,p}^{(agv)}$) are determined by summing all effects:

$$x_{l,b,p}^{(agv)} = \mathbf{N}\{x,p\} + x_{l,b,p}^{(trans)} + x_{l,b,p}^{(rot)}, \tag{9}$$

$$y_{l,b,p}^{(agv)} = \mathbf{N}\{y,p\} + y_{l,b,p}^{(trans)} + y_{l,b,p}^{(rot)} + y_{l,b,p}^{(sqee)}, \tag{10}$$

where $\mathbf{N}\{x,p\}$ and $\mathbf{N}\{y,p\}$ are the nominal value for the offset-X and offset-Y of the p-th pad.

3.2. Common Cause Variation Affecting Height

The height of solder paste in each pad is affected by the solder mask and the squeegee's action. The effect of the solder mask translates into systematic deviations across all PCBs in the same lot (because the same mask is used in each production lot). The allowed deviation due to the solder mask is here defined by $\Delta_{h,sold}$. Furthermore, its impact on the height is simulated through (Figure 5):

(i) An overall component that affects all PCBs in a lot—inter-lot variation: $\alpha_h^{(inter)} z_{l,:,:}^{(inter)} \Delta_{h,sold}$;
(ii) A specific component for each PCB—intra-lot variation: $\alpha_h^{(intra)} z_{l,b,:}^{(intra)} \Delta_{h,sold}$.

For both cases, $z_{l,:,:}^{(inter)}$ and $z_{l,b,:}^{(intra)}$ are generated from the standard normal distribution. The $\alpha_h^{(inter)}$ and $\alpha_h^{(intra)}$ factors scale the contribution of each component. These factors are between 0 and 1 and conditioned to:

$$\left(\alpha_h^{(inter)}\right)^2 + \left(\alpha_h^{(intra)}\right)^2 = 1. \tag{11}$$

As the total variability of the height is greater than that caused by the solder mask, the remaining variability is added as a specific component for each pad, $z_{l,b,p}^{(pad)} \sqrt{\sigma_{tol}^2\{h,p\}\varphi_h^2 - \Delta_{h,sold}^2}$. Following these considerations, the solder mark effect on the height of the p-th pad of the b-th PCB of the l-th lot ($h_{l,b,p}^{(sold)}$) is computed as follows (see also Table 6):

$$h_{l,b,p}^{(sold)} = \left(\alpha_h^{(inter)} z_{l,:,:}^{(inter)} + \alpha_h^{(intra)} z_{l,b,:}^{(intra)}\right)\Delta_{h,sold} + z_{l,b,p}^{(pad)} \sqrt{\sigma_{tol}^2\{h,p\}\varphi_h^2 - \Delta_{h,sold}^2}. \tag{12}$$

Table 6. Pseudo-code for generating the solder mask effect on height.

1. For each lot, $l = 1, \ldots, L$:
   - 1.1. Generate inter-lot variability: $z_{l,:,:}^{(inter)} \sim N(0,1)$;
   - 1.2. For each PCB, $b = 1, \ldots, B$:
     - 1.2.1. Generate intra-lot variability: $z_{l,b,:}^{(intra)} \sim N(0,1)$;
     - 1.2.2. For each pad, $p = 1, \ldots, P$:
       - 1.2.2.1. Compute solder mask effect on height:
         $h_{l,b,p}^{(sold)} = \left(\alpha_h^{(inter)} z_{l,:,:}^{(inter)} + \alpha_h^{(intra)} z_{l,b,:}^{(intra)}\right)\Delta_{h,sold} + z_{l,b,p}^{(pad)} \sqrt{\sigma_{tol}^2\{h,p\}\varphi_h^2 - \Delta_{h,sold}^2}$.

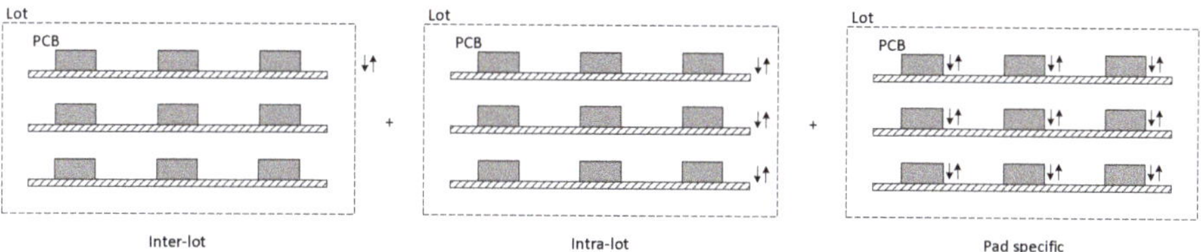

Figure 5. Schematic representation of inter-lot, intra-lot and pad specific components used to generate the solder mask effect on height.

As in the case of the offset-Y, the height is also affected by the printing direction of the squeegee. This effect relates to a progressive increase in the pressure of the squeegee. At the beginning of the printing process, the pressure is low and thus the heights are typically lower than intended. As the printing progresses, the heights reach the intended nominal values. The same applies when the printing direction is reversed, but now affecting the pads on the other extreme of the PCB (Figure 6). To be consistent with the squeegee effect on the offset-Y (see Section 3.1), it is considered that PCBs with an odd index b are printed in the direction of increasing y-coordinates, while PCBs with an even index are printed in the reversed direction.

To systematically address both printing directions, the original y-coordinate of the pads is transformed into an equivalent distance, related to the point where the printing beginnings, by:

$$d = (-1)^{b+1}\left(\mathbf{c}_y\{p\} - \frac{\max(\mathbf{c}_y) + \min(\mathbf{c}_y)}{2}\right) + \frac{\max(\mathbf{c}_y) - \min(\mathbf{c}_y)}{2}, \tag{13}$$

where $\mathbf{c}_y$ is a vector with the y-coordinates of each pad, p is the pad of interest and b is the PCB index. Based on this, the influence of the squeegee in the height ($h_{l,b,p}^{(squee)}$) is modeled through an exponential decay function (Table 7):

$$h_{l,b,p}^{(squee)} = -\Delta_{h,squee} u_{l,b,:} e^{-\frac{d}{\tau}}, \tag{14}$$

where $\Delta_{h,squee}$ is the maximum allowance deviation, τ is a length constant and $u_{l,b,:}$ is a random variable following $U(0,1)$. Process knowledge indicates that the deviation in $h_{l,b,p}^{(squee)}$ is more pronounced in the first third of the PCB (i.e., until $d = \left(\max(\mathbf{c}_y) - \min(\mathbf{c}_y)\right)/3$). Thus, by considering that for $d = 2\tau$ the deviation in $h_{l,b,p}^{(squee)}$ is reduced to 13.5% of the allowance deviation, the length constant is set to:

$$\tau = \frac{1}{6}\left(\max(\mathbf{c}_y) - \min(\mathbf{c}_y)\right). \tag{15}$$

Table 7. Pseudo-code for generating the squeegee effect on height.

1. For each lot, $l = 1, \ldots, L$:
    1.1. For each PCB, $b = 1, \ldots, B$:
        1.1.1. Generate squeegee variability: $u_{l,b,:} \sim U(0,1)$;
        1.1.2. For each pad, $p = 1, \ldots, P$:
            1.1.2.1. Compute squeegee effect on height:

$$h_{l,b,p}^{(squee)} = -\Delta_{h,squee} u_{l,b,:} e^{-\frac{d}{\tau}}, \text{ where}$$

$$d = (-1)^{b+1}\left(\mathbf{c}_y\{p\} - \frac{\max(\mathbf{c}_y)+\min(\mathbf{c}_y)}{2}\right) + \frac{\max(\mathbf{c}_y)-\min(\mathbf{c}_y)}{2}.$$

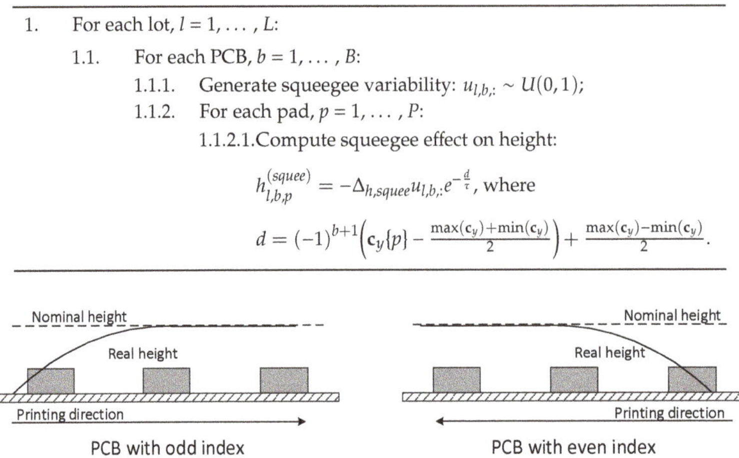

Figure 6. Schematic representation of the squeegee effect on height. Dimensions are not to scale.

The final value for the height of each pad ($h_{l,b,p}^{(agv)}$) is determined by summing the solder mask and squeegee effects:

$$h_{l,b,p}^{(agv)} = \mathbf{N}\{h,p\} + h_{l,b,p}^{(sold)} + h_{l,b,p}^{(squee)}. \tag{16}$$

3.3. Common Cause Variation Affecting Area

The area common cause variation can be conceptually subdivided into three components representing (the structure of each component is analogous to Figure 3):

(i). Inter-lot deviation that impacts all pads of all PCBs in the same lot—inter-lot variation: $\alpha_a^{(inter)} z_{l,:,:}^{(inter)} \boldsymbol{\sigma}_{tol}\{a,p\}\varphi_a$;

(ii). Intra-lot deviation that affects all pads in each PCB—intra-lot variation: $\alpha_a^{(intra)} z_{l,b,:}^{(intra)} \boldsymbol{\sigma}_{tol}\{a,p\}\varphi_a$;

(iii). Local variability that affects each pad—pad specific variability: $\alpha_a^{(pad)} z_{l,b,p}^{(pad)} \boldsymbol{\sigma}_{tol}\{a,p\}\varphi_a$.

The random part of each component ($z_{l,:,:}^{(inter)}$, $z_{l,b,:}^{(intra)}$ and $z_{l,b,p}^{(pad)}$) is drawn from the standard normal distribution and scaled by the allowed standard deviation of each pad's area ($\boldsymbol{\sigma}_{tol}\{a,p\}$) using a scaling factor for the total variance (φ_a). Furthermore, the contribution of each component is codified by the scaling factors $\alpha_a^{(inter)}$, $\alpha_a^{(intra)}$ and $\alpha_a^{(pad)}$, which are between 0 and 1 and are subject to:

$$\left(\alpha_a^{(inter)}\right)^2 + \left(\alpha_a^{(intra)}\right)^2 + \left(\alpha_a^{(pad)}\right)^2 = 1. \tag{17}$$

Therefore, the final variance for the area of each pad is $\boldsymbol{\sigma}_{tol}^2\{a,p\}\varphi_a^2$. Following these definitions, the values for the area ($a_{l,b,p}^{(agv)}$) are obtained by (Table 8):

$$a_{l,b,p}^{(agv)} = \mathbf{N}\{a,p\} + \left(\alpha_a^{(inter)} z_{l,:,:}^{(inter)} + \alpha_a^{(intra)} z_{l,b,:}^{(intra)} + \alpha_a^{(pad)} z_{l,b,p}^{(pad)}\right)\boldsymbol{\sigma}_{tol}\{a,p\}\varphi_a. \tag{18}$$

Table 8. Pseudo-code for generating the area.

1. For each lot, $l = 1, \ldots, L$:
   1.1. Generate inter-lot variability: $z_{l,:,:}^{(inter)} \sim N(0,1)$;
   1.2. For each PCB, $b = 1, \ldots, B$:
      1.2.1. Generate intra-lot variability: $z_{l,b,:}^{(intra)} \sim N(0,1)$;
      1.2.2. For each pad, $p = 1, \ldots, P$:
         1.2.2.1. Generate random variability: $z_{l,b,p}^{(pad)} \sim N(0,1)$;
         1.2.2.2. Compute area:
         $a_{l,b,p}^{(agv)} = \mathbf{N}\{a,p\} + \left(\alpha_a^{(inter)} z_{l,:,:}^{(inter)} + \alpha_a^{(intra)} z_{l,b,:}^{(intra)} + \alpha_a^{(pad)} z_{l,b,p}^{(pad)}\right) \boldsymbol{\sigma}_{tol}\{a,p\}\varphi_a.$

3.4. Common Cause Variation Affecting Volume

The volume of solder paste deposit in each pad ($v_{l,b,p}^{(agv)}$) is considered to be proportional to the product of area ($a_{l,b,p}^{(agv)}$) and height ($h_{l,b,p}^{(agv)}$):

$$v_{l,b,p}^{(agv)} = a_{l,b,p}^{(agv)} h_{l,b,p}^{(agv)} \mathbf{f}\{p\}, \tag{19}$$

where $\mathbf{f}\{p\}$ is a correction constant (Table 9). For the case studies considered in this article, three different values for $\mathbf{f}\{p\}$ were observed. However, it was not possible to establish a relationship between this parameter and the pads' location or geometry. Therefore, in the absence of further prior information, it is suggested that $\mathbf{f}\{p\}$ is estimated using the nominal values for volume, area and height such that:

$$\mathbf{f}\{p\} = \frac{\mathbf{N}\{v,p\}}{\mathbf{N}\{a,p\}\mathbf{N}\{h,p\}}. \tag{20}$$

Table 9. Pseudo-code for generating the volume.

1. For each lot, $l = 1, \ldots, L$:
   1.1. For each PCB, $b = 1, \ldots, B$:
      1.1.1. For each pad, $p = 1, \ldots, P$:
         1.1.1.1. Compute volume:
         $v_{l,b,p}^{(agv)} = a_{l,b,p}^{(agv)} h_{l,b,p}^{(agv)} \frac{\mathbf{N}\{v,p\}}{\mathbf{N}\{a,p\}\mathbf{N}\{h,p\}}.$

4. Multivariate Statistical Process Monitoring Based on Principal Component Analysis (MSPM-PCA)

In the previous section, details were given regarding the artificial generation of common cause variability for the SPD printing process in a Surface Mount Technology (SMT) production line. This is done in the AGV module, and the simulation outputs for area (a), height (h), volume (v), offset-X (x) and offset-Y (y), are used to augment the reference NOC dataset, enriching it with a wide overage of common cause variation modes. The augmented NOC dataset can then be used to set up a multivariate statistical process monitoring scheme, which in the present case is based on principal component analysis (PCA).

PCA is a latent variable methodology that decomposes the original data matrix, $\mathbf{X}$, with n observations and m variables, as:

$$\mathbf{X} = \mathbf{T}\mathbf{P}^{\mathrm{T}} + \mathbf{E}, \tag{21}$$

where $\mathbf{T}$ is a $(n \times k)$ matrix of PCA scores, $\mathbf{P}$ is a $(m \times k)$ matrix with the PCA loadings, $\mathbf{E}$ is a $(n \times m)$ matrix of residuals and k is the number of retained principal components (PC). As PCA is scale-dependent, the data matrix **X** is usually standardized or "autoscaled" to zero mean and unit variance.

The principal components of PCA are monitored by the Hotelling's T^2 as proposed by [52,53],

$$T^2 = \sum_{i=1}^{k} \frac{t_i^2}{\lambda_i} = \mathbf{x}^{\mathrm{T}}\mathbf{P}\mathbf{\Lambda}_k^{-1}\mathbf{P}^{\mathrm{T}}\mathbf{x}, \tag{22}$$

where $\mathbf{x}$ is a $(m \times 1)$ vector with the current observations, $\mathbf{\Lambda}_k$ is a diagonal matrix with the first k eigenvalues in the main diagonal. If the process follows a multivariate normal distribution, then the upper control limit (UCL) for T^2 can be computed as [54,55]:

$$UCL_{T^2} = \frac{k(n-1)(n+1)}{n^2 - nk} F_{\alpha,k,n-k}, \tag{23}$$

where $F_{\alpha,k,n-k}$ is the upper α-percentile of the F distribution, with k and $n-k$ degrees of freedom.

The complementary residuals' subspace is monitored through the Q-statistic based on the squared prediction error (SPE) of the residuals, $\mathbf{e}$ $(m \times 1)$ [53]:

$$Q = \mathbf{e}^{\mathrm{T}}\mathbf{e} = \left(\mathbf{x} - \hat{\mathbf{x}}\right)^{\mathrm{T}}\left(\mathbf{x} - \hat{\mathbf{x}}\right) = \mathbf{x}^{\mathrm{T}}\left(\mathbf{I} - \mathbf{P}\mathbf{P}^{\mathrm{T}}\right)\mathbf{x}, \tag{24}$$

where $\hat{\mathbf{x}}$ $(m \times 1)$ is the projection of $\mathbf{x}$ $(m \times 1)$ onto the PCA subspace. The UCL for this statistic is usually determined by [53,56]:

$$UCL_Q = \theta_1 \left(\frac{z_\alpha \sqrt{2\theta_2 h_0^2}}{\theta_1} + 1 + \frac{\theta_2 h_0 (h_0 - 1)}{\theta_1^2} \right)^{1/h_0}, \tag{25}$$

with,

$$\theta_i = \sum_{j=k+1}^{m} \lambda_j^i,\ i = 1,\ 2,\ 3, \tag{26}$$

$$h_0 = 1 - \frac{2\theta_1\theta_3}{3\theta_2^2}, \tag{27}$$

where, z_α is the upper $1-\alpha$ percentile of the standard normal distribution.

Due to deviations in the model assumptions, the theoretical control limits for the PCA monitoring statistics are often found to be inaccurate. Therefore, it is often preferable to set the control limits by adjusting a scaled χ^2 distribution to the empirical distributions of T^2 and Q [57–59], leading to:

$$UCL_{T^2} = g_{T^2} \cdot \chi^2\left(\alpha, h_{T^2}\right) \tag{28}$$

$$UCL_Q = g_Q \cdot \chi^2\left(\alpha, h_Q\right) \tag{29}$$

where g is a weighting factor and h is the effective number of degrees of freedom for the χ^2 distribution. Both values are obtained by matching the moments of the χ^2 distribution with those of the empirical distributions obtained from NOC data, leading to $g = v/(2u)$ and $h = 2u^2/v$, where v is the sample variance and u the sample mean of the monitoring statistic.

Under the PCA framework, fault diagnosis can be performed by resort to contribution plots [60–63]. In this context, the contribution of the i-th variable for the T^2 and Q statistics are computed as [62,64],

$$T_i^2 = \left\|\mathbf{\Lambda}^{-1/2}\mathbf{P}_i{}^{\mathrm{T}}\mathbf{x}_i\right\|^2, \tag{30}$$

$$Q_i = ||\mathbf{x}_i - \hat{\mathbf{x}}_i||^2, \tag{31}$$

where $\mathbf{P}_i$ is the i-th row of $\mathbf{P}$ and $\mathbf{x}_i$ is the i-th element of $\mathbf{x}$. The theoretical control limits for each T_i^2 and Q_i can be found in Refs. [62,64]. Alternatively, the scaled χ^2 approximation described before can also be used.

5. Results for the Implementation of AGV in an Industrial SMT Process

The results presented next regard the printing of solder paste deposits in PCBs composed by 3517 pads. For each pad, measurements for five parameters are collected: (i) area; (ii) height; (iii) volume; (iv) offset-X; (v) and offset-Y. For monitoring purposes, only 3507 pads are considered since the remaining pads are associated with atypical shapes for which the accumulated historic data is not yet enough to set up the AGV module parameters; they are also known to produce unusual offsets in x- and y-coordinates even in NOC. Consequently, the total number of variables under monitoring is $m = 3507\ pads \times 5\ parameters = 17{,}535\ variables$.

To analyze the characteristics of the proposed AGV module, three real datasets are considered in this study. The first dataset (CS1; Section 5.1) corresponds to NOC data and is composed by 337 PCBs. The second dataset (CS2; Section 5.2) contains 2080 PCBs taken from a different production run. Similarly, the third dataset (CS3; Section 5.3) has 1330 PCBs taken from a third production run. The CS2 and CS3 datasets are mostly composed by normal PCBs, but also contain some faulty PCBs. Furthermore, due to structured common cause variation in the process, the operational state of the three case studies is typically distinct.

In this study, the PCA model is trained using only simulated data generated by the AGV module with the values for the tuning parameters presented in Table 2. No data from the process was employed for building the PCA model or setting up the monitoring charts (they can be also considered, for this reason, pre-control charts). 20 lots with 300 PCBs in each lot were generated in the AGV module. Ten of these lots were used to build the PCA model (training dataset). The other ten lots were used to set the control limits by adjusting the empirical distributions of the monitoring statistics (validation dataset). The control limits were set to a false alarm rate of 0.01. Although we only present here the results for one set of the 20 lots, several replicates were analyzed leading to similar results.

From analysis of the scree plot in Figure 7, the number of retained principal components was set to 5. By further inspection of the PCA loadings (Figure 8) it is verified that the first five components have a close relationship with known phenomena. PC1 describes the relationship between heights and volumes. This PC also shows that the variation in the SPD's height and volume is related to the pads' location in the PCB. PC2 and PC3 explain the rigid body translation effects in the PCBs, while PC4 describes the rotation effects. PC5 is related to the squeegee effect in the heights and its propagation to the volumes. For this reason, PC5 is also related to the printing direction. As for the remaining PCs, they are mostly associated with relationships between areas and volumes of different SPDs. However, since the variation in the area is assumed to be independent from pad to pad, each one of them tend to generates a different PC. Therefore, many PCs are generated by this effect (note that there are over three thousands of pads involved in this mechanism). This justifies why the first 5 PCs explaining the most relevant structured phenomena for process monitoring contain less than 5% of the total variance in the simulated data. However, the monitoring procedure still comprises all the process variation: both the variation captured by the first 5 PCs (through the Hotelling's T^2 statistic for the selected PC scores) and the complementary residual variation (using the Q-statistic for the PCA residuals). Therefore, this asymmetrical distribution of common cause variability is consistent with the physics of the process and does not pose any problem or limitation for the monitoring scheme.

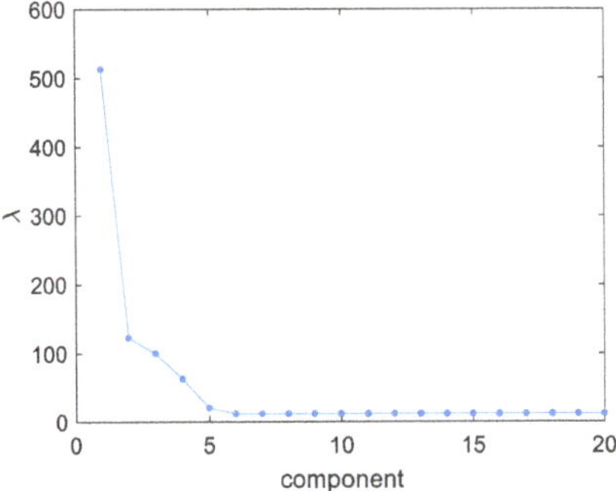

Figure 7. Scree plot of the first 20 eigenvalues of the PCA model.

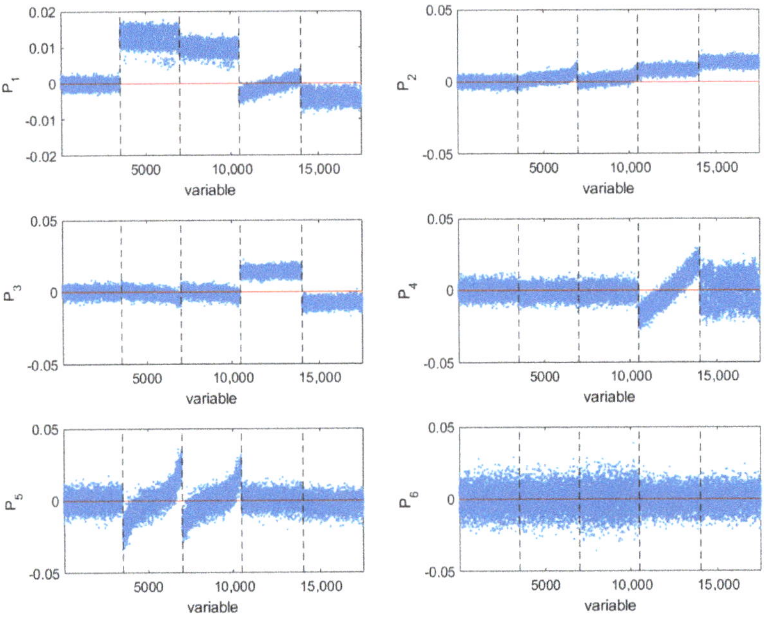

Figure 8. Loadings of the first six principal components. Each block of loadings corresponds to (from left to right) areas, heights, volumes, offset-X and offset-Y.

5.1. Analysis of Dataset CS1: NOC Data

The CS1 is composed by NOC data, and therefore it will be used to validate the AGV simulation module. By applying the PCA model to this dataset the score plots in Figure 9 are obtained. The results show that the scores from CS1 do not fully overlap the scores from the simulated lots. This is visible in the plot of PC3 vs. PC4 (see Figure 9c), and is an expected behavior since the inter-lot variability causes the lot to fall in a different region of the scores space. In this case, CS1 has lower scores in PC3 than the simulated lots. PC3 is related to translation effects in the PCBs, which implies that the PCBs in CS1 have a systematic translation in a given direction. Note however that some of the simulated lots have translations with similar magnitudes but on the opposite direction. As these inter-lot translations occur randomly and in either direction, we consider that the simulated data is still representative of real operation conditions. The different behavior across the lots is also visible for PC1 and PC4. Since both PCs are related with changes in offset-X and offset-Y, this indicates that the alignment of the PCBs in each lot is a critical factor in the final quality of the products. This information will be further explored in Sections 5.2 and 5.3 to identify possible problems in the different datasets under analysis. As for the intra-lot variability, it is verified that the dispersion of the real data is similar to that obtained with the AGV module.

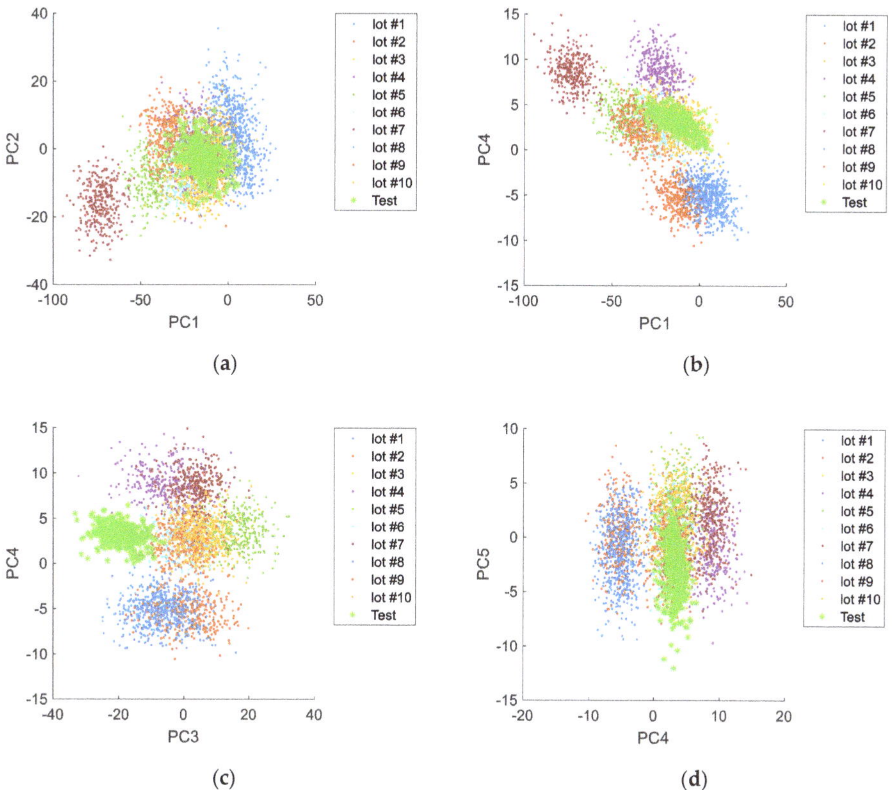

Figure 9. Relation between the principal components of the simulated validation lots and the data Figure 1. (**a**) first and second principal components; (**b**) first and fourth principal components; (**c**) third and fourth principal components; (**d**) fourth and fifth principal components.

As for the control charts produced by the PCA model (Figure 10), it is noticeable that the monitoring statistics are relatively low compared to the control limits. This happens because the PCA model was built using multiple lots, while the monitored data respects to a single lot operating in a well-defined and localized region, spanning only a small fraction of the operational NOC space.

One word of caution should be referred about the proposed approach, resulting from our accumulated experience on its application to real data. Although the proposed methodology decreases the chance of declaring a false alarm due to a change to another lot and solves the issue of false alarms that render classical monitoring methods useless, it also decreases the sensitivity to detect small deviations within a lot since a slightly deviating PCB may be treated as belonging to a different lot rather than being abnormal. This is a consequence of incorporating inter-lot variability in the AGV simulation scenarios.

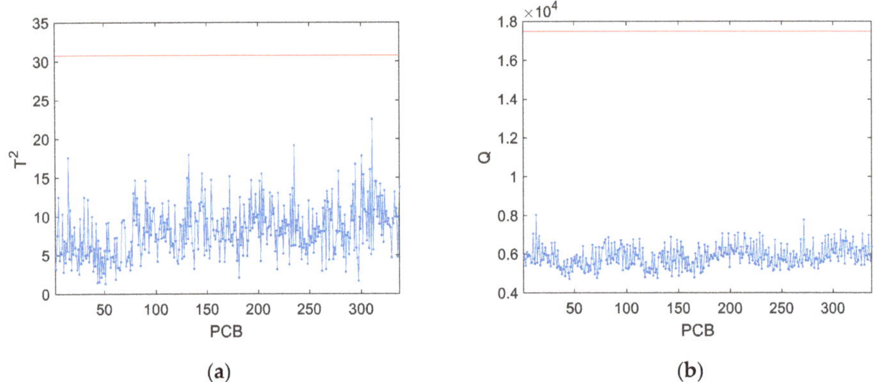

Figure 10. Control charts for the PCBs in CS1: (**a**) T^2-statisctic; (**b**) Q-statistic.

Related to this issue, it was also noted that the Q-statistic is affected by the large number of monitored variables (m). In this case, there are 17,535 variables, but only 5 PCs are retained in the PCA model. These 5 PCs also explain a small fraction of the total variance (less than 5%), meaning that most of the variability of the data is monitored by the Q-statistic. In other words, the Q-statistic cumulates a large number of squared residuals with rather large variances. Thus, when a fault occurs, even with a large magnitude, it can suffer from dilution effect from the contribution of all residuals.

From the analysis of the dataset CS1, we can confirm that real NOC data fall inside the expected NOC envelop obtained from the AGV module. The dispersion characteristics of the CS1 lot in the PCA subspace are also similar to the simulated lots, which occupy different regions in the PCA subspace due to the expected (and simulated) inter-lot variability. These observations consubstantiate the rigor placed on the model development process, and provide confidence to the analysis of the remaining datasets using the PCA model developed in this section based on the simulated AGV data.

5.2. Analysis of Dataset CS2: Test Data (NOC and Faulty)

From analysis of the scores obtained for this case study (Figures 11 and 12) it is apparent that this dataset is composed by at least five distinct operation periods. By comparison with the previously simulated validation lots, it is verified that all periods still fall within the typical operation conditions described by the PCA model applied to AGV data. It is also visible that each period resembles individual lots with specific characteristics. In this case study, a clear distinction in the printing direction is observed for some of the production periods. This phenomenon is more noticeable when PC5 is plotted against PC2 (Figure 13).

By inspecting the control charts for this case study (Figure 14), it is verified that all PCBs are marked as in-control. Nevertheless, it is noticeable that PCBs #159 and #1596 have atypical Q-statistics. While for PCB #1596 most of the variables are within their tolerance specifications, PCB #159 present significant deviations in 151 variables. Therefore, PCB #159 is clearly a false negative. The deviations in the measurements of PCB #159 are passed to the residual subspace of PCA. However, since the critical residuals are only a small part of the sum of squared residual over all variables, they are not large enough to trigger an alarm. In spite of this, the contribution plots of PCA are able to identify the abnormal pads and parameters (Figure 15). This result suggests that a monitoring statistic focused solely on the relevant residuals (i.e., those with a significant contribution in the residual subspace) could lead to a more sensitive fault detection.

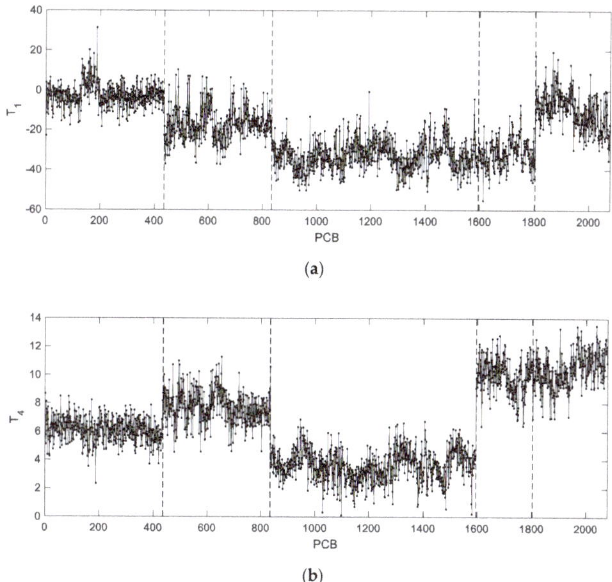

Figure 11. Principal components of CS2: (**a**) first principal component; (**b**) fourth principal component.

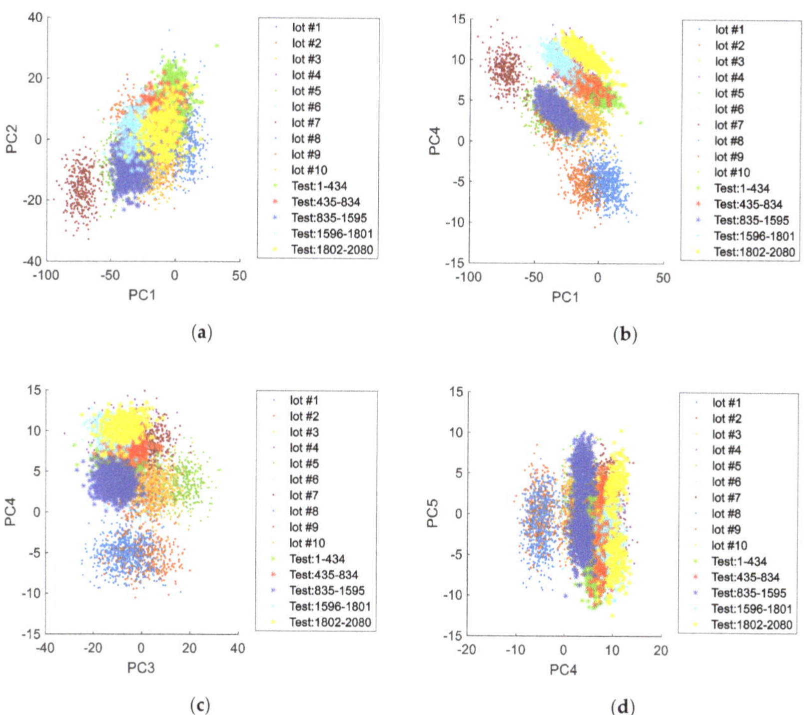

Figure 12. Relation between the principal components of the simulated validation lots and the data from CS2: (**a**) first and second principal components; (**b**) first and fourth principal components; (**c**) third and fourth principal components; (**d**) fourth and fifth principal components.

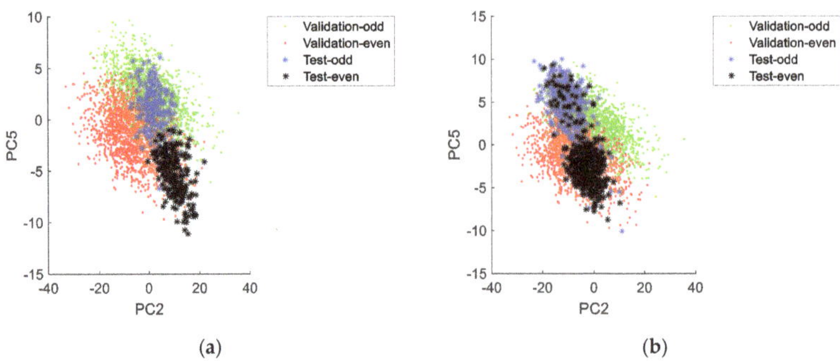

Figure 13. Second and fifth principal components of CS2 stratified by printing direction: (**a**) PCBs #435 to #834. (**b**) PCBs #835 to #1595.

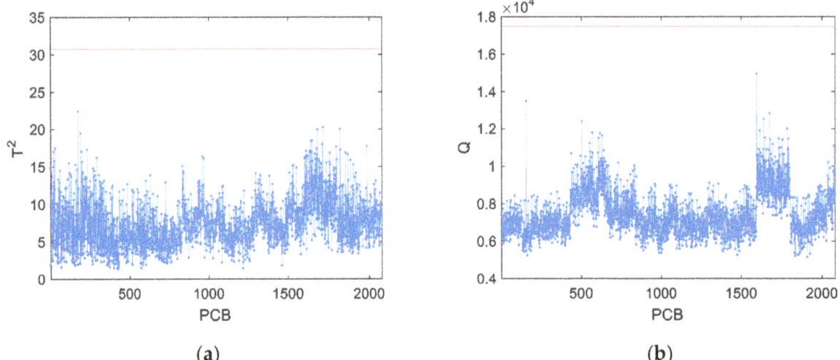

Figure 14. Control charts for the PCBs in CS2: (**a**) T^2-statisctic; (**b**) Q-statistic.

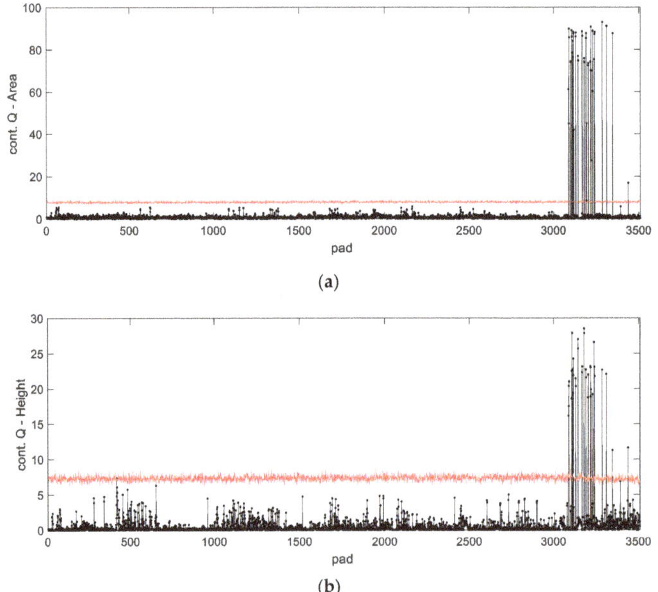

Figure 15. Contribution of each pads' parameter to the Q-statistic of PCB #159 of CS2: (**a**) contributions for the pads' area; (**b**) contributions for the pads' height.

5.3. Analysis of Dataset CS3: Test Data (NOC and Faulty)

This dataset comprises at least six distinct operation periods (Figure 16). Once again, these periods are mostly identified by PC1 and PC4, meaning that they are related to lots with different characteristic rotations in the PCBs during the SPD printing process. This result also highlights that the rotation effects on the offsets is mostly driven by inter-lot variation. From Figure 17 it is also verified that the selected values for the tuning parameters of the AGV module led to simulated lots resembling real data.

Although the AGV module generates data with realistic features, the PCA-based monitoring methodology presents limitations in its detection capabilities due to the incorporation of inter-lot variability (Figure 18). For instance, from the Q-statistic, it is observed that PCBs #248, #401 and #1191 have larger deviations than the others. Nevertheless, these deviations are not large enough to overcome the sum of squared residuals over a large number of variables. By inspecting the contribution plots for these PCBs, it is verified that some of the pads have unusual heights (Figure 19) and therefore these PCBs should have been signaled as potentially faulty. Once again, this issue may be solved in the future if the squared residuals below a given threshold are not included in the Q-statistic, which will cause the monitoring scheme to be more sensitive to critical deviations.

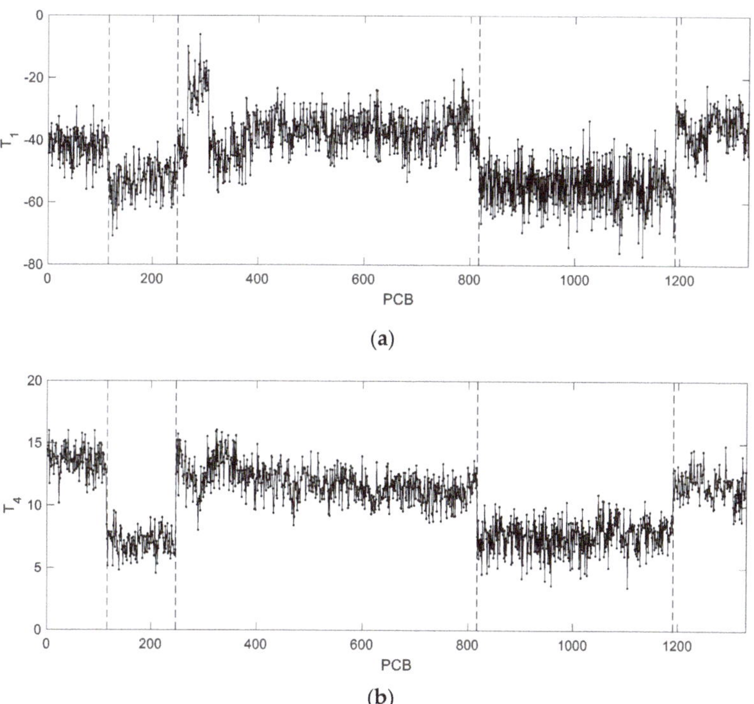

Figure 16. Principal components of CS3: (**a**) first principal component; (**b**) fourth principal component.

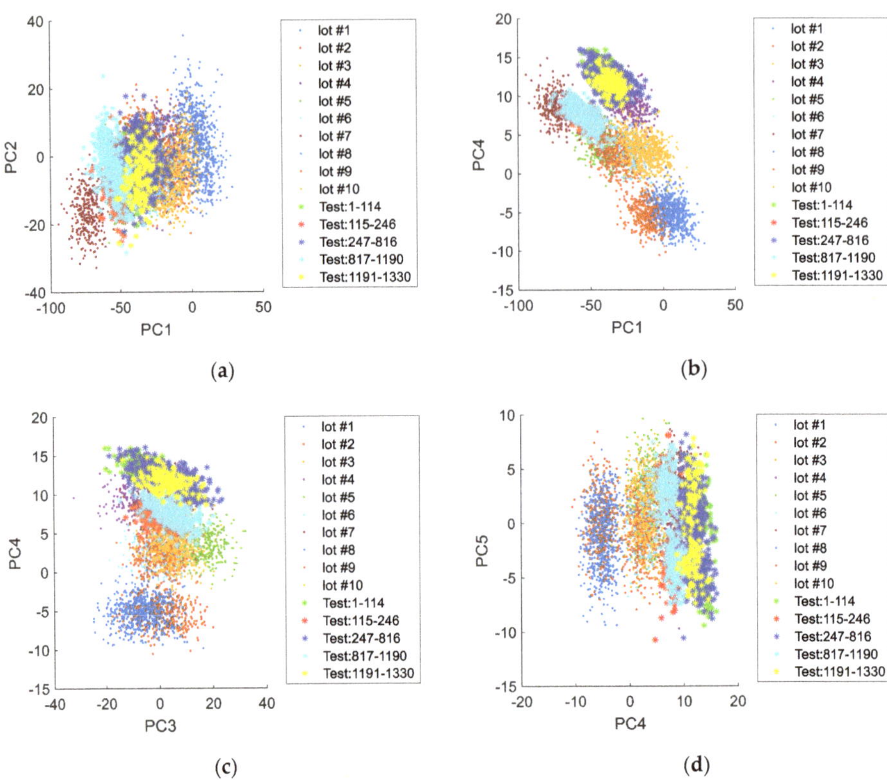

Figure 17. Relation between the principal components of the simulated validation lots and the data from CS3: (**a**) first and second principal components; (**b**) first and fourth principal components; (**c**) third and fourth principal components; (**d**) fourth and fifth principal components.

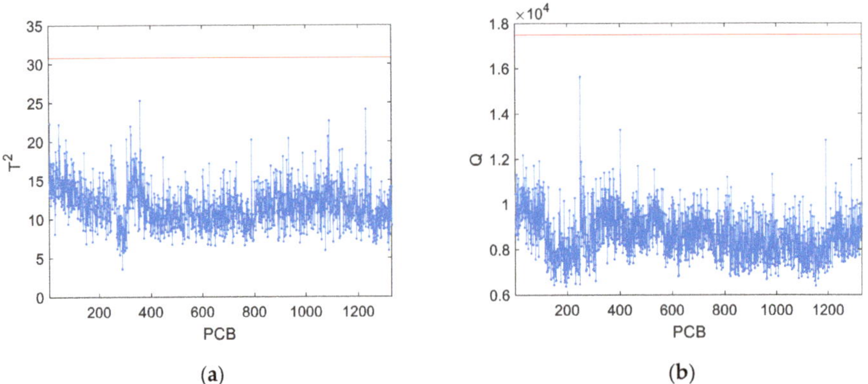

Figure 18. Control charts for the PCBs in CS3: (**a**) T^2-statisctic; (**b**) Q-statistic.

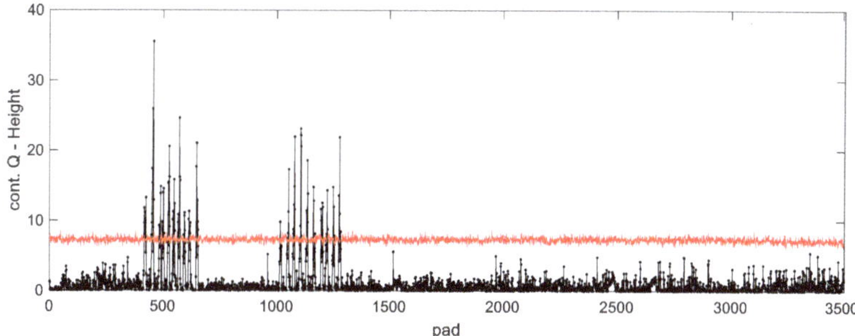

Figure 19. Contribution of the pads' height to the Q-statistic of PCB #248 of CS3.

6. Discussion

A suitable monitoring scheme requires an adequate model that describes the normal operations conditions of the process. One way to construct such NOC model is through the use of data-driven methods based on historical records of normal operation. However, when a new product enters the production line, such past information is not available. To overcome this situation, one can simply run the process without monitoring until sufficient data is acquired. However, this approach can lead to inadequate models since the data collected in a short period of time might not represent the overall variability of the process. This is particularly more relevant when the process has non-stationary characteristics, such as those caused by distinct production lots. As an alternative to the current data-driven modelling approaches, it is here proposed to generate artificial data based on accumulated process knowledge and then build a monitoring scheme using the augmented NOC dataset, enriched with information about inter-lot variation sources.

In this work, the proposed AGV module aims to simulate the structural components of common cause variability of a solder past printing process, accounting for (i) translation and rotation effects on offset-X and offset-Y, (ii) squeegee effects on offset-Y and height (iii) and solder mask effects on height. Furthermore, these effects are structured into multiple components in order to accommodate for inter-lot (common to all PCBs in a lot), intra-lot (common to all pads in a PCB) and pad specific (unique to each pad) variability. The amount of variability attributed to each component, as well as the magnitude of the effects is established by a set of tuning parameters handily defined by the user.

Throughout this work, it was observed that the tuning parameters can define a large variety of operation conditions. While this gives a high degree of flexibility to the AGV module, it also implies that selecting the best set of tuning parameters is a critical task in order to achieve realistic simulations. In this case, we had accesses to historical data and the tuning parameters were chosen to resemble the real data of the product under study. For a new product with distinct characteristics, the best combination of tuning parameters might be different. We expect the tuning parameters to be related to external factors such as operators and equipment operation and therefore they should not vary much from product to product. Nevertheless, further analysis is required to confirm this conjecture and the applicability of the similar tuning parameters to PCBs with different layouts.

By use of the PCA model built on the simulated data, it was verified that the simulated data was distributed in clusters related to the simulated lots. This behavior was also observed for the real data. Real data was dispersed over the same range of values as simulated data, which confirms the reliability of the AGV module as a realistic data augmentation tool. The loadings of the PCA model trained on the simulated data also showed to be related with known phenomena of the printing process. Therefore, the PCA model can provide very useful insights on the operation conditions of each PCB, as well as fault diagnosis. For instance, the PCA model can be used to identify distinct lots and the printing direction.

In this study the number of simulated lots was selected to have a reasonable representation of the process variability while also allowing for an easy visualization of the results. For a more accurate representation of the process the number of lots should be selected as the point where the PCA parameters and control limits stabilize. Throughout this work several realizations of 20 lots were simulated, and it was confirmed that they lead to consistent results of the PCA models and control limits for the statistics.

Following the good results in terms of process modelling, the PCA model was used for process monitoring. In this regard, the monitoring scheme showed to solve the excessive false alarms problem of current methodologies, at the cost of presenting low sensitivity to fault detection when faults are located on a relatively small number of SPDs. The reason for this is twofold. Firstly, the PCA model was built to account for the variability across multiple lots. However, for a given lot, the PCBs are likely to vary within a localized region of the principal components subspace. Furthermore, to detect abnormal PCBs within a lot, their measurements must differ not only from the typical measurements on their lot, but also typical measurements on other lots. Secondly, deviations from the PCA model can be masked by the sum of squared residuals over a very large number of variables. To overcome this limitation, in analogy to the cumulative sum (CUSUM) control chart [6,11], we propose to screen for the relevant residuals (i.e., those above a given allowance threshold) and then build a truncated Q-statistic using only the residuals that were found relevant. This approach is already under development and preliminary results show that it reduces the number of variables/residuals included in the Q-statistic and therefore puts more focus on the most critical deviations.

Although none of the case studies led to alarms, fault diagnosis was run for PCBs with atypical monitoring statistics. For these cases, the contribution plots were able to identify abnormal measurements as well as the affected pads. Therefore, the resort to simulated data can provide useful information for fault diagnosis. These results also support the development of an alternative monitoring statistic to enhance the detection of deviations in the residual subspace.

Finally, it is noted that real data can also be included in the PCA model as the process progresses. Another improvement that can be made as soon as real data is available is to replace the correction constants for the volume ($\mathbf{f}$, Equation (20)) by those computed from real data. Consequently, the relationships between volume and the product of area and height will be closer to reality. Nevertheless, even without the use of real data, the current study already demonstrates the usefulness of the AGV module for process monitoring when historical data is not available.

7. Conclusions

In this work, a data augmentation framework for Artificial Generation of common cause Variability in SMT industrial assembling lines was proposed. The aim of the AGV module is to allow for a prompt application of monitoring procedures in new products for which extensive historical data is not available, without incurring in prohibitive rates of false alarms. The AGV module is based on process knowledge and codifies the dominant modes of common cause variation due to known physical phenomena. The proposed approach can be even implemented in the absence of an historical dataset, as long as there is accumulated information from related processes. This particular aspect was also covered in our analysis in this paper.

The data generated by the AGV module proved to reproduce well the long term variability of real process data. The critical issue of high false alarm rates was solved. On the other hand, we also report that the monitoring statistics show limited detection capabilities. This happens because the PCA model accounts for inter-lot variation, while the process will typically operate in localized regions at a given time. Furthermore, the sensitivity of the PCA monitoring scheme is also affected by the very large number of variables under monitoring.

Future work should therefore consider the application of the AGV module to other case studies. Furthermore, novel monitoring approaches should be development in order to account for the presence of different lots (improve the sensitivity of the T^2-statistic) and reduce the impact of irrelevant residuals when the number of monitored variables is very large (improve the sensitivity of the Q-statistic).

Author Contributions: T.J.R. and M.S.R. developed the methodology and wrote the paper; P.D. and C.M. provided the data and process knowledge. All authors have read and agreed to the published version of the manuscript.

Funding: This research received no external funding.

Conflicts of Interest: The authors declare no conflict of interest.

References

1. Reis, M.S.; Gins, G.; Rato, T.J. Incorporation of process-specific structure in statistical process monitoring: A review. *J. Qual. Technol.* **2019**, *51*, 407–421. [CrossRef]
2. Del Castillo, E. *Statistical Process Adjustment for Quality Control*; Wiley: Hoboken, NJ, USA, 2002.
3. Box, G.E.P.; Kramer, T. Statistical process control and feedback adjustments—A discussion. *Technometrics* **1992**, *34*, 251–285. [CrossRef]
4. Ge, Z.; Song, Z.; Gao, F. Review of recent research on data-based process monitoring. *Ind. Eng. Chem. Res.* **2013**, *52*, 3543–3562. [CrossRef]
5. Montgomery, D.C.; Mastrangelo, C.M. Some statistical process control methods for autocorrelated data. *J. Qual. Technol.* **1991**, *23*, 179–193. [CrossRef]
6. Lu, C.-W.; Reynolds, M.R., Jr. Cusum charts for monitoring an autocorrelated process. *J. Qual. Technol.* **2001**, *33*, 316–334. [CrossRef]
7. Runger, G.C.; Willemain, T.R. Model-based and model-free control of autocorrelated processes. *J. Qual. Technol.* **1995**, *27*, 283–292. [CrossRef]
8. Ku, W.; Storer, R.H.; Georgakis, C. Disturbance detection and isolation by dynamic principal component analysis. *Chemom. Intell. Lab. Syst.* **1995**, *30*, 179–196. [CrossRef]
9. Rato, T.J.; Reis, M.S. Fault detection in the tennessee eastman process using dynamic principal components analysis with decorrelated residuals (dpca-dr). *Chemom. Intell. Lab. Syst.* **2013**, *125*, 101–108. [CrossRef]
10. Rato, T.J.; Reis, M.S. Advantage of using decorrelated residuals in dynamic principal component analysis for monitoring large-scale systems. *Ind. Eng. Chem. Res.* **2013**, *52*, 13685–13698. [CrossRef]
11. Nomikos, P.; MacGregor, J.F. Multivariate spc charts for monitoring batch processes. *Technometrics* **1995**, *37*, 41–59. [CrossRef]
12. Rendall, R.; Chiang, L.H.; Reis, M.S. Data-driven methods for batch data analysis—A critical overview and mapping on the complexity scale. *Comput. Chem. Eng.* **2019**, *124*, 1–13. [CrossRef]
13. Wold, S.; Kettaneh, N.; Friden, H.; Holmberg, A. Modelling and diagnostics of batch processes and analogous kinetic experiments. *Chemom. Intell. Lab. Syst.* **1998**, *44*, 331–340. [CrossRef]
14. Wold, S.; Kettaneh-Wold, N.; MacGregor, J.F.; Dunn, K.G. Batch process modeling and mspc. In *Comprehensive Chemometrics*; Elsevier: Oxford, UK, 2009; pp. 163–197.
15. Undey, C.; Cinar, A. Statistical monitoring of multistage, multiphase batch processes. *IEEE Control Syst.* **2002**, *22*, 40–52.
16. Camacho, J.; Picó, J. Multi-phase principal component analysis for batch processes modelling. *Chemom. Intell. Lab. Syst.* **2006**, *81*, 127–136. [CrossRef]
17. Yao, Y.; Gao, F. A survey on multistage/multiphase statistical modeling methods for batch processes. *Annu. Rev. Control* **2009**, *33*, 172–183. [CrossRef]
18. Chiang, L.H.; Braatz, R.D. Process monitoring using the causal map and multivariate statistics: Fault detection and identification. *Chemom. Intell. Lab. Syst.* **2003**, *65*, 159–178. [CrossRef]
19. Bauer, M.; Thornhill, N.F. A practical method for identifying the propagation path of plant-wide disturbances. *J. Process Control* **2008**, *18*, 707–719. [CrossRef]
20. Rato, T.J.; Reis, M.S. On-line process monitoring using local measures of association. Part i: Detection performance. *Chemom. Intell. Lab. Syst.* **2015**, *142*, 255–264. [CrossRef]
21. Rato, T.J.; Reis, M.S. On-line process monitoring using local measures of association. Part ii: Design issues and fault diagnosis. *Chemom. Intell. Lab. Syst.* **2015**, *142*, 265–275. [CrossRef]
22. Rato, T.J.; Reis, M.S. Markovian and non-markovian sensitivity enhancing transformations for process monitoring. *Chem. Eng. Sci.* **2017**, *163*, 223–233. [CrossRef]

23. Bakshi, B.R. Multiscale pca with application to multivariate statistical process control. *Aiche J.* **1998**, *44*, 1596–1610. [CrossRef]
24. Fourie, S.H.; de Vaal, P. Advanced process monitoring using an on-line non-linear multiscale principal component analysis methodology. *Comput. Chem. Eng.* **2000**, *24*, 755–760. [CrossRef]
25. Reis, M.S. Multiscale and multi-granularity process analytics: A review. *Processes* **2019**, *61*, 1–21. [CrossRef]
26. Reis, M.S.; Bakshi, B.R.; Saraiva, P.M. Multiscale statistical process control using wavelet packets. *Aiche J.* **2008**, *54*, 2366–2378. [CrossRef]
27. Reis, M.S.; Saraiva, P.M. Multiscale statistical process control with multiresolution data. *Aiche J.* **2006**, *52*, 2107–2119. [CrossRef]
28. Woodall, W.H. Controversies and contradictions in statistical process control. *J. Qual. Technol.* **2000**, *32*, 341–350. [CrossRef]
29. Page, E.S. Continuous inspection schemes. *Biometrics* **1954**, *41*, 100–115. [CrossRef]
30. Roberts, S.W. Control charts tests based on geometric moving averages. *Technometrics* **1959**, *1*, 239–250. [CrossRef]
31. Shewhart, W.A. *Economic Control of Quality of Manufactured Product*; D. Van Nostrand Company, Inc.: New York, NY, USA, 1931.
32. Crosier, R.B. Multivariate generalizations of cumulative sum quality-control schemes. *Technometrics* **1988**, *30*, 291–303. [CrossRef]
33. Hotelling, H. Multivariate quality control, illustrated by the air testing of sample bombsights. In *Selected Techniques of Statistical Analysis*; Eisenhart, C., Hastay, M.W., Wallis, W.A., Eds.; McGraw-Hill: NewYork, NY, USA, 1947.
34. Kourti, T.; MacGregor, J.F. Multivariate spc methods for process and product monitoring. *J. Qual. Technol.* **1996**, *28*, 409–428. [CrossRef]
35. Lowry, C.A.; Woodall, W.H.; Champ, C.W.; Rigdon, C.E. A multivariate exponentially weighted moving average control chart. *Technometrics* **1992**, *34*, 46–53. [CrossRef]
36. Kim, K.; Mahmoud, M.A.; Woodall, W.H. On the monitoring of linear profiles. *J. Qual. Technol.* **2003**, *35*, 317–328. [CrossRef]
37. Reis, M.S.; Saraiva, P.M. Multiscale statistical process control of paper surface profiles. *Qual. Technol. Quant. Manag.* **2006**, *3*, 263–282. [CrossRef]
38. Woodall, W.H.; Spitzner, D.J.; Montgomery, D.C.; Gupta, S. Using control charts to monitor process and product quality profiles. *J. Qual. Technol.* **2004**, *36*, 309–320. [CrossRef]
39. Trygg, J.; Kettaneh-Wold, N.; Wallbäcks, L. 2d wavelet analysis and compression of on-line industrial process data. *J. Chemom.* **2001**, *15*, 299–319. [CrossRef]
40. Brink, M.; Mandenius, C.-F.; Skoglund, A. On-line predictions of the aspen fibre and birch bark content in unbleached hardwood pulp, using nir spectroscopy and multivariate data analysis. *Chemom. Intell. Lab. Syst.* **2010**, *103*, 53–58. [CrossRef]
41. Sahni, N.S.; Aastveit, A.H.; Naes, T. In-line process and product control using spectroscopy and multivariate calibration. *J. Qual. Technol.* **2005**, *37*, 1–20. [CrossRef]
42. Reis, M.S.; Bauer, A. Wavelet texture analysis of on-line acquired images for paper formation assessment and monitoring. *Chemom. Intell. Lab. Syst.* **2009**, *95*, 129–137. [CrossRef]
43. Bharati, M.H.; MacGregor, J.F. Multivariate image analysis for real-time process monitoring and control. *Ind. Eng. Chem. Res.* **1998**, *37*, 4715–4724. [CrossRef]
44. Pereira, A.C.; Reis, M.S.; Saraiva, P.M. Quality control of food products using image analysis and multivariate statistical tools. *Ind. Eng. Chem. Res.* **2009**, *48*, 988–998. [CrossRef]
45. Geladi, P.; Grahn, H. *Multivariate Image Analysis*; Wiley: Chichester, UK, 1996.
46. Yu, H.; MacGregor, J.F. Monitoring flames in an industrial boiler using multivariate image analysis. *Aiche J.* **2004**, *50*, 1474–1483. [CrossRef]
47. Yu, H.; MacGregor, J.F.; Haarsma, G.; Bourg, W. Digital imaging for online monitoring and control of industrial snack food processes. *Ind. Eng. Chem. Res.* **2003**, *42*, 3036–3044. [CrossRef]
48. Smilde, A.K.; Bro, R.; Geladi, P. *Multi-Way Analysis with Applications in the Chemical Sciences*; Wiley: Chichester, UK, 2004.
49. Reis, M.S.; Delgado, P. A large-scale statistical process control approach for the monitoring of electronic devices assemblage. *Comput. Chem. Eng.* **2012**, *39*, 163–169. [CrossRef]

50. Jackson, J.E. *A User's Guide to Principal Components*; Wiley: New York, NY, USA, 1991.
51. Jolliffe, I.T. *Principal Component Analysis*, 2nd ed.; Springer: New York, NY, USA, 2002.
52. Jackson, J.E. Quality control methods for several related variables. *Technometrics* **1959**, *1*, 359–377. [CrossRef]
53. Jackson, J.E.; Mudholkar, G.S. Control procedures for residuals associated with principal component analysis. *Technometrics* **1979**, *21*, 341–349. [CrossRef]
54. MacGregor, J.F.; Kourti, T. Statistical process control of multivariate processes. *Control Eng. Pract.* **1995**, *3*, 403–414. [CrossRef]
55. Tracy, N.D.; Young, J.C.; Mason, R.L. Multivariate control charts for individual observations. *J. Qual. Technol.* **1992**, *24*, 88–95. [CrossRef]
56. Qin, S.J. Statistical process monitoring: Basics and beyond. *J. Chemom.* **2003**, *17*, 480–502. [CrossRef]
57. Cinar, A.; Palazoglu, A.; Kayihan, F. *Chemical Process Performance Evaluation*; CRC Press: Boca Raton, FL, USA, 2007.
58. Nomikos, P.; MacGregor, J.F. Monitoring batch processes using multiway principal component analysis. *Aiche J.* **1994**, *40*, 1361–1375. [CrossRef]
59. Box, G.E.P. Some theorems on quadratic forms applied in the study of analysis of variance problems, i. Effect of inequality of variance in the one-way classification. *Ann. Math. Stat.* **1954**, *25*, 290–302. [CrossRef]
60. Conlin, A.K.; Martin, E.B.; Morris, A.J. Confidence limits for contribution plots. *J. Chemom.* **2000**, *14*, 725–736. [CrossRef]
61. Westerhuis, J.A.; Gurden, S.P.; Smilde, A.K. Generalized contribution plots in multivariate statistical process monitoring. *Chemom. Intell. Lab. Syst.* **2000**, *51*, 95–114. [CrossRef]
62. Qin, S.J.; Valle, S.; Piovoso, M.J. On unifying multiblock analysis with application to decentralized process monitoring. *J. Chemom.* **2001**, *15*, 715–742. [CrossRef]
63. Van den Kerkhof, P.; Vanlaer, J.; Gins, G.; Van Impe, J.F.M. Analysis of smearing-out in contribution plot based fault isolation for statistical process control. *Chem. Eng. Sci.* **2013**, *104*, 285–293. [CrossRef]
64. Rato, T.J.; Blue, J.; Pinaton, J.; Reis, M.S. Translation-invariant multiscale energy-based pca for monitoring batch processes in semiconductor manufacturing. *IEEE Trans. Autom. Sci. Eng.* **2017**, *14*, 894–904. [CrossRef]

Publisher's Note: MDPI stays neutral with regard to jurisdictional claims in published maps and institutional affiliations.

Article

Multiscale Convolutional and Recurrent Neural Network for Quality Prediction of Continuous Casting Slabs

Xing Wu [1,2,*], Hanlu Jin [1], Xueming Ye [1], Jianjia Wang [1,2], Zuosheng Lei [1], Ying Liu [3], Jie Wang [4] and Yike Guo [5]

1 School of Computer Engineering and Science, Shanghai University, Shanghai 200444, China; Lucyjinhanlu@shu.edu.cn (H.J.); yuneming@shu.edu.cn (X.Y.); jianjiawang@shu.edu.cn (J.W.); lei_zsh@staff.shu.edu.cn (Z.L.)
2 Shanghai Institute for Advanced Communication and Data Science, Shanghai University, Shanghai 200444, China
3 School of Economics and Management, University of Chinese Academy of Sciences, Beijing 100190, China; liuy@ucas.ac.cn
4 Center for Sustainable Development and Global Competitiveness, Stanford University, Stanford, CA 94305, USA; jiewang@stanford.edu
5 Department of Computing, Imperial College London, London SW7 2AZ, UK; yikeguo@hkbu.edu.hk
* Correspondence: xingwu@shu.edu.cn

Abstract: Quality prediction in the continuous casting process is of great significance to the quality improvement of casting slabs. Due to the uncertainty and nonlinear relationship between the quality of continuous casting slabs (CCSs) and various factors, reliable prediction of CCS quality poses a challenge to the steel industry. However, traditional prediction models based on domain knowledge and expertise are difficult to adapt to the changes in multiple operating conditions and raw materials from various enterprises. To meet the challenge, we propose a framework with a multiscale convolutional and recurrent neural network (MCRNN) for reliable CCS quality prediction. The proposed framework outperforms conventional time series classification methods with better feature representation since the input is transformed at different scales and frequencies, which captures both long-term trends and short-term changes in time series. Moreover, we generate different category distributions based on the random undersampling (RUS) method to mitigate the impact of the skewed data distribution due to the natural imbalance of continuous casting data. The experimental results and comprehensive comparison with the state-of-the-art methods show the superiority of the proposed MCRNN framework, which has not only satisfactory prediction performance but also good potential to improve continuous casting process understanding and CCS quality.

Keywords: quality prediction; continuous casting; multiscale; convolutional neural network; time series classification; imbalanced data

Citation: Wu, X.; Jin, H.; Ye, X.; Wang, J.; Lei, Z.; Liu, Y.; Wang, J.; Guo, Y. Multiscale Convolutional and Recurrent Neural Network for Quality Prediction of Continuous Casting Slabs. *Processes* **2021**, *9*, 33. https://dx.doi.org/10.3390/pr9010033

Received: 15 October 2020
Accepted: 21 December 2020
Published: 25 December 2020

1. Introduction

At present, the steel industry is facing unprecedented challenges including resource consumption, serious environmental pollution, substandard process and product stability, and low productivity [1]. Steelmaking is a typical process industry, with long production processes, complicated manufacturing processes, and many process control factors involved [2]. The changes in product types and raw materials of different companies will be different, and it is difficult for knowledge-based models to adapt to all changes, which makes the migration and maintenance of models difficult. Therefore, the deep integration of information technology and the steel manufacturing industry, as the entry point for industrial upgrading, is of great significance to the realization of intelligent and green steel production.

Continuous casting is the most critical part of steelmaking [3]. Stable and high-quality continuous casting production is the top priority of iron and steel enterprises. Continuous casting is the process of solidifying molten metal into semifinished slabs and rolling them in a finishing mill [4]. As shown in Figure 1, the molten metal is transferred from the ladle to a tundish and slowly injected into the continuous caster. Then, the crystallizer in the continuous caster shapes the casting and rapidly solidifies and crystallizes. In this process, the mold level fluctuation will greatly affect the quality of continuous casting slabs (CCSs). With the sharp fluctuation of the liquid level in the mold, the content of oxide inclusions under the slabs will increase significantly [5]. However, mold level fluctuation is likely to cause slag entrapment of molten steel, which further leads to the deterioration of slab quality.

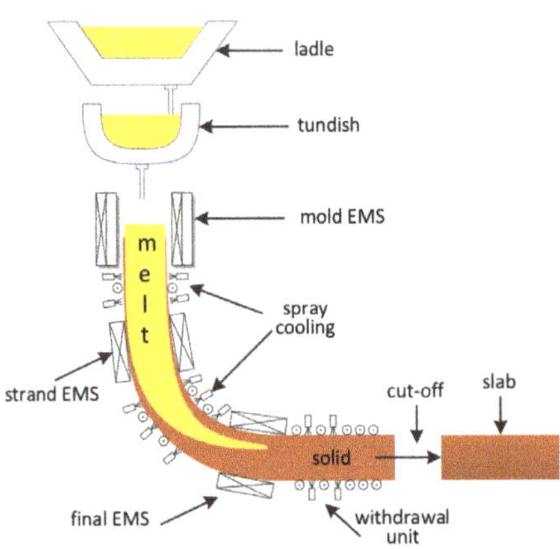

Figure 1. A schematic diagram of the continuous casting process.

Major steel producers are leveraging information technology such as the Internet of Things (IoT) and embracing big data to change the current state of the steel industry [6]. The use of sensor-based data acquisition systems in factories and the explosive growth of steel data make data modeling and analysis possible [7]. Furthermore, over the last decade, intelligent technologies, represented by data mining [8] and neural networks [9], have been developed from the theoretical research into their industrial applications. In the field of steelmaking, numerous scholars focus on the classification of steel surface defects [10,11]. Although continuous casting is the main process phase affecting the final quality of the steel products, the continuous casting system has a large number of complex input parameters; thus it is well adapted for big data analysis. Lei et al. have used machine learning methods to develop an offline system for continuous casting data collection and data mining [12], a small amount of research work involves the classification and prediction of continuous casting slabs quality. Nandkumar et al. [13] predicted and improved the quality of iron casting with the Six Sigma approach. A two-layer feedforward backpropagation neural network model was developed to predict the possibility of defects in foundry products [14]. The feedforward backpropagation neural net is out of practice currently, and the vanilla recurrent neural net performs poorly in engineering. Artur et al. designed a specific convolutional neural network (CNN) to detect stickers during continuous casting [15]. Although their method can reduce false alarms, when CNN is used alone for detection, the effect is not respectable. Indeed, we have incorporated two neural net architectures into our multiscale convolutional and recurrent neural network (MCRNN) to build one more robust and better network.

In this work, based on the process data acquisition system, a real-time prediction closed-loop control system was constructed to predict and improve the quality of CCS. In the system, a framework composed of an MCRNN is proposed for real-time quality prediction of CCS. Various conversions are made at different times and frequencies to obtain time series data for fluctuations in the level of the original mold. The CNN can apply to time series analysis of sensor data well, and it can also be used to analyze signal data with a fixed-length period. Feature extractors based on the fully convolutional network (FCN) and long short-term memory (LSTM) are used to capture long-term dependencies and extract local features of time series, respectively, and we use the advantages of CNN to automatically learn features [16] in the downsampling transformation representation and frequency domain, extracting features of different time scales and frequencies and solving the limitations of many previous features that can only be extracted at a single time scale [17,18]. As a result, the proposed MCRNN enhances feature representation and improves the performance of quality prediction compared to traditional time series classification models. Moreover, the number of normal samples is much larger than the number of abnormal samples. Average production is 100 slabs, with production of only 5 abnormal slabs. We use the random undersampling (RUS) method to reduce the number of majority classes to address the class imbalance. We introduced expert knowledge into the system. When the predictive model detects an abnormal slab, the continuous casting process adjusts in real-time based on expert knowledge, which improves steelmaking efficiency and slab quality.

The organizational structure is as follows: In Section 2, we review the work related to time series classification. In Section 3, we describe our proposed MCRNN and established system in detail, which is the core section of the paper. In Section 4, we present the detailed process and experimental results of the method. Finally, in Section 5, we draw the main conclusions of this work.

2. Related Work

In our real world, time series data are ubiquitous; examples include temperature, click volume, stock prices, and sensor data. They are sequential data of real value type with a large amount of data, high data dimensions, and constant updating of data. In the data-driven era, there is an increasing demand for information extracted from time series, the main task of which is time series classification (TSC). It is a long-standing problem involving a wide range of practical applications, such as the classification of financial time series [19], the judgment of individual agricultural land-cover types [20], and early churn detection [21].

Traditional time series classification methods are mostly based on distance measurement. Lines and Bagnall [22] proposed nearest neighbor classifiers with elastic distance measures to improve classification accuracy. In particular, the dynamic time warping (DTW) distance combined with the nearest neighbor classifier has proved to be a strong baseline [23]. Nevertheless, the performance could be rarely acceptable when it was applied to the engineering field with big data. There are other methods of distance measurement and spatial transformation for time series, such as information entropy [24], weighted dynamic time warping (WDTW) [25], and shapelet transformation [26]. Moreover, enhanced weighted dynamic time warping [27] and distributed fast-shapelet transform [28] were proposed to improve the performance of times series classification. Based on ensemble schemes and data conversion, Bagnall et al. not only aggregated different classifiers on the same transformation but also collected different classifiers in different time series representations [29]. However, these methods only have linear separability.

In recent years, deep learning has developed rapidly and achieved excellent results in classification tasks. Convolutional neural networks and recurrent neural networks are widely used in image recognition [30], video classification [31], machine translation [32], information extraction [33], and other fields. CNN can use convolutional layers to learn complex feature representations automatically, with the advantage of absorbing a large amount of data to learn feature representations. In recent years, many neural networks for time series classification, such as multilayer perceptron (MLP), fully convolutional network (FCN), and residual network (ResNet) [34], emerged. Convolutional neural networks (CNN) have been applied to time series applications, though CNN is mainly for the image field [35,36]. In the classification of high-dimensional time series, Zheng et al. proposed to use a multichannel convolutional neural network for modeling [37]. The echo state network (ESN) is a time-warping invariant, limited to static patterns rather than temporal patterns, and was applied to time series classification tasks [38]. Joan et al. studied the use of a time series encoder and established a hybrid deep CNN with an attention mechanism [39]. For the quality prediction system, however, these present methods cannot meet the demands of overall continuous casting slab production pipelines.

3. Methodology

Given a series of mold level fluctuations, our goal is to predict the quality of the continuous casting slab (CCS) in production. The quality of CCS will also change under different production conditions, such as different raw materials and technological parameters. In addition, it is worth noting that the quality of CCS is normal in most cases, while only a few are abnormal. Unbalanced time series classification is a challenging task when using only FCN or LSTM to extract time series on a single scale. We consider that time series should be represented comprehensively in multiscale and multifrequency dimensions to improve the classification performance and obtain a robust model. To address these problems for quality prediction of the CCS, we propose a new MCRNN architecture, where the input is the time series of mold level fluctuation to be predicted and the output is its quality label, as shown in Figure 2. The more details of layouts of each network are tabulated in Table 1. We use the grid search to obtain hyperparameters and iteratively find the best hyperparameters. This architecture mainly includes three sequential stages: the input representation stage, the feature learning stage, and the classification stage.

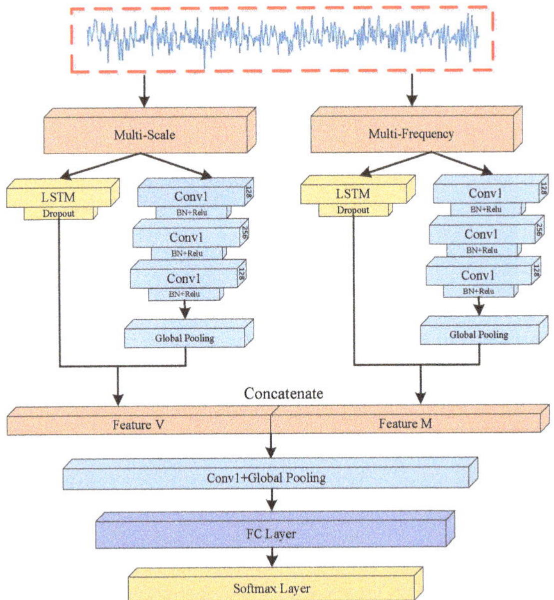

Figure 2. The proposed multiscale convolutional and recurrent neural network (MCRNN) framework.

Table 1. Details of the the MCRNN structure.

Layer Type	Feature Maps	Kernel Size	Stride
Convolution	128	8	1
BN + ReLU	128	-	-
Convolution	256	5	1
BN + ReLU	256	-	-
Convolution	128	3	1
BN + ReLU	128	-	-
AvgPooling	128	243	0
Concate	768	-	-
Convolution	64	4	1
AvgPooling	64	765	0
Full-connected	2	-	-

3.1. *Class Imbalance*

In the process of quality prediction, the number of abnormal and normal samples is extremely unbalanced, and the imbalance ratio is about 20:1. Class imbalance can have a negative impact on classification performance, because the classifier trained on unbalanced data favor major classes. We utilize the RUS method to achieve a more balanced class distribution, which improves the classification performance.

The RUS method is a form of data sampling that randomly selects major class instances and removes them from the dataset until the desired class distribution is achieved. Based on the original unbalanced dataset, RUS is used to generate the training dataset of three sample ratios, which are 1:1, 1:2, and 1:3. The normal sample ratio is followed by the abnormal sample ratio. We try to see how different sampling ratios affect the classification performance of the trained neural network and select the best sampling dataset. However, the test set is generated from unbalanced raw data without RUS because of realistic prediction requirements. As shown in Figure 3, in the original dataset of continuous casting slabs, the number of abnormal continuous casting slabs is far less than the number of normal continuous casting slabs. The desired class distribution is achieved by randomly removing the normal CCS and retaining the entire abnormal CCS, which can cause the loss of majority class information.

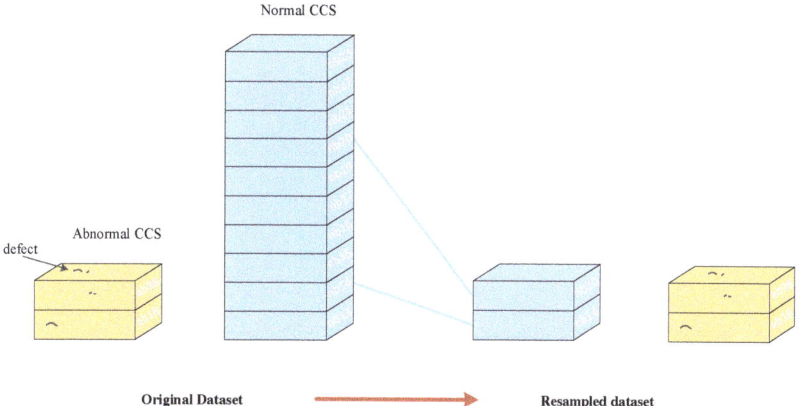

Figure 3. The random undersampling process of continuous casting slabs (CCSs).

3.2. *MCRNN Architecture*

3.2.1. Input Representation

Consideration should be given to using multiscale time series to build an accurate and reliable time series model. The long-term temporal pattern shows general trend changes, and the short-term temporal pattern reflects fine-grained fluctuations. Both patterns are critical to the performance of TSC. In our research work, we transform the original input space to obtain representation at different time scales and frequencies inspired by Cui et al. [40]. The transformation includes two stages: downsampling transformation in the time domain and smoothing transformation in the frequency domain. In the first stage, we downsample from the sequence $X = [x_1, x_2, ..., x_T]$ of mold level fluctuation and the downsampling rate is r. Then, new time series X^r is generated from the original sequence by retaining every r^{th} data points.

$$X^r = \{x_{1+r*i}\}, i = 0, 1, ..., \lfloor \frac{T-1}{r} \rfloor \tag{1}$$

Due to the influence of high-frequency disturbances and random noise, we carry out the moving average of the time series in the second stage to solve the problem. Given an original sequence $X = [x_1, x_2, ..., x_T]$ of mold level fluctuation, a new time series can be defined as X_w according to different degrees of smoothness.

$$X^w = \{\frac{1}{w} \sum_{i=(j-1)w+1}^{jw} x_i\}, j = 1, 2, ..., \frac{T}{w} \tag{2}$$

where w is the window size.

As shown in Figure 4, a sequence of the mold level fluctuation values in the production time of one slab transforms in time and frequency dimensions. For different downsampling rates and degrees of smoothness, we can get multiple time sequences, each of which corresponds to different scale representations of original sequence input. With the multiscale transformation of input, long-term temporal patterns and short-term temporal patterns can be employed to build a robust model. At the same time, the new time series based on the moving average of different windows reduces the noise of the original sequence. After two stages of transformation, the input is divided into two modules and fed into the neural network. For r and w, it is related to the sampling size. Sampling size is the sample points for each slab. We compared the sampling size values when the sampling rate is 1:2. As shown in Table 2, the model trained well when the sampling size was equal to 256, so we use 256 in our model.

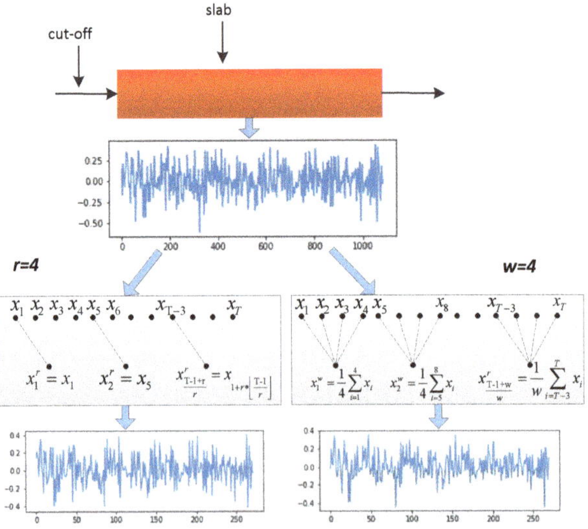

Figure 4. Illustration of the input transformations when r = 4 and w = 4.

Table 2. Comparison of sampling size with sampling ratios = 1:2.

Sampling Size	Accuracy
128	0.5176
256	0.6250
512	0.4778

3.2.2. Feature Learning

The feature extractor architecture is composed of the LSTM module and a fully convolutional module. The goal of this phase is to learn effective time series features in a parallel manner through multiple pairs of recurrent layers and convolutional layers in advance.

1. LSTM module: This module contains an LSTM layer, followed by a dropout layer. We employ an LSTM feature extractor to capture temporal patterns of CCS time series with multiscale and multifrequency dimensions. Specifically, the mold level fluctuation input $X = [x_1, x_2, ..., x_T]$ and the hidden state H_{t-1} of the previous time step given for the time step t. The definition of input gate i_t, forget gate f_t, and output gate o_t is as follows. The input gate controls the extent to which a new value flows into the cell.

$$i_t = \sigma(X_t W_{xi} + H_{t-1} W_{hi} + b_i) \quad (3)$$

The forget gate decides what information should be dropped.

$$f_t = \sigma(X_t W_{xf} + H_{t-1} W_{hf} + b_f) \quad (4)$$

The output gate determines which parts are useful.

$$o_t = \sigma(X_t W_{xo} + H_{t-1} W_{ho} + b_o) \quad (5)$$

The candidate memory cells $\tilde{C}_t$ at time step t are calculated as

$$\tilde{C}_t = \tanh(X_t W_{xc} + H_{t-1} W_{hc} + b_c) \quad (6)$$

The calculation of the current time step memory cell C_t combines the information of the last time step memory cell and the current time step candidate memory cell, and controls the flow of information through the forgetting gate and the input gate.

$$C_t = f_t \odot C_{t-1} + i_t \odot \tilde{C}_t \quad (7)$$

The output gate controls the flow of information from memory cells to the hidden state H_t, which can be calculated as:

$$H_t = o_t \odot tanh(C_t) \quad (8)$$

We feed the raw or transformed mold level fluctuation to LSTM and get output vector $O_v = [H_1, H_2, ..., H_T]$ from the last layer of the LSTM. We use output at time step t as feature $O_v^T = H_T$ extracted by LSTM. To prevent overfitting, the output of the LSTM layer is followed by the dropout layer with a dropout rate of 0.8 as shown in Figure 2. With dropout, final feature vector F_v can denote as:

$$F_v = \mathbf{r} * O_v^T \quad (9)$$

$$r_i \sim Bernoulli(p) \quad (10)$$

Here, $*$ denotes an element-wise product. For output vector at time step t, $\mathbf{r}$ is a vector of independent *Bernoulli* random variables, each of which has probability p of being 1.

2. Fully convolutional module: The core component of fully convolutional module is a convolutional block that contains:
   - Convolutional layer with a filter size of 128 or 256, the kernel with a size of 8, 5, 3 and stride of 1.
   - Batch normalization layer with a momentum of 0.99 and epsilon of 0.001.
   - A ReLU activation at the end of the module.

   In this module, we utilize convolution kernel $w \in \mathbb{R}^m$ to slide over the input sequence and extract local features. The output c_i of the i-node in the feature map is defined by

$$c_i = \sigma(w^T * x_{i:i+m-1} + b) \tag{11}$$

   where $x_{i:i+m-1}$ represents m-length subsequence from the ith time step to the $(i+m-1)$th time step of input sequence, $*$ denotes the convolution operator, b denotes the bias term, and $\sigma(.)$ is a nonlinear activation function.
   Accordingly, the convolution kernel is slid from the beginning time step to the end and we get the feature map of the jth kernel as

$$\mathbf{c}_j = [c_1, c_2, \ldots, c_{T-m+1}] \tag{12}$$

After convolution, batch normalization followed by a ReLU activation function accelerates fast training speed and improves model generalization ability. The fully convolutional module contains three convolutional blocks which are used as a feature extractor. Then, it performs a one-dimensional global average pooling operation on the feature map of the last block to obtain the vector, which reduces feature dimensions while increasing the receptive field of the kernel. The vector obtained by global average pooling on the final output channel can be expressed as

$$F_c = [a_1, a_2, \ldots, a_k] \tag{13}$$

$$a_j = \frac{1}{T-m+1} \sum \mathbf{c}_j \tag{14}$$

where k represents the filter size of the last convolutional block. We concatenate the features extracted by LSTM with a fully convolutional module. As mentioned in the previous section, the original input is transformed at different time scales and frequencies, so we use feature extractors on different input expressions and feed the final features into the next stage as input.

3.2.3. Classification

Finally, the concatenated feature vector obtained in the feature learning stage is directly fed to the classification module, which is composed of a convolution and global average pooling layer, a fully connected layer, and a softmax layer. As a result, it outputs conditional probability for each class. The softmax function rescales the n-dimensional vector of the FC layer output so that the output value is in the range [0, 1] and the sum is 1, which is defined by the following:

$$s(v_i) = \frac{e^{v_i}}{\sum_{j=1}^{n} e^{v_j}} \tag{15}$$

The full convolution module and LSTM module process the same time series input in two different fields of view. The full convolution is a fixed-size perception field to extract local features of time series. On the contrary, LSTM effectively captures time dependencies. The method of combining with convolutional and recurrent neural networks is crucial to enhance the performance of the proposed framework.

3.3. *Quality Prediction System Based on MCRNN*

Based on a large amount of process information collected by sensors, a quality prediction and control system is established for intelligent decision-making and control. To elaborate on the infrastructure of an established system, the framework of the system based on MCRNN is described in Figure 5. It mainly consists of three parts: data acquisition, quality prediction, and dynamic control. Data acquisition module based on various sensor networks collects massive real-time production data about the continuous casting process, such as temperature, water volume, and casting speed. The real-time collected process data will be sent to the quality prediction module and stored as historical data for visualizing the display and training of the model. Moreover, the quality information of each rolled slab is collected to label continuous casting data.

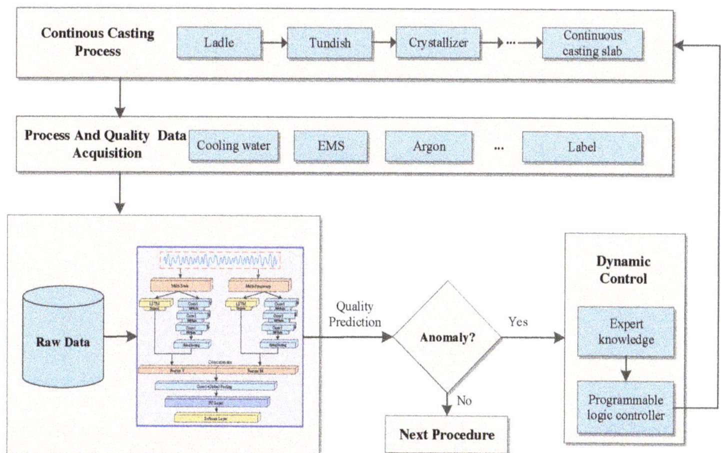

Figure 5. The framework of quality prediction system based on MCRNN.

With production process parameters and slab labels, a quality prediction model based on the proposed MCRNN is built. In the real-time production process, the original time series data are entered into the model and transformed with different time scales and frequencies. The output of the model is the quality label of CCS. Once the slab in producing is judged to be abnormal by the prediction model, the knowledge of domain experts will be employed to dynamically adjust the production process. The dynamic control module adjusts the process and equipment parameters in time through the programmable logic controller to avoid affecting the next rolling process and causing waste. Abnormal CCS produced will be sorted into the cleaning process of the machine to eliminate defects. The workflow improves efficiency, reduces costs, and enhances yield greatly.

4. Experiments and Results

In this section, we first describe the dataset and the evaluation metrics. Then, the effects of the RUS method and multiscale transformations are discussed in our studies. Finally, the proposed MCRNN model compares with different baseline models.

4.1. Dataset

Based on the installed data collector, the mold level fluctuation of the continuous casting production is recorded every 0.5 s in time series. In this way, we obtain a one-year continuous casting real-time process (CCRP) dataset which is not labeled. The continuous casting slab is rolled, and then the label information is generated by the inspection machine. Therefore, we get slightly delayed slab quality information, called the slab label dataset, from another system.

The slab label dataset contains abnormal reasons to be used as anomaly labels. We cannot obtain the quality information of CCS in the production process immediately, and can only get feedback results after hot rolling. The only connection to the CCRP dataset and the slab label dataset is the time of continuous casting. We map the anomaly labels in the slab label dataset to the CCRP dataset through casting time. Each slab corresponds to a large amount of real-time information during the continuous casting period. With the help of the start and end times in the slab label dataset, we match quality labels to the time series data during this period.

After marking the CCRP dataset with the slab label dataset, we obtained 9628 time series of slabs with the label. Among them, 9073 time-series were labeled as normal samples, and 555 time series were labeled as abnormal samples. In all experiments, we used a leave-one-out approach to train and test the classifier, divided the sample into two, 70% of the samples for training and 30% of the samples for testing, and used k-fold cross-validation to ensure the robustness of the model; cross-validation was repeated 5 times. However, normal and abnormal samples were extremely unbalanced. We utilized the RUS method described in Section 3.3 on the training set to ensure sample balance.

4.2. Evaluation Metrics

The confusion matrix is used to evaluate the quality of the algorithm in the classification task. In particular, we focus on three important metrics, the average accuracy of the classifier, the recall value for each class, and F_1 score. Our goal is to find a balance between false negatives and false positives, and find as many abnormal slabs as possible for good judgment. Specifically, if our model does not detect a CCS with abnormal quality, the abnormal slab will move on to the next process, and the final result is that the produced steel plate cannot be sold. If a CCS of normal quality is predicted to be abnormal by the model, it will undergo further processing attempts to change the quality status, which will increase costs. The most important point is that the cost of sending defective products to customers can be much higher than that of inspecting the products. Therefore, we want to maximize recall rates of exception class and sacrifice as few normal samples as possible.

$$Recall = \frac{TP}{TP + FN} \tag{16}$$

$$Precision = \frac{TP}{TP + FP} \tag{17}$$

$$F_1 = \sum_i 2 \times w_i \frac{Precision_i \times Recall_i}{Precision_i + Recall_i} \tag{18}$$

where i refers to class index and $w_i = \frac{n_i}{N}$ represents the proportion of samples of class i, with n_i being the number of samples of the ith class and N being the total number of samples.

4.3. Effect of Random Undersampling

The training errors of different sampling rates (1:1, 1:2, 1:3) shows in the form of loss curves in Figure 6. When the sampling rate is 1:2, the curve drops more smoothly, so the sampling effect is better.

Tables 3–5 show the results of k-fold cross-validation of the proposed MCRNN method at different sampling rates, k = 5. The result of the proposed MCRNN method at different sampling ratios is shown in Table 6. From the results, we can see the effect of sampling on the predictive performance of the model, and our model has a certain degree of robustness. Without sampling, recall for abnormal class and normal class is 0 and 1, respectively. Obviously, the trained models predicted all the slabs as normal to acquire the highest accuracy, without any ability to detect abnormal slabs. As the proportion of abnormal samples in the training sample increases, the recall of abnormal class increases. The SMOTE sampling algorithm has a certain effect on solving the problem of imbalanced data [41]. We also compared the SMOTE sampling algorithm with RUS in Table 6, and it was obvious that the RUS algorithm we proposed has a better effect on our data set. However, when the sampling ratio is 1:1, although more than 50% of abnormal slabs can be identified, a large number of normal slabs are misjudged at the same time. It is reflected in the low F_1 score and accuracy.

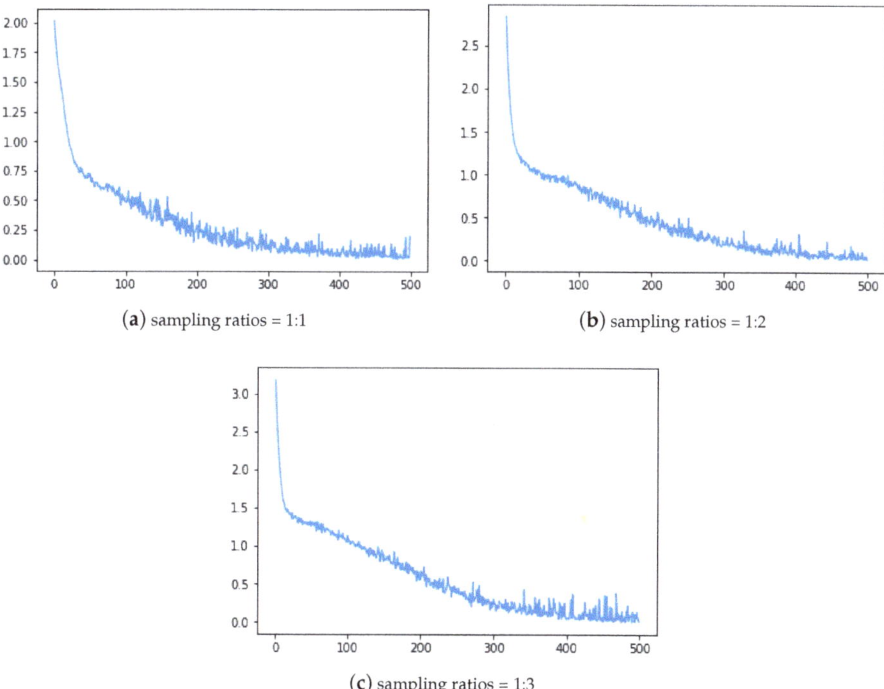

(a) sampling ratios = 1:1 (b) sampling ratios = 1:2 (c) sampling ratios = 1:3

Figure 6. The MCRNN training loss curve with different sampling ratios.

Table 3. Results for sampling ratios = 1:1 with k = 5.

	Accuracy	F1	Recall-Abnormal	Recall-Normal
1	0.3978	0.5164	0.6165	0.6071
2	0.4147	0.5338	0.5987	0.6122
3	0.4549	0.5724	0.5553	0.5714
4	0.4551	0.5747	0.5525	0.5663
5	0.4531	0.5737	0.4531	0.5306

Table 4. Results for sampling ratios = 1:2 with k = 5.

	Accuracy	F1	Recall-Abnormal	Recall-Normal
1	0.6247	0.7206	0.3588	0.3827
2	0.6227	0.7190	0.3651	0.4439
3	0.6058	0.7062	0.3797	0.3929
4	0.6393	0.7315	0.3404	0.3418
5	0.6325	0.7264	0.3512	0.3929

Through the sampling of training samples, the prediction ability of the model for abnormal slab can be improved, but the best proportion is one that is not completely balanced. When the sampling ratio is 1:2 or 1:3, the trained model has a certain ability to detect abnormal slabs without misjudging a large number of normal slabs. In the actual quality prediction of CCS, we adopt the sampling strategy with a sampling ratio of 1:2 because sending defective slabs to customers based on prediction can be more expensive than misjudgment, and we want to detect as many abnormal slabs as possible to avoid inferior products.

Table 5. Results for sampling ratios = 1:3 with $k = 5$.

	Accuracy	F1	Recall-Abnormal	Recall-Normal
1	0.7142	0.7843	0.2590	0.3214
2	0.7292	0.7940	0.2406	0.2857
3	0.6875	0.7659	0.2876	0.3214
4	0.6940	0.7705	0.2816	0.3367
5	0.7181	0.7871	0.2563	0.3418

Table 6. Results for different sampling ratios.

Sampling Ratio	Accuracy	F1	Recall-Abnormal	Recall-Normal
1:1	0.4351	0.5542	0.5552	0.5776
1:2	0.6250	0.7207	0.3590	0.3908
1:3	0.7086	0.7804	0.2650	0.3214
SMOTE	0.4566	0.5274	0.4942	0.5272
No sampling	0.9445	0.9277	0	1

4.4. Effect of Multiscale Transformations

In order to validate the effectiveness of multiscale input transformations, we performed experiments with transformed and untransformed inputs. The results are shown in Figure 7. We can see that the F_1 score with input transformations is higher than that without input transformations when the sampling ratio is 1:2 and 1:3. When the sampling ratio is 1:1, the F_1 score of the two scenarios are almost identical. However, input transformations have a positive effect on the recall for abnormal class. It can be concluded from the right part of the figure that more abnormal slabs can be detected with input transformations. In most cases, performing input transformations will help greatly improve classification performance. The effectiveness of the multiscale transformations is demonstrated in the recall rate of the abnormal class and F_1 score.

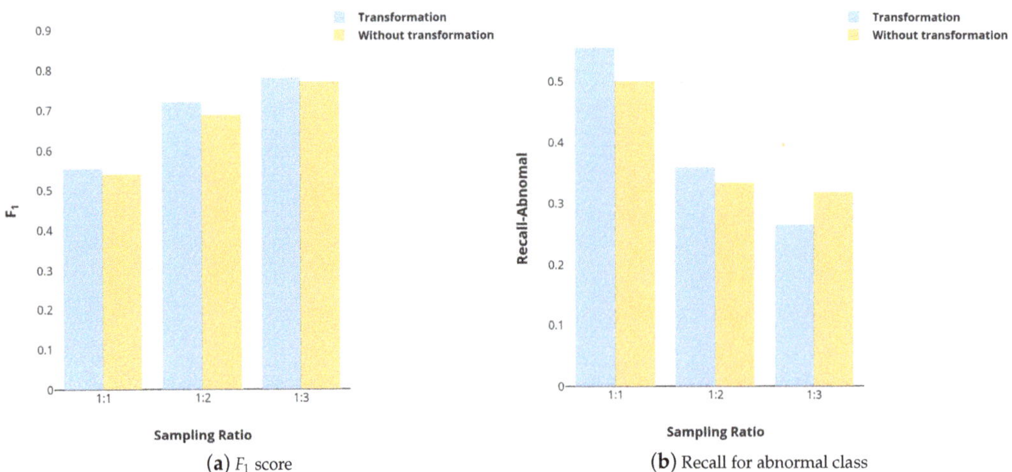

(**a**) F_1 score (**b**) Recall for abnormal class

Figure 7. Effects of multiscale transformation on classification performance.

4.5. Comparison

We conducted experiments on our dataset using two baseline methods from the publication of Wang et al. [34] for comparison to our developed approach: fully convolutional network (FCN) and residual network (ResNet), which have been proved to be useful as standard benchmarks for end-to-end time series classification networks. The FCN basic block is a convolutional layer, followed by a batch of normalization layer and a ReLU activation layer, and the final output comes from the softmax layer. The convolution operation is completed by three 1-D kernels of size 8, 5, 3. The final network is constructed by stacking three convolution blocks. The filter size of each convolution block is 128, 256, 128. ResNet uses the convolution block in FCN to construct each residual block, and finally stacks three residual blocks, followed by a global average pooling layer and a softmax layer. The number of filters for each residual block is 64, 128, 128. Furthermore, long short-term memory (LSTM) is used to compare with our proposed method, which has been proved to apply to periodic time series data. We have optimized the parameters of all networks participating in the comparison experiment to achieve the best results in this problem domain.

Table 4 shows the recall rate for the abnormal class of the proposed model and the other methods of baselines. Table 5 compares the F_1 score of our proposed model with other models. The results illustrate that our proposed model achieves the highest recall for abnormal class at different sampling ratios. According to Tables 7 and 8, the proposed model achieves the highest recall for abnormal class while maintaining a high F_1 score. When the sampling ratio is 1:2, the proposed model obtains the recall for an abnormal class of 0.3590 and the F_1 of 0.7207. It is best for our task. We hope that the model can detect more abnormal slabs and minimize misjudgment, which is a cost consideration.

Table 7. Recall-Abnormal comparison between the proposed model and the other baseline methods.

Methods	Sampling Ratio		
	1:1	1:2	1:3
FCN	0.5303	0.3485	0.2576
ResNet	0.5455	0.3536	0.2272
LSTM	0.5303	0.0606	0.0151
MCRNN	**0.5552**	**0.3590**	**0.2650**

Table 8. F_1 score comparison between the proposed model and the other baseline methods.

Methods	Sampling Ratio		
	1:1	1:2	1:3
FCN	0.5249	0.6778	0.8155
ResNet	0.5137	0.6751	0.8246
LSTM	0.6244	0.8962	0.9445
MCRNN	0.5542	0.7207	0.7804

By comparison of the three methods, LSTM is bad in comparison to ResNet and FCN for Recall-Abnormal and MCRNN is not superior to ResNet and FCN in the F_1 score. However, the MCRNN is superior to LSTM in the Recall-Abnormal score, though the MCRNN shows inferior slightly to LSTM in the F_1 score. Considering the engineering scenario of steel production prediction, the Recall-Abnormal is more important than the F_1 score to prevent low-level steel slabs from escaping check. FCN and ResNet, though slightly inferior to our model, also achieved good classification performance. However, LSTM performs unsatisfactorily in most cases except for the 1:1 sampling ratio. LSTM can easily deal with periodic time series data, but there are still some challenges with cluttered sensor data. Compared with FCN and ResNet, the MCRNN extracts features at different time scales and frequencies. Inputs of different transformations capture long-term trends and short-term changes, which is essential for classification. It can explain that the traditional methods simply perform a large number of convolutions over the same time scale.

5. Conclusions

We proposed a novel MCRNN architecture for the quality prediction of CCS. The major contributions of the new architecture are the transformations of time series input and feature extraction with LSTM and FCN. The proposed architecture can automatically extract the long-term trend and short-term change of time series, which greatly enhances feature learning ability and abnormal slab detecting performance. Extensive experimental results show that traditional methods are more incapable when dealing with messy and unbalanced data, and multiscale convolution and recurrent neural networks outperform other state-of-the-art baseline methods in quality prediction. Accordingly, a real-time quality prediction system based on MCRNN architecture has also been developed. The mold level fluctuation collected by the data module in the system is fed into the trained model. The continuous casting process will be adjusted in real-time based on expert knowledge if there is a high probability of prediction that it is an abnormal slab. The system greatly enhances steelmaking efficiency, improves slab quality, and reduces costs. Due to class imbalance caused by a few abnormal slabs, we use a random sampling method to generate training sets with three different sampling ratios to help mitigate class imbalance. Experimental results demonstrated that the proposed method can detect more abnormal slabs and reduce the misjudgment of normal slabs when the sampling ratio is 1:2.

For future research, although the established quality system has achieved certain results, it is still insufficient in several aspects such as interpretability of prediction and root cause analysis, the sampling method of dealing with the problem of unbalanced data is still worthy of our continued study. In recent years, the interpretability of deep learning is an important research field. In the future, we will utilize the interpretable method and root cause analysis to find out the cause of the abnormal slab, which will further improve the performance of intelligent steelmaking.

Author Contributions: X.W., H.J., X.Y. and J.W. (Jianjia Wang) are the main authors of this manuscript. All the authors contributed to this manuscript. Conceptualization, X.W.; data curation, J.W. (Jianjia Wang); methodology, X.W.; software, H.J. and X.Y.; validation, X.Y. and J.W. (Jianjia Wang); writing—original draft preparation, H.J. and X.Y.; writing—review and editing, X.W., H.J., X.Y., J.W. (Jianjia Wang), Z.L., Y.L., J.W. (Jie Wang) and Y.G.; supervision, X.W. All authors have read and agreed to the published version of the manuscript.

Funding: This work was supported by the Natural Science Foundation of Shanghai, China (Grant No. 20ZR1420400), the State Key Program of National Natural Sc hasience Foundation of China (Grant No. 61936001).

Institutional Review Board Statement: Not applicable.

Informed Consent Statement: Not applicable.

Data Availability Statement: Data is contained within the article.

Acknowledgments: We appreciate the High Performance Computing Center of Shanghai University, and Shanghai Engineering Research Center of Intelligent Computing System (No. 19DZ2252600) for providing the computing resources.

Conflicts of Interest: The authors declare no conflict of interest.

Abbreviations

The following abbreviations are used in this manuscript:

CCS	Continuous Casting Slabs
MCRNN	Multiscale Convolutional and Recurrent Neural Network
RUS	Random Undersampling
IoT	Internet of Things
CNN	Convolutional Neural Network
LSTM	Long Short-Term Memory
TSC	Time Series Classification
DTW	Dynamic Time Warping
WDTW	Weighted Dynamic Time Warping
FCN	Fully Convolutional Network
MLP	Multilayer Perceptron
ESNs	Echo State Networks
CCRP	Continuous Casting Real-time Process
ResNet	Residual Network
SMOTE	Synthetic Minority Oversampling Technique

References

1. Nikiforova, V.A. World steel industry: Current challenges and development trends (analytical overview). *Econ. Ind.* **2018**, *1*, 86–114. [CrossRef]
2. Xiang, F.; Zhi, Z.; Jiang, G. Digital Twins technolgy and its data fusion in iron and steel product life cycle. In Proceedings of the 2018 IEEE 15th International Conference on Networking, Sensing and Control (ICNSC), Zhuhai, China, 27–29 March 2018; pp. 1–5.
3. Mazumdar, D. Review, Analysis, and Modeling of Continuous Casting Tundish Systems. *Steel Res. Int.* **2019**, *90*, 1800279. [CrossRef]
4. Louhenkilpi, S. Continuous casting of steel. In *Treatise on Process Metallurgy*; Elsevier: Boston, MA, USA, 2014; pp. 373–434.
5. Smirnov, A.; Kuberskii, S.; Smirnov, E.; Verzilov, A.; Maksaev, E. Influence of meniscus fluctuations in the mold on crust formation in slab casting. *Steel Transl.* **2017**, *47*, 478–482. [CrossRef]
6. Peters, H. How could industry 4.0 transform the steel industry? In Proceedings of the Future Steel Forum, Warsaw, Poland, 14–15 June 2017.
7. Kuo, Y.H.; Kusiak, A. From data to big data in production research: The past and future trends. *Int. J. Prod. Res.* **2019**, *57*, 4828–4853. [CrossRef]
8. Ge, Z.; Song, Z.; Ding, S.X.; Huang, B. Data mining and analytics in the process industry: The role of machine learning. *IEEE Access* **2017**, *5*, 20590–20616. [CrossRef]
9. Xing, S.; Ju, J.; Xing, J. Research on hot-rolling steel products quality control based on BP neural network inverse model. *Neural Comput. Appl.* **2019**, *31*, 1577–1584. [CrossRef]
10. Liu, Y.; Geng, J.; Su, Z.; Zhang, W.; Li, J. Real-time classification of steel strip surface defects based on deep CNNs. In *Proceedings of 2018 Chinese Intelligent Systems Conference*; Springer: Singapore, 2019; pp. 257–266.
11. He, D.; Xu, K.; Wang, D. Design of multi-scale receptive field convolutional neural network for surface inspection of hot rolled steels. *Image Vis. Comput.* **2019**, *89*, 12–20. [CrossRef]
12. Lei, Z.; Li, B.; Zhou, Y.; Wu, X.; Zhong, Y.; Ren, Z. Two Paradigms on Study Slab Continuous Casting Process with Mold Electromagnetic Stirring. *MS&E* **2018**, *424*, 012035.
13. Mishra, N.; Rane, S.B. Prediction and improvement of iron casting quality through analytics and Six Sigma approach. *Int. J. Lean Six Sigma* **2019**, *10*, 189–210. [CrossRef]
14. Hore, S.; Das, S.K.; Humane, M.M.; Peethala, A.K. Neural Network Modelling to Characterize Steel Continuous Casting Process Parameters and Prediction of Casting Defects. *Trans. Indian Inst. Met.* **2019**, *72*, 3015–3025. [CrossRef]
15. Faizullin, A.; Zymbler, M.; Lieftucht, D.; Fanghänel, F. Use of Deep Learning for Sticker Detection During Continuous Casting. In Proceedings of the 2018 Global Smart Industry Conference (GloSIC), Chelyabinsk, Russia, 13–15 November 2018; pp. 1–6.
16. Domingos, P. A few useful things to know about machine learning. *Commun. ACM* **2012**, *55*, 78–87. [CrossRef]
17. Murad, A.; Pyun, J.Y. Deep recurrent neural networks for human activity recognition. *Sensors* **2017**, *17*, 2556. [CrossRef] [PubMed]
18. Minar, M.R.; Naher, J. Recent advances in deep learning: An overview. *arXiv* **2018**, arXiv:1807.08169.
19. Chao, L.; Zhipeng, J.; Yuanjie, Z. A novel reconstructed training-set SVM with roulette cooperative coevolution for financial time series classification. *Expert Syst. Appl.* **2019**, *123*, 283–298. [CrossRef]
20. Whelen, T.; Siqueira, P. Time-series classification of Sentinel-1 agricultural data over North Dakota. *Remote Sens. Lett.* **2018**, *9*, 411–420. [CrossRef]
21. Óskarsdóttir, M.; Van Calster, T.; Baesens, B.; Lemahieu, W.; Vanthienen, J. Time series for early churn detection: Using similarity based classification for dynamic networks. *Expert Syst. Appl.* **2018**, *106*, 55–65. [CrossRef]
22. Lines, J.; Bagnall, A. Time series classification with ensembles of elastic distance measures. *Data Min. Knowl. Discov.* **2015**, *29*, 565–592. [CrossRef]

23. Bagnall, A.; Lines, J.; Bostrom, A.; Large, J.; Keogh, E. The great time series classification bake off: A review and experimental evaluation of recent algorithmic advances. *Data Min. Knowl. Discov.* **2017**, *31*, 606–660. [CrossRef]
24. Deng, H.; Runger, G.; Tuv, E.; Vladimir, M. A time series forest for classification and feature extraction. *Inf. Sci.* **2013**, *239*, 142–153. [CrossRef]
25. Jeong, Y.S.; Jeong, M.K.; Omitaomu, O.A. Weighted dynamic time warping for time series classification. *Pattern Recognit.* **2011**, *44*, 2231–2240. [CrossRef]
26. Hills, J.; Lines, J.; Baranauskas, E.; Mapp, J.; Bagnall, A. Classification of time series by shapelet transformation. *Data Min. Knowl. Discov.* **2014**, *28*, 851–881. [CrossRef]
27. Anantasech, P.; Ratanamahatana, C.A. Enhanced Weighted Dynamic Time Warping for Time Series Classification. In *Third International Congress on Information and Communication Technology*; Springer: Singapore, 2019; pp. 655–664.
28. Baldán, F.J.; Benítez, J.M. Distributed FastShapelet Transform: A Big Data time series classification algorithm. *Inf. Sci.* **2019**, *496*, 451–463. [CrossRef]
29. Bagnall, A.; Lines, J.; Hills, J.; Bostrom, A. Time-series classification with COTE: The collective of transformation-based ensembles. *IEEE Trans. Knowl. Data Eng.* **2015**, *27*, 2522–2535. [CrossRef]
30. Pang, L.; Lan, Y.; Guo, J.; Xu, J.; Wan, S.; Cheng, X. Text matching as image recognition. In Proceedings of the Thirtieth AAAI Conference on Artificial Intelligence, Phoenix, AZ, USA, 12–17 February 2016.
31. Karpathy, A.; Toderici, G.; Shetty, S.; Leung, T.; Sukthankar, R.; Fei-Fei, L. Large-scale video classification with convolutional neural networks. In Proceedings of the IEEE Conference on Computer Vision and Pattern Recognition, Columbus, OH, USA, 23–28 June 2014; pp. 1725–1732.
32. Gehring, J.; Auli, M.; Grangier, D.; Yarats, D.; Dauphin, Y.N. Convolutional sequence to sequence learning. In Proceedings of the 34th International Conference on Machine Learning, Sydney, Australia, 6–11 August 2017; Volume 70, pp. 1243–1252.
33. Stanovsky, G.; Michael, J.; Zettlemoyer, L.; Dagan, I. Supervised open information extraction. In *Proceedings of the 2018 Conference of the North American Chapter of the Association for Computational Linguistics: Human Language Technologies, New Orleans, Louisiana*; Association for Computational Linguistics (ACL): Stroudsburg, PA, USA, 2018; Volume 1, pp. 885–895, long papers.
34. Wang, Z.; Yan, W.; Oates, T. Time series classification from scratch with deep neural networks: A strong baseline. In Proceedings of the 2017 International Joint Conference on nEural Networks (IJCNN), Anchorage, AK, USA, 14–19 May 2017; pp. 1578–1585.
35. Le Guennec, A.; Malinowski, S.; Tavenard, R. Data Augmentation for Time Series Classification Using Convolutional Neural Networks. In *ECML/PKDD Workshop on Advanced Analytics and Learning on Temporal Data*; halshs: Riva Del Garda, Italy, 2016.
36. Zhao, B.; Lu, H.; Chen, S.; Liu, J.; Wu, D. Convolutional neural networks for time series classification. *J. Syst. Eng. Electron.* **2017**, *28*, 162–169. [CrossRef]
37. Zheng, Y.; Liu, Q.; Chen, E.; Ge, Y.; Zhao, J.L. Exploiting multi-channels deep convolutional neural networks for multivariate time series classification. *Front. Comput. Sci.* **2016**, *10*, 96–112. [CrossRef]
38. Tanisaro, P.; Heidemann, G. Time series classification using time warping invariant echo state networks. In Proceedings of the 2016 15th IEEE International Conference on Machine Learning and Applications (ICMLA), Anaheim, CA, USA, 18–20 December 2016; pp. 831–836.
39. Serrà, J.; Pascual, S.; Karatzoglou, A. Towards a Universal Neural Network Encoder for Time Series. *arXiv* **2018**, arXiv:1805.03908.
40. Cui, Z.; Chen, W.; Chen, Y. Multi-scale convolutional neural networks for time series classification. *arXiv* **2016**, arXiv:1603.06995.
41. Chawla, N.V.; Bowyer, K.W.; Hall, L.O.; Kegelmeyer, W.P. SMOTE: Synthetic minority over-sampling technique. *J. Artif. Intell. Res.* **2002**, *16*, 321–357. [CrossRef]

Article

Identification of Abnormal Processes with Spatial-Temporal Data Using Convolutional Neural Networks

Yumin Liu [1], Zheyun Zhao [1,*], Shuai Zhang [1] and Uk Jung [2]

1 Business School, Zhengzhou University, Zhengzhou 450001, China; yuminliu@zzu.edu.cn (Y.L.); zhang1227zzu@163.com (S.Z.)

2 Department of Management, School of Business, Dongguk University-Seoul, Seoul 04620, Korea; ukjung@dongguk.edu

* Correspondence: zzu_zzy@126.com

Received: 2 December 2019; Accepted: 2 January 2020; Published: 6 January 2020

Abstract: Identifying abnormal process operation with spatial-temporal data remains an important and challenging work in many practical situations. Although spatial-temporal data identification has been extensively studied in some domains, such as public health, geological condition, and environment pollution, the challenge associated with designing accurate and convenient recognition schemes is very rarely addressed in modern manufacturing processes. This paper proposes a general recognition framework for identifying abnormal process with spatial-temporal data by employing a convolutional neural network (CNN) model. Firstly, motivated by the pasting case study, the spatial-temporal data are transformed into process images for capturing spatial and temporal interrelationship. Then, the CNN recognition model is presented for identifying different types of these process images, leading to the identification of abnormal process with spatial-temporal data. The specific architecture parameters of CNN are determined step by step. According to the performance comparison with alternative methods, the proposed method is able to accurately identify the abnormal process with spatial-temporal data.

Keywords: spatial-temporal data; pasting process; process image; convolutional neural network

1. Introduction

Advanced sensing technologies are being increasingly applied in data collection systems for the areas including public health, geological condition, environment pollution, and manufacturing process. If the output of sensors is represented by the data with space and time structure, it can be termed as spatial-temporal data [1]. A lot of research focuses on the abnormality identification of abnormal-spatial temporal data, such as identifying outliers of the hourly air quality [2], detecting abnormal ozone measurements caused by air pollution or correlation among neighbor sensors [3], and diagnosing whether a disease is randomly distributed over space and time [4]. With the development of manufacturing technology, many sensors have been installed in the production lines, and a large number of spatial-temporal data can be collected from such processes. In order to improve the quality of manufacturing process, the abnormality identification of such spatial-temporal data has attracted much attention. Wang et al. [5] proposed a spatial-temporal data modeling method to identify the abnormality of a wafer production process. The identification scheme developed by Megahed et al. [6] can quickly detect the emergence of a fault in the nonwoven textile production process. Yu et al. [1] presented a rapid spatial-temporal quality control procedure for detecting systematic and random outliers. Current research is being conducted on identifying whether the process with spatial-temporal data is normal or not. Their common objective is to accurately detect the time and location of changes in the occurrence rate as soon as possible [7]. In other words, existing research only focuses on the process monitoring of spatial-temporal data, but it is very rare to consider the root cause of abnormal data. In a real process, however, both normal and abnormal data can be collected from the process, and process monitoring and fault diagnosis can be applied simultaneously with spatial-temporal data. Normal and abnormal data display different variation patterns, which can be observed in process. Hence, how to identify such patterns precisely is the key problem in modern manufacturing process control.

Generally speaking, spatial-temporal data collected from the production process has many observations over time and location, and the adjacent observations are highly correlated. Thus, the high volume and their correlation of spatial-temporal data provide a considerable challenge for the identification of abnormal process. Moreover, the curse of dimensionality and complex data structure makes it difficult to build an identification model. Therefore, dimension reduction techniques are required before using the identification model. To capture intrinsic spatial- and temporal- correlations in an abnormal process state, principal component analysis (PCA), a widely used technique in dimension reduction, can be applied to extract features from spatial-temporal data via unfolding the original data set [8]. However, PCA cannot be directly applied to two or higher dimensional tensor data, unless such data are reshaped into a vector. Because the vectorization operation breaks the spatial and temporal correlation structure, it would lose potentially useful information contained in original data [9]. It is known that analyzing spatial-temporal data is more challenging than analyzing one-dimensional data. To overcome this issue, multilinear PCA (MPCA) and uncorrelated multilinear PCA (UMPCA) are proposed as alternatives to PCA [10]. In these methods, the tensor structure of spatial-temporal data are considered and more effective representation can be extracted [11,12]. Although these methods have desirable performance in processing spatial-temporal data, they focus separately on how to extract useful features from raw data and construct an effective identification model of an abnormal process [13]. If extracted features cannot interpret abnormal processes sufficiently or the identification model does not understand the extracted features, the performance is not robust. Hence, how to propose an effective approach integrating features self-learning and the identification of abnormal process with spatial-temporal data are still a challenge to be overcome.

Convolutional neural networks (CNN), one of the most effective deep learning models for tensor data processing, has been widely applied in natural language processing [14], image recognition [15], electrocardiogram (ECG) analysis [16], and fault diagnosis [17]. Benefiting from the mechanism of CNN in tensor data processing, the correlation structure of spatial-temporal data can be well-preserved by CNN. Meanwhile, CNN does not need to extract abnormal features manually, as the features can be learned from spatial-temporal data hierarchically and automatically. Taking the advantages of CNN, a novel method for identifying abnormal production process with spatial-temporal data is proposed in this paper. The case study considered is a pasting process, which is a critical process in lead-acid battery production and the output of sensors constitute a typical example of spatial-temporal data. Motivated by this process, a CNN-based identification approach for abnormal process with spatial-temporal data are presented. In order to show the recognition accuracy and effect of this approach, UMPCA is used as a benchmark in our study.

This paper is organized as follows. In Section 2, the pasting process as a motivating example is introduced and its spatial-temporal data are acquired. Section 3 develops a general CNN framework for identifying abnormal process with spatial-temporal data. We investigate the validation of the CNN recognition model in Section 4. In Section 5, the CNN method is applied to online identify the abnormal pasting process, and then the performance of the proposed method is evaluated. Suggestions and directions for further research are discussed in conclusions.

2. Case Study: Pasting Process

2.1. Spatial-Temporal Data Acquisition

A lead-acid battery consists of basic cell blocks, and each cell block contains several plates. Plates are the basic components of lead-acid batteries, and unqualified plates will directly affect the initial capacity and cycle life of batteries. In general, plate production includes five processes: ball-milling, paste mixing, grid casting, pasting, and plate curing. Pasting is a critical process in plate production [18,19], and most components of poor quality batteries can be traced back to this process. The identification of abnormal pasting process is the key to ensure the quality of batteries. In the pasting process, the lead oxide pastes are squeezed into the gap between two sides of the grid, and then it is turned into a plate. The mechanism of the whole pasting process is shown in Figure 1. The uniformity of the lead oxide paste in the plate surface is a critical quality characteristic, which can be measured by plate thickness. Therefore, the change of uniformity will directly reflect the abnormal state of the pasting process.

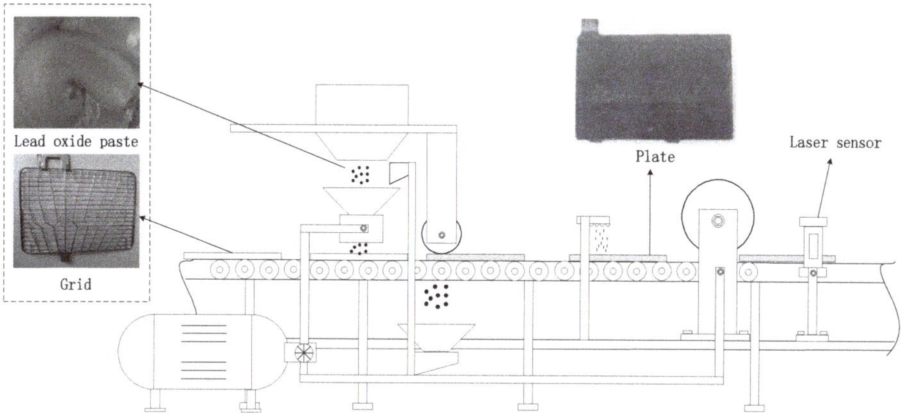

Figure 1. Pasting process.

To obtain plate thickness data, a laser sensor is installed at the end of pasting equipment to collect the observations of the uniformity as shown in Figure 1. When a plate moves in the pasting process, the laser sensor records its thickness values at different locations over time as shown in Figure 2. In the pasting example, there are m locations to measure the uniformity of the plate. When a plate moves through the sensor, the data observed over m locations can be collected at one time. In other words, the uniformity of the plate can be described by the observations measured in the m locations, and the uniformity of different plates can be observed over time to indicate the condition of the current process. Actually, the observation of uniformity is represented as a vector, which can be collected at time t. The vector will become a matrix over time, which can indicate the stability of the pasting process. The matrix collected from the pasting process is bi-dimensional, including space and time dimension. The matrix visualized by a surface is shown in Figure 2.

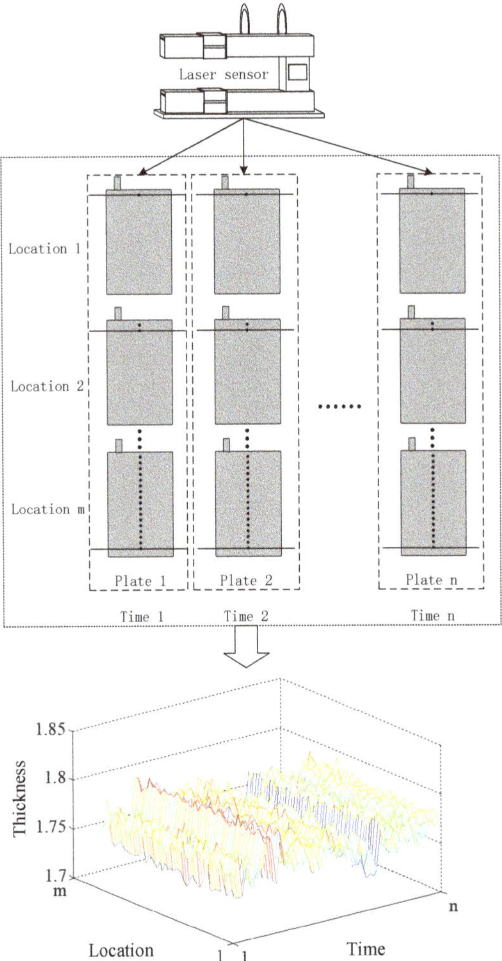

Figure 2. Spatial-temporal data collection in the pasting process.

The abnormal changes of plate thickness in the pasting process often result from unexpected causes, such as the failure of the pasting machine and unqualified grids, which are the root causes of the abnormal process. Once these root causes are identified and removed, the pasting process will return to normal. The plate thickness data change randomly in space and time when the pasting process is running normally, which is referred to as a normal process pattern, F_0, seen in Figure 3a. In general, different causes will lead to various abnormal process patterns, which can be reflected by the changes of plate thickness in space and time domains. For example, when the plate thicknesses have an upward shift, it is usually caused by the wear of the parts in the pasting machine.

According to the changes of spatial-temporal data from pasting process and engineering experience, we find out seven common abnormal process patterns. When the uniformity of lead oxide paste becomes worse, the plate thickness data will change unevenly in time and space, which is denoted by abnormal process pattern F_1, seen in Figure 3b. This abnormal process pattern results from the failure of the compression roller or acid spouting system, such as the blockage of the sprayer in the acid spouting system and aging spring in a compression roller. When the plate thickness is not uniform on both sides, that is, one side is thick and another is thin, the spatial-temporal data of plate thickness at different locations will gradually become steeper with time. This abnormal process pattern is labeled as F_2, and its corresponding causes are unusual clearance between the pasting machine and conveyor belt. When the plate thickness becomes thicker gradually, there is a steady rise over time in the spatial-temporal data, seen in Figure 3d. This pattern, denoted by abnormal process pattern F_3, is caused by the insufficiency of conveyor belt tension under the pasting machine. When the thickness uniformity of plates becomes worse in a sudden way, the spatial-temporal data will become nonuniform suddenly, which is denoted by abnormal process pattern F_4, as seen in Figure 3e. In general, this situation could be attributed to the low strength of the steel in a new batch of grids. When the thickness between the two sides of plates becomes nonuniform suddenly, the spatial-temporal data of plate thickness at one side will step up and become steep suddenly, which can be seen in Figure 3f. This situation is denoted by abnormal process pattern F_5 and its corresponding cause is usually that the roller in the pasting machine slants to one side. When the plate thickness becomes thicker suddenly, the overall upward step will be shown in the spatial-temporal data of plate thickness, seen in Figure 3d. The abnormal process pattern F_6 could be used as the label of this situation, which results from the electromagnetic fault in the pressure machine of compressed air. When the thickness of plates become thicker in a periodic manner, a periodical change will be observed in spatial-temporal data, seen in Figure 3h. This type of situation is labeled as abnormal process pattern F_7, and the corrosion status of pasting conveyor belt is usually required to be checked by engineers. All types of abnormal process patterns are shown in Figure 3.

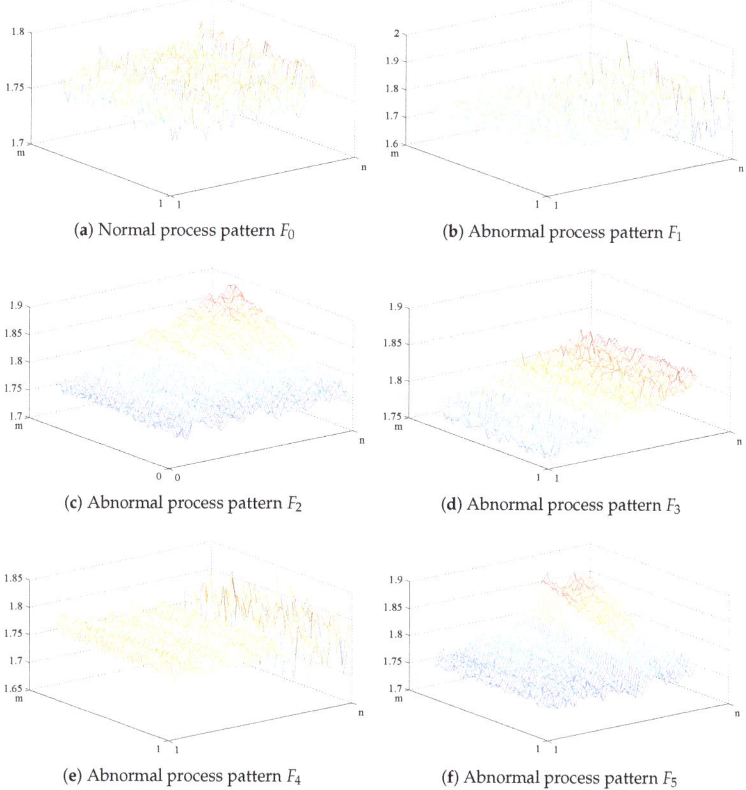

(**a**) Normal process pattern F_0

(**b**) Abnormal process pattern F_1

(**c**) Abnormal process pattern F_2

(**d**) Abnormal process pattern F_3

(**e**) Abnormal process pattern F_4

(**f**) Abnormal process pattern F_5

Figure 3. *Cont.*

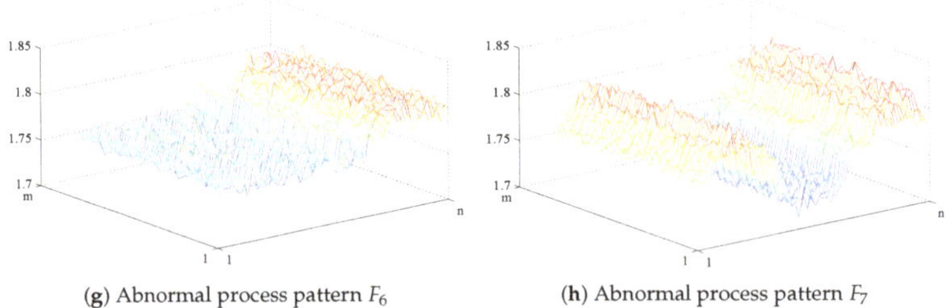

(g) Abnormal process pattern F_6 (h) Abnormal process pattern F_7

Figure 3. Normal and abnormal process patterns in the pasting process.

From Figure 3, it can be observed that the obvious shape differences among the spatial-temporal data of plate thicknesses show different abnormal process patterns. Therefore, the identification of abnormal process with spatial-temporal data are converted into the problem of identifying these abnormal process patterns. Once a certain abnormal process pattern is identified, its corresponding root causes could be found out simultaneously.

2.2. Abnormal Process Image Collection

Although the abnormal process patterns describe the abnormal states of the process with spatial-temporal data very well, high dimensionality, spatial-temporal correlation, and a high amount of noise make it difficult to identify them directly. Grayscale image can visually capture important abnormal patterns of observation data without any parameters predefined by users' experiences [20]. In order to effectively distinguish normal and seven abnormal process patterns, spatial-temporal data from the pasting process can be transformed into grayscale images, which is called process image. Given a spatial-temporal data matrix X, $x_{i,j}$ refers to a measured value, where i and j represent the location and time of the spatial-temporal data matrix, respectively. For generating the process image of the spatial-temporal data, each grayscale can be calculated from data matrix X by normalizing, multiplying 255 and taking integer. The transformation formula is given by:

$$y_{i,j}^{(0)} = rounding(\frac{x_{i,j} - Min(X)}{Max(X) - Min(X)} \times 255), \quad (1)$$

where $y_{i,j}^{(0)}$ is the grayscale value corresponding to $x_{i,j}$, $rounding$ is the function for taking integer, and $Min(X)$ and $Max(X)$ are used for extracting maximum and minimal elements from matrix X. Thus, the spatial-temporal data have been represented as a process image to visualizing the operation status of the manufacturing process. Process images of normal and seven abnormal process patterns discussed above are shown in Figure 4. For the convenience, they are still denoted as F_0, F_1, F_2, F_3, F_4, F_5, F_6, and F_7.

In the grayscale image of a normal process pattern, the pixels are arranged randomly without obvious change. Because the spatial and temporal data of plate thickness varies unevenly, in the grayscale image F_1, some pixels are white and some are dark. When the spatial-temporal data of plate thickness becomes steeper gradually in F_2, the pixels of the steep data gradually appear to be white in the process image F_2. The important features of other abnormal process patterns can be directly represented in their process images, as shown in Figure 4. From the above discussion, it can be observed that each process image can capture the trend and variation features of its corresponding pattern shown in Figure 3. Therefore, the problem of identifying abnormal process patterns is converted into detecting abnormal process images. How to use an effective recognition method of the abnormal process image is a challenge in this paper.

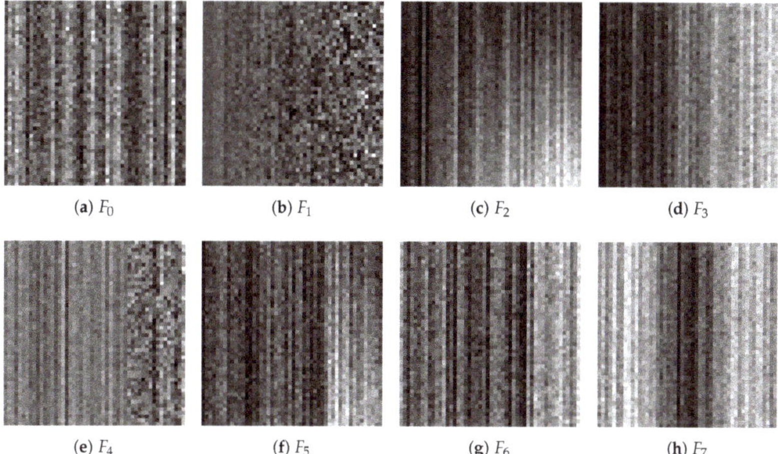

(**a**) F_0 (**b**) F_1 (**c**) F_2 (**d**) F_3

(**e**) F_4 (**f**) F_5 (**g**) F_6 (**h**) F_7

Figure 4. Normal and abnormal process images in the pasting process.

3. CNN Framework for Process Images Recognition

3.1. Architecture of the CNN Model

Convolutional neural network (CNN) is a kind of feed-forward artificial neural network (ANN), which is inspired by biological vision from pixel to abstract feature [21]. Unlike the traditional neural networks that need to concatenate raw data into a vector, CNN can directly deal with image data. Taking this advantage of CNN, the recognition model for process image will be constructed in this paper. A CNN recognition model is comprised of an input layer, several convolution layers and pooling layers, fully-connected (FC) layers, and an output layer, as illustrated in Figure 5. The input layer is to import process images for detection. The convolution layer and pooling layer are used to extract abnormal information from process images. The fully-connected layer serves as the integration of abnormal process information. The output layer is to provide the categories of abnormal process images.

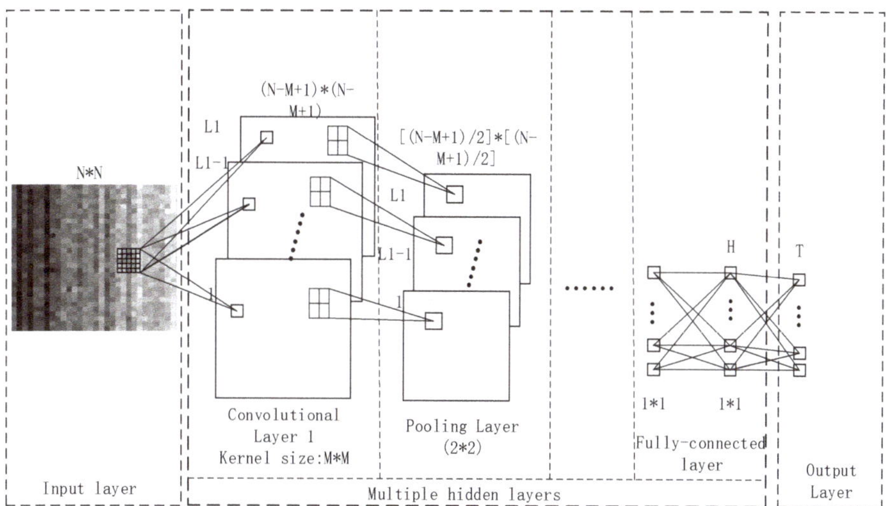

Figure 5. The general architecture of convolutional neural network (CNN) model for the process image.

A CNN model usually consists of several convolution and pooling layers. For convenience, suppose that there are R convolution and pool layers in CNN model and a process image is a square with $N \times N$ size, seen in Figure 5. Convolution layer is a most important component in the CNN model, which assigns weights to the grayscales of the input image by a convolution kernel, so as to extract process abnormal features.

3.2. Underlying Mechanism of the CNN Model

In the first convolution layer, suppose that there are L_1 convolution kernels. The $(w_{k,i,j}^{(1)})_{M\times M}$, $(k = 1, 2, \ldots L_1; i = 1, 2, \ldots, M; j = 1, 2, \ldots, M)$ is used to represent the kth convolution kernel with $M \times M$ size $(M < N)$, where $w_{k,i,j}^{(1)}$ means the weight value at row i and column j of the kth kernel. In order to get the convolution results, a moving window with $M \times M$ size should be set up. The window moves one stride at a time, where the area formed is called a receptive field. When all pixels in the process image are covered by the moving window, $(N - M + 1) \times (N - M + 1)$ receptive fields can be obtained. For all the receptive fields and convolution kernels in the first layer, the convolution result can be obtained by the following:

$$\begin{aligned} & y_{k,i,j}^{(1)} = f(\sum_{s=1}^{M}\sum_{t=1}^{M} w_{k,s,t}^{(1)} \cdot y_{i+s-1,j+t-1}^{(0)} + b_k^{(1)}), \\ & i \; and \; j = 1, \ldots, (N - M + 1), \\ & s \; and \; t = 1, \ldots, M, \\ & k = 1, \ldots, L_1, \end{aligned} \tag{2}$$

where $y_{k,i,j}^{(1)}$ is the convolution output of the current receptive field for the kth convolution kernel, $w_{k,s,t}^{(1)}$ is the weight element at row s and column t of the kth convolution kernel, $y_{i+s-1,j+t-1}^{(0)}$ is the grayscale in current receptive field, and $b_k^{(1)}$ $(k = 1, 2, \ldots L_1)$ represents the bias value. $f()$ is RuLU activation function [22]. For the kth convolution kernel, all of the $y_{k,i,j}^{(1)}$ values from receptive fields formed a matrix, which is called a feature map corresponding to the kth convolution kernel $(k = 1, 2, \ldots L_1)$. The feature maps of the first convolutional layer will be input to the pooling layer.

The pooling layer is mainly used to compress feature maps and obtain condensed feature maps via pooling function. In the first pooling layer, L_1 feature maps are entered respectively. Each feature map is partitioned into several non-overlapping square areas, which is known as pooling fields. In general, every pooling field with 2×2 size is preferred [23]. To get the pooling results, maximum pooling and average pooling are widely used pooling functions [24]. For the kth feature map, the pooling results, also known as condensed feature map, are obtained. The pooling value at row i and column j of the kth condensed feature map, $y_{k,i,j}^{(2)}$, can be computed by following equation:

$$\begin{aligned} & y_{k,i,j}^{(2)} = pooling(y_{k,2i-1,2j-1}^{(1)}, y_{k,2i-1,2j}^{(1)}, y_{k,2i,2j-1}^{(1)}, y_{k,2i,2j}^{(1)}), \\ & i = 1, \ldots, \frac{N - M + 1}{2}, j = 1, \ldots, \frac{N - M + 1}{2}, \end{aligned} \tag{3}$$

where $y_{k,2i-1,2j-1}^{(1)}, y_{k,2i-1,2j}^{(1)}, y_{k,2i,2j-1}^{(1)}, y_{k,2i,2j}^{(1)}$ are four values in the current pooling field connecting to $y_{k,i,j}^{(2)}$, and $pooling$ refers to pooling function. After pooling operation, L_1 condensed feature maps are obtained, which will be input into the next convolution layer.

Similarly, the above work is repeated on several alternative convolution and pooling layers until the last pool layer. The condensed feature maps of the last pooling layer will be unfolded and imported to the Fully-connected (FC) layer. Before fully-connected operation, L_R condensed feature maps are unfolded to a vector, y_i^{FC}, $i = 1, \ldots, L_R \times q \times q$. Suppose there are H nodes in the FC layer, and, for the hth node, the weights connected to the ith input nodes can be denoted as $w_{i,h}^{(FC)}$ $(i = 1, 2, \ldots, L_R \times q \times q, h = 1, 2, \ldots, H)$. The value of the hth node in the FC layer can be computed as follows:

$$y_h^F = f(\sum_{i=1}^{L_R \times q \times q} y_i^{(FC)} \times w_{i,h}^{(FC)} + b_h^{(F)}), h = 1, \ldots, H, i = 1, \ldots, L_R \times q \times q, \tag{4}$$

where y_h^F is the output the hth node in FC layer, $b_h^{(F)} (h = 1, 2, \ldots H)$ represents the bias of the hth node, and $f()$ is ReLU activation function. In the output layer, the connection between the hth FC node and the jth output node is represented by $w_{hj}^{(O)}$, $(h = 1, 2, \ldots H; j = 1, 2, ..T)$. To classify the input data, the probability output of the jth output node is required to be computed as follows:

$$P_j = f(\sum_{h=1}^{H} y_h^{(F)} \times w_{hj}^{(O)} + b_j), h = 1, \ldots, H, j = 1, \ldots, T, \tag{5}$$

where T is the number of output nodes, P_j is the result of the jth node in the output layer, and b_j is the bias value of the jth output nodes. $f()$ is the normalized exponential function, through which the probability can be obtained:

$$f(y_j) = \frac{e^{y_j}}{\sum_{j=1}^{T} e^{y_j}}, j = 1, \ldots, T. \tag{6}$$

As discussed above, the weights and biases in each layer of the CNN model can directly affect the final results of the output layer. If the difference between the actual output and the expected output is too large to be accepted, the weights and biases are required to be updated. The difference can be measured by the following loss function:

$$F = -\frac{1}{T} \sum_{j=1}^{T} [P_j^* ln(P_j) + (1 - P_j^*) ln(1 - P_j)], \tag{7}$$

where T refers to the total number of trained categories, P_j^* and P_j are the expected output and the actual output, respectively. A small loss value means an accurate probability output. To reduce the loss value, The back-propagation (BP) algorithm [25] is utilized. For convenience, the weights and biases in all layers are referred to as w and b, respectively. Thus, F is the loss function of w and b. The partial derivatives of F for w and b can be expressed as follows:

$$\begin{aligned} w_t &= w_{t-1} - \eta \frac{\partial F}{\partial w}, \\ b_t &= b_{t-1} - \eta \frac{\partial F}{\partial b}, \end{aligned} \tag{8}$$

where w_t and b_t represent the updated weights and biases after the tth iteration. η is the learning rate, and $\eta = 0.01$ is a common selection [26]. The CNN architecture parameters consist of the numbers of convolution layers and convolution kernels, the size of convolution kernels per layer, the selection of pooling function, and the number of output nodes in the FC layer. These parameters are required to be stepwise determined, which will be discussed in the next section.

3.3. CNN Identification Framework

A CNN framework for the identifying abnormal process with spatial-temporal data is presented in this paper. This framework can be divided into two phases: offline learning and online identifying. The offline learning phase aims to train a CNN model from offline collected process images and establish an appropriate CNN recognition model. In the online identifying phase, this CNN model trained in the offline phase can be applied to identify the abnormal process with real-time spatial-temporal data. The CNN identification framework is shown in Figure 6, and the details are introduced in the following.

There are two main steps in offline learning: The first step is to obtain training samples, which consists of process images and their corresponding categories. Process images are converted by spatial-temporal data and their corresponding categories can be obtained according to engineering experience. The second step is to determine the architecture parameters of CNN, and update the weights and biases from the convolution, pooling, and FC layer of the CNN using the BP algorithm. After offline learning, the CNN recognition model will be obtained.

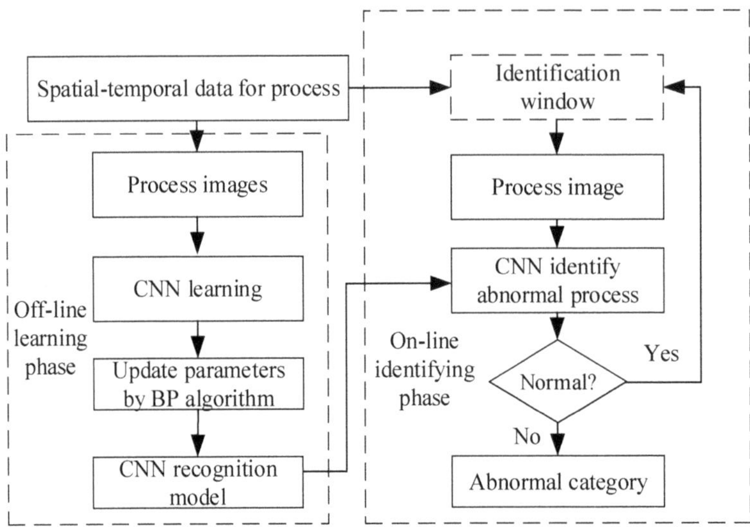

Figure 6. The CNN identification framework.

After the offline learning phase, the CNN recognition model is established and applied to online recognition of the abnormal process with spatial-temporal data. Two main steps in the online identifying phase are as follows: The first step is to collect the real-time spatial-temporal data from a process through the moving identification window. The size of the window should be determined by the product processing time, so that the spatial-temporal data in the current window can be mapped into a suitable process image. The second step is to identify the abnormal process images. The process images in the current identification window will be recognized by the CNN recognition model. If the decision result made by the CNN recognition model is a normal process image, the sliding window will move forward to collect new observations until an abnormal process image is identified.

According to the above two phases, the abnormal process with spatial-temporal data can be identified in a real-time way. It benefits from the powerful learning ability of the CNN recognition model.

4. Validation Results

The validation of the proposed approach depends on the architecture parameters of the CNN model, which will be investigated by the spatial-temporal data collected from the pasting process in this section. To evaluate the ability of detecting process images under various architectures, the recognition accuracy of test data (ATD) is used for the evaluation index, which is a ratio of the number of samples recognized correctly and the total number of samples. When an abnormal pattern happens in the process, the CNN model is expected to identify the abnormal type as precisely as possible. Meanwhile, when the process is normal, the CNN model is required to recognize the normal pattern more accurately, so that the higher ATD indicates the better validity of the CNN model. The following validation analysis consists of the determination of layer and node number, and selection of the kernel size and pooling function. In addition, 700 process images with 50 × 50 size are collected from the pasting process under each process pattern separately, where 200 and 500 process images are selected randomly as the training and test samples, respectively. Training data are applied to learn the CNN recognition model with different architectures, and test data are used to compare their recognition accuracy to find the best one, which is conducted in Caffe, Ubuntu 16.04 with a Tesla K80 GPU.

4.1. Determination of Layer and Node Numbers

The number of layers and nodes discussed here contains the numbers of convolutional layers, convolution kernels per layer, and FC nodes. To achieve the optimal performance of CNN model, the number of convolution layers and kernels per layer are important parameters required to be specified in advance [27]. Adding more convolutional layers and kernels to the network could capture high-level features of input data at the price of making the model complex to train [28]. Thus, to obtain an optimal network architecture, the numbers of convolutional layers and convolution kernels will be determined layer by layer.

For the process image size of 50 × 50 size, the size of a condensed feature map will be 2 × 2 after four convolution and pooling operations, which cannot be further convolved. Thus, here we consider the CNN model with 4 convolution layers. Twenty scenarios for the numbers of convolution kernels from 5 to 100 are tested via recognition accuracy. The average accuracy and standard deviation are used to measure the performance of the proposed CNN model. To evaluate recognition accuracy, all the results of proposed method are replicated at least 100 times, and the results are shown in Table 1. By comparing different scenarios in the first convolution layer, the highest average recognition accuracy can reach 95.18% when the number of convolution kernels is 65. Considering the same scenarios in the second convolution layer, the number of convolution kernels is set to 70 because the highest average accuracy is 96.36%. In a similar way, the optimal numbers of convolution kernels can be determined layer by layer, shown in Table 1. It is expected that the recognition accuracy will increase until the optimal number of convolution layer is found out.

Generally, the recognition accuracy will be improved with the increase of convolution layers. However, we can see that the recognition accuracy begins to decrease from the third convolution layer in Table 1. In this situation, it can be inferred that the optimal number of convolutional layer is 3, and the corresponding number of convolution kernels is 65, 70, and 80 respectively, which is denoted as 65-70-80.

Under the 65–70–80 structure of the convolutional layer, the number of nodes in the FC layer needs to be determined. Ten scenarios for the numbers of FC nodes from 100 to 1000 are shown in Figure 7. By comparing the average recognition accuracy and standard deviation in different scenarios, we find that the accuracy increases to 98.08% at 800 nodes, and, after that, the accuracy does not exceed 98.08% by adding more nodes. Additionally, we observe that the variation of recognition accuracy becomes lower as the number of nodes in FC layer increases. From Figure 7, it is noted that the difference of recognition accuracy using various FC node numbers is not obvious enough, especially using 600 and 800. In this research, the node number corresponding to higher recognition accuracy and lower variation is preferred. Because the recognition accuracy using 800 nodes is slightly better than using 600, and the optimal number of FC nodes is set to 800. For the above discussion, the optimal architecture of CNN model is denoted as 65–70–80–800.

Table 1. The recognition accuracy comparison among different number of layers and convolution kernels. The boldface entries represent the highest average accuracy under the current layer. Numbers in parentheses are standard deviation of recognition accuracy.

The Number of Convolution Kernels	Accuracy of Test Data (%)			
	1st Convolution Layer	2nd Convolution Layer	3rd Convolution Layer	4th Convolution Layer
5	94.16 (0.232)	95.56 (0.253)	96.45 (0.225)	96.32 (0.336)
10	94.37 (0.145)	95.90 (0.238)	96.96 (0.169)	96.68 (0.278)
15	94.67 (0.117)	96.01 (0.146)	97.12 (0.187)	96.70 (0.168)
20	94.76 (0.156)	96.21 (0.154)	97.18 (0.119)	96.83 (0.165)
25	94.78 (0.110)	96.12 (0.162)	97.21 (0.172)	96.97 (0.157)
30	94.86 (0.118)	96.12 (0.102)	97.28 (0.123)	97.03 (0.146)
35	94.80 (0.114)	96.23 (0.162)	97.31 (0.168)	97.08 (0.137)
40	94.92 (0.147)	96.25 (0.166)	97.35 (0.158)	**97.25 (0.148)**
45	94.98 (0.120)	96.22 (0.116)	97.40 (0.162)	97.10 (0.151)
50	94.97 (0.116)	96.24 (0.168)	97.01 (0.128)	97.02 (0.139)
55	95.05 (0.100)	96.25 (0.081)	97.42 (0.144)	96.89 (0.155)
60	95.10 (0.115)	96.30 (0.104)	97.66 (0.201)	96.82 (0.187)
65	**95.18 (0.112)**	96.28 (0.336)	97.61 (0.159)	96.82 (0.103)
70	95.12 (0.126)	**96.36 (0.120)**	97.83 (0.173)	96.69 (0.098)
75	95.09 (0.098)	96.14 (0.064)	97.91 (0.139)	96.64 (0.111)
80	95.10 (0.102)	96.32 (0.149)	**98.00 (0.106)**	96.69 (0.132)
85	95.11 (0.121)	96.33 (0.105)	97.78 (0.092)	96.76 (0.139)
90	95.14 (0.133)	96.30 (0.071)	97.43 (0.094)	96.53 (0.167)
95	95.16 (0.075)	96.33 (0.090)	97.39 (0.122)	96.50 (0.143)
100	95.14 (0.116)	96.34 (0.110)	97.36 (0.118)	96.31 (0.121)

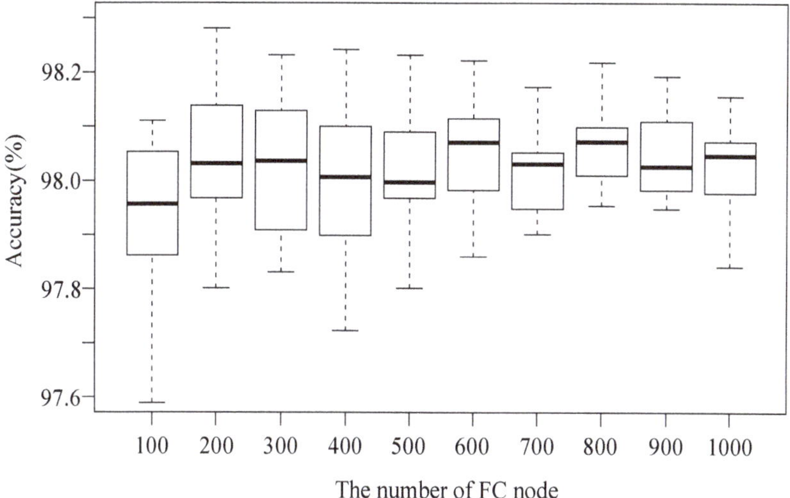

Figure 7. The accuracy comparison among fully-connected (FC) node numbers.

4.2. Selection of Kernel Size and Pooling Function

The sizes of convolution kernels and pooling function are also important architecture parameters of a CNN model. The size of convolution kernels will be first discussed. In general, a convolution kernel with a small size can capture details of abnormal information from process images. Thus, under the optimal architecture, 65–70–80–800, the CNN recognition models with kernel sizes 3×3, 5×5, and 7×7 are considered respectively, seen in Table 2. The results of the CNN model with three kernel sizes are obtained and the optimal kernel size can be determined as 3×3. Then, the optimal pooling function will be selected in the following.

Table 2. The recognition accuracy comparison for kernel sizes. The boldface entries represent the highest accuracy.

Architecture	Kernel Size	Average Accuracy of Test Data (%)
65-70-80-800	3×3	**98.08**
65-70-80-800	5×5	97.27
65-70-80-800	7×7	96.58

Like the above discussion, the max-pooling and average-pooling shown as Equations (4) and (5) are widely used pooling functions. The comparison of recognition accuracy for two pooling functions is shown in Table 3. The max-pooling function is the best one, which can be used in the CNN recognition model.

Table 3. The recognition accuracy comparison for pooling functions. The boldface entries represent the highest accuracy.

Architecture	Kernel Size	Pooling Function	Accuracy of Test Data (%)
65-70-80-800	3×3	Max pooling function	**98.08**
65-70-80-800	3×3	Average pooling function	92.29

To summarize, the CNN recognition model for abnormal pasting process images shown in Table 4 has been constructed, and its validation has been proved.

Table 4. The architecture parameters of the proposed convolutional neural network (CNN) recognition model for the pasting process.

The Number of Convolution Layers	The Number of Kernels	Kernel Size	Pooling Function	The Number of Nodes in FC Layer
3	65-70-80	3×3	Max pooling function	800

5. Performance Comparison

In order to demonstrate the performance of the proposed method, the abnormal pasting process with spatial-temporal data will be identified in this section. The plate thickness data used for performance comparison are obtained from the pasting machine. After data cleaning and clustering analysis, the data of normal pattern and abnormal patterns is extracted and converted to process images. Each type of process patterns includes 200 and 500 process images with 50×50 size for training and testing respectively.

The abnormal pasting process images are online recognized using the constructed CNN recognition model. In the real pasting process, because the processing time of a plate is 0.5 s, the plate thickness observations at 50 locations in 25 s can be mapped to a process image with 50×50. Thus, the size of the moving identification window is set to 50×50. Taking 0.5 s as a stride length, once a new observation is obtained, the sliding window will move one step. As the window slides, the process images formed by the first 64 observations change smoothly and steadily, in which the output results of the CNN model are normal process images, shown in Figure 8. When the window moves to observation 115, the probability output is (0, 0, 0.005, 0.012, 0.003, 0.007, **0.973**, 0), which indicates that the abnormal process pattern F_6 happens in the pasting process. The corresponding cause is insufficient conveyor belt tension caused by the electromagnetic fault of the air pressure machine. When the root cause is removed, the plate thickness comes back to the target value nearby. When the identification window moves to the 199th observation, the probability output of CNN model is (0.011, 0.027, 0.02, 0, **0.934**, 0.008, 0, 0), which means that the abnormal pattern F_4 is detected. After checking the current process, we find that the plate flatness changes suddenly, which results from the low steel strength of new grids. After replacing this batch of grids with other qualified grids, the process returns to normal. When observation 281 enters the moving window, the probability output is (0.002, 0.008, 0, **0.9825**, 0, 0.003, 0.0045, 0). Adjusting the levelness of the pasting machine makes the process go back to normal.

From the above application, we can conclude that the proposed method has practicality to the online identification of abnormal process with spatial-temporal data. To further evaluate the performance of proposed CNN recognition model, the UMPCA based recognition method [11] as a benchmark method is considered for comparison. In this UMPCA based method, Bayes Classifier (BC) is utilized to achieve identification, which is denoted as UMPCA-BC. After 100 experiments, the comparison results are obtained and shown in Figure 9.

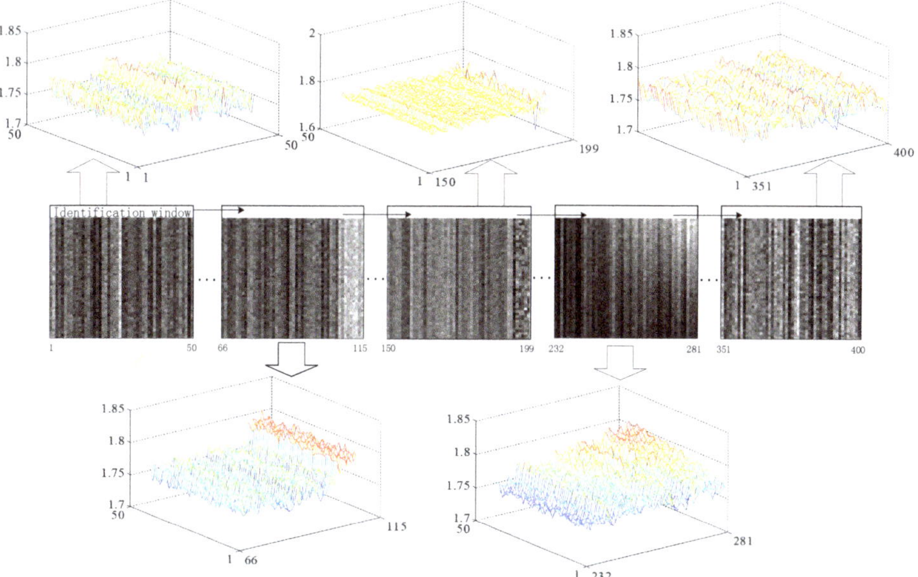

Figure 8. Online identification for the pasting process.

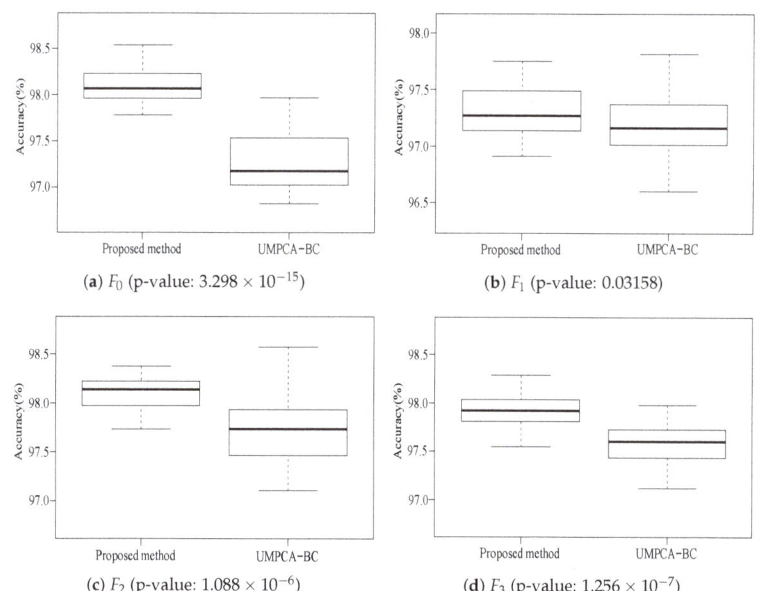

(**a**) F_0 (p-value: 3.298×10^{-15}) (**b**) F_1 (p-value: 0.03158)

(**c**) F_2 (p-value: 1.088×10^{-6}) (**d**) F_3 (p-value: 1.256×10^{-7})

Figure 9. *Cont.*

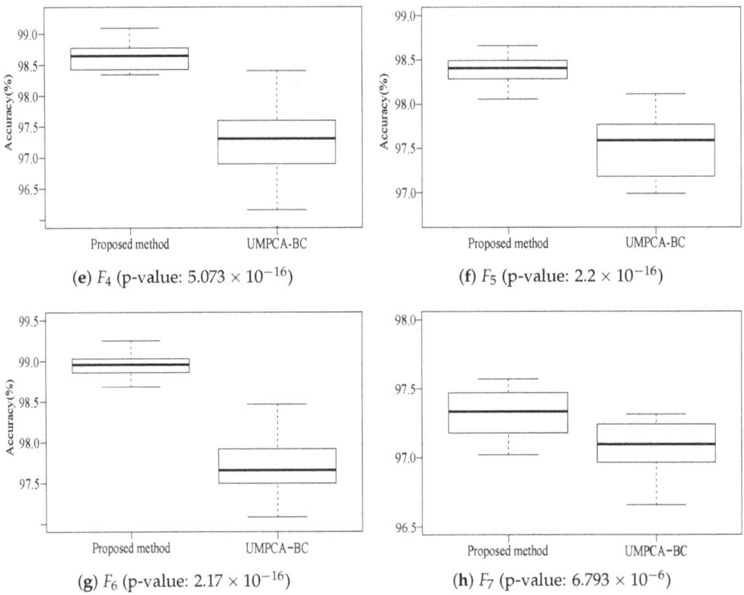

(**e**) F_4 (p-value: 5.073×10^{-16}) (**f**) F_5 (p-value: 2.2×10^{-16})

(**g**) F_6 (p-value: 2.17×10^{-16}) (**h**) F_7 (p-value: 6.793×10^{-6})

Figure 9. Performance comparison. Numbers in parentheses are the *p*-values of the Mann–Whitney U test.

From Figure 9, in all process patterns, the proposed method is better than UMPCA-BC, though only slightly better at F_1. To further test whether the proposed method outperforms the alternative method significantly in all process patterns, Mann–Whitney U tests are conducted, and their *p*-values are also presented in Figure 9. If the 1% level of significance is taken, the superiority of the proposed model is significant except F_1. There are two reasons for this situation. Firstly, the rounding operation of the grayscale image makes the F_1 process lose detail information, which further makes our model not sensitive to a small shift in abnormal process pattern F_1. Secondly, in practical production, the failure of the acid spouting system does have a great impact on the thickness uniformity of plates in a short time. In other words, the variation of thickness uniformity in F_1 is usually small in a limited size of identification window, which increases the similarity between normal pattern and F_1 pattern. As a result, the performance of the proposed method has a limited improvement at F_1. However, in general, the CNN method proposed in this paper can identify the abnormal process with spatial-temporal data more accurately than the UMPCA-BC method.

To further verify the reliability of the proposed model, sensitivity analysis is carried out and the effect of noise on performance is studied. The white Gaussian noise is generated and added to the original test data as follows:

$$x_{i,j}^{noise} = x_{i,j} + g, g \sim N(0, \sigma^2), \quad (9)$$

where $x_{i,j}$ is the value at row i and column j of original data matrix, and g refers to the noise that obeys normal distribution with a mean of zero and a standard deviation of σ. In this example, the specification limits of plate thickness are 1.75 ± 0.02, and only the values of σ at 0 to 0.02 are considered. For convenience, five scenarios, including 0, 0.005, 0.01, 0.015, and 0.002 are implemented, shown in Table 5.

Table 5. The performance comparison for various noise levels.

Noise Level (σ)	Average Recognition of Test Data (%)	
	Proposed Method	Uncorrelated Multilinear Principal Component Analysis and Bayes Classifier (UMPCA-BC)
0	98.08	97.41
0.005	97.17	95.98
0.01	96.01	95.13
0.015	94.24	93.46
0.02	93.44	92.73

The recognition accuracy of the proposed method and UMPCA-BC is inevitably decreased with an increase of σ. However, from Table 5, the proposed method still outperforms UMPCA-BC with an increasing noise level. Therefore, the CNN method proposed in this paper can identify the abnormal process with spatial-temporal data with better results.

6. Limitations of the Proposed Methodology

Some aspects may limit the application and assessment of the proposed framework, such as the following ones:

- All the process images are generated by normalizing, multiplying by 255, and rounding, which may result in some information loss. How to measure the impact of missing information on CNN model performance is worthy of attention.
- In this study, the recognition accuracy as an evaluation criterion of performance is not very comprehensive. In practice, there are other metrics and representations used to evaluate the performance of CNN model, such as training time and test data loss. It is an interesting topic to evaluate and optimize CNN model based on other metrics.
- Generally, the operation state of manufacturing process is also affected by other unexpected factors, such as the new abnormal mode caused by an unknown fault. This situation will affect the performance of the current recognition model. Therefore, it is necessary to update the current data and build the recognition model again to consider the new pattern.

7. Conclusions and Further Work

This paper develops a general framework based on the CNN model to detect the abnormal pattern and diagnose the causes in the pasting process with spatial-temporal data. Different from traditional schemes, the main contribution of our proposed framework makes full use of both normal and abnormal information from historical data, and it overcomes the dilemma of multiple data types in real applications. The proposed model is tested on the example of the pasting process and achieved a better recognition performance than the alternative method. Experimental results demonstrated that better performance can be achieved at all abnormal process patterns in the pasting process. In addition, the sensitivity analysis of noise is also provided to verify the superiority of the proposed method. In addition, the procedure for constructing the recognition model is convenient. Our proposed CNN recognition model shows the good potential of online monitoring and tracing the root cause simultaneously. Benefiting from the CNN model, the spatial and temporal interrelationship of abnormal information can be captured and all the historical information can be utilized by the proposed CNN model.

However, there are two outstanding issues on this topic. First, although this paper focuses on the pasting process, the CNN recognition framework we proposed could be applied to any other abnormal process monitoring and diagnosis, where the observations are spatial-temporal data. In order to improve the performance of the CNN recognition model, the CNN model should be investigated further to make the proposed framework more suitable for other general situations.

Second, the parameter optimization of the CNN model is a challenging work in the deep learning domain. However, it is not our concern in current work, thus the parameters of the CNN model only for the pasting process are determined in our paper. In fact, the architecture parameters are related to the data type, shape of the abnormal patterns, and the number of the data sample, thus more suggestions for determining proper parameters are needed. Some other advanced parameter optimization techniques should be added to the CNN framework to improve the recognition accuracy further. Therefore, future improvements can be conducted in the following ways. First, this framework can be modified to identify the process of other general situations by using the transfer learning method. Second, other advanced techniques for hyper-parameters optimization can be studied further to replace the manual method, such as heuristic search algorithms and design of experiments' techniques.

Author Contributions: Conceptualization, Z.Z. and Y.L.; methodology, Z.Z.; software, S.Z.; validation, Y.L.; formal analysis, Y.L. and Z.Z.; investigate Z.Z.; resources, Y.L.; data curation, Z.Z.; writing—original draft preparation, Z.Z.; writing—review and editing, Y.L.; visualization, Z.Z.; supervision, U.J.; project administration, Y.L.; funding acquisition, Y.L. All authors have read and agreed to the published version of the manuscript.

Funding: This research was funded by the National Natural Science Foundation of China Grant No. 71672182, U1604262, U1904211, and 71672209.

Conflicts of Interest: The authors declare no conflict of interest.

References

1. Yu, Y.; Workman, A.; Grasmick, J.G.; Mooney, M.A.; Hering, A.S. Space-time outlier identification in a large ground deformation data set. *J. Qual. Technol.* **2018**, *50*, 431–445. [CrossRef]
2. Michel, B.; Michel, M.; Yves, M.; Jean-Michel, P.; Bruno, P. Spatial outlier detection in the PM10 monitoring network of Normandy (France). *Atmos. Pollut. Res.* **2015**, *6*, 476–483. [CrossRef]
3. Harrou, F.; Kadri, F.; Khadraoui, S.; Sun, Y. Ozone measurements monitoring using data-based approach. *Process Saf. Environ. Prot.* **2016**, *100*, 220–231. [CrossRef]
4. Kulldorff, M. Prospective time periodic geographical disease surveillance using a scan statistic. *J. R. Stat. Soc. Ser. A Statistics Soc.* **2001**, *164*, 61–72. [CrossRef]
5. Wang, A.; Wang, K.; Tsung, F. Statistical surface monitoring by spatial-structure modeling. *J. Qual. Technol.* **2014**, *46*, 359–376. [CrossRef]
6. Megahed, F.M.; Wells, L.J.; Camelio, J.A.; Woodall, W.H. A spatiotemporal method for the monitoring of image data. *Qual. Reliab. Eng. Int.* **2012**, *28*, 967–980. [CrossRef]
7. Tsui, K.L.; Wong, S.Y.; Jiang, W.; Lin, C.J. Recent research and developments in temporal and spatiotemporal surveillance for public health. *IEEE Trans. Reliab.* **2011**, *60*, 49–58. [CrossRef]
8. Colosimo, B.M.; Pacella, M. On the use of principal component analysis to identify systematic patterns in roundness profiles. *Qual. Reliab. Eng. Int.* **2007**, *23*, 707–725. [CrossRef]
9. Ye, J.; Janardan, R.; Li, Q. GPCA: An efficient dimension reduction scheme for image compression and retrieval. In Proceedings of the Tenth ACM SIGKDD International Conference on Knowledge Discovery and Data Mining, Seattle, WA, USA, 22–25 August 2004; pp. 354–363.
10. Lu, H.; Plataniotis, K.N.; Venetsanopoulos, A.N. Uncorrelated multilinear principal component analysis for unsupervised multilinear subspace learning. *IEEE Trans. Neural Netw.* **2009**, *20*, 1820–1836.
11. Paynabar, K.; Jin, J.; Pacella, M. Monitoring and diagnosis of multichannel nonlinear profile variations using uncorrelated multilinear principal component analysis. *Iie Trans.* **2013**, *45*, 1235–1247. [CrossRef]
12. Pacella, M. Unsupervised classification of multichannel profile data using PCA: An application to an emission control system. *Comput. Ind. Eng.* **2018**, *122*, 161–169. [CrossRef]
13. Zhang, L.; Gao, H.; Wen, J.; Li, S.; Liu, Q. A deep learning-based recognition method for degradation monitoring of ball screw with multi-sensor data fusion. *Microelectron. Reliab.* **2017**, *75*, 215–222. [CrossRef]
14. Dahl, G.E.; Yu, D.; Deng, L.; Acero, A. Context-dependent pre-trained deep neural networks for large-vocabulary speech recognition. *IEEE Trans. Audio Speech Lang. Process.* **2011**, *20*, 30–42. [CrossRef]
15. He, K.; Zhang, X.; Ren, S.; Sun, J. Spatial pyramid pooling in deep convolutional networks for visual recognition. *IEEE Trans. Pattern Anal. Mach. Intell.* **2015**, *37*, 1904–1916. [CrossRef]

16. Kiranyaz, S.; Ince, T.; Gabbouj, M. Real-time patient-specific ECG classification by 1D convolutional neural networks. *IEEE Trans. Biomed. Eng.* **2015**, *63*, 664–675. [CrossRef]
17. Wen, L.; Li, X.; Gao, L.; Zhang, Y. A new convolutional neural network-based data-driven fault diagnosis method. *IEEE Trans. Ind. Electron.* **2017**, *65*, 5990–5998. [CrossRef]
18. Brik, K.; ben Ammar, F. Causal tree analysis of depth degradation of the lead acid battery. *J. Power Sources* **2013**, *228*, 39–46. [CrossRef]
19. Schiffer, J.; Sauer, D.U.; Bindner, H.; Cronin, T.; Lundsager, P.; Kaiser, R. Model prediction for ranking lead-acid batteries according to expected lifetime in renewable energy systems and autonomous power-supply systems. *J. Power Sources* **2007**, *168*, 66–78. [CrossRef]
20. Wang, Y.; Zhou, H.; Feng, H.; Ye, M. Network traffic classification method basing on CNN. *J. Commun.* **2018**, *1*, 14–23.
21. Krizhevsky, A.; Sutskever, I.; Hinton, G.E. Imagenet classification with deep convolutional neural networks. In Proceedings of the Advances in Neural Information Processing Systems, Stateline, NV, USA, 3–8 December 2012; pp. 1097–1105.
22. Zhao, H.; Liu, F.; Li, L.; Luo, C. A novel softplus linear unit for deep convolutional neural networks. *Appl. Intell.* **2018**, *48*, 1707–1720. [CrossRef]
23. Feiyan, Z.; Linpeng, J.; Jun, D. Review of convolutional neural network. *Chin. J. Comput.* **2017**, *40*, 1229–1251.
24. Boureau, Y.L.; Ponce, J.; LeCun, Y. A theoretical analysis of feature pooling in visual recognition. In Proceedings of the 27th International Conference on Machine Learning (ICML-10), Haifa, Israel, 21–24 June 2010; pp. 111–118.
25. Khaw, H.Y.; Soon, F.C.; Chuah, J.H.; Chow, C.O. Image noise types recognition using convolutional neural network with principal components analysis. *IET Image Process.* **2017**, *11*, 1238–1245. [CrossRef]
26. Zou, J.; Wu, Q.; Tan, Y.; Wu, F.; Wang, W. Analysis Range of Coefficients in Learning Rate Methods of Convolution Neural Network. In Proceedings of the 2015 14th International Symposium on Distributed Computing and Applications for Business Engineering and Science (DCABES), Guiyang, China, 18–24 August 2015; pp. 513–517.
27. Gu, J.; Wang, Z.; Kuen, J.; Ma, L.; Shahroudy, A.; Shuai, B.; Liu, T.; Wang, X.; Wang, G.; Cai, J.; et al. Recent advances in convolutional neural networks. *Pattern Recognit.* **2018**, *77*, 354–377. [CrossRef]
28. Zeng, H.; Edwards, M.D.; Liu, G.; Gifford, D.K. Convolutional neural network architectures for predicting DNA–protein binding. *Bioinformatics* **2016**, *32*, i121–i127. [CrossRef]

Article

Automatic Implementation of a Self-Adaption Non-Intrusive Load Monitoring Method Based on the Convolutional Neural Network

Xin Wu *, Dian Jiao and Yu Du

School of Electric and Electronic Engineering, North China Electric Power University, Changping District, Beijing 102206, China; a1321907679@163.com (D.J.); duyu120192201603@163.com (Y.D.)

* Correspondence: wuxin07@ncepu.edu.cn; Tel.: +86-186-1130-8185

Received: 9 May 2020; Accepted: 14 June 2020; Published: 18 June 2020

Abstract: Non-intrusive load monitoring (NILM) is an effective way to achieve demand-side measurement and energy efficiency optimization. This paper studies a method of non-intrusive on-line load monitoring under a high-frequency mode of electric data acquisition, which enables the NILM to be automated and in real-time, including the short-term construction of a dynamic signature library and continuous on-line load identification. Firstly, in the short initial operation phase, load separation and category determination are carried out to construct the load waveform library of the monitoring user. Then, the continuous load monitoring phase begins. Based on the data of each user's signature library, the decomposition waveforms are classified by convolutional neural network models that are constructed to be suitable for each signature library in order to realize load identification. The real-time power consumption status of the load can be obtained continuously. In this paper, the electricity data of actual users are collected and used to perform the experiments, which show that the proposed method can construct the load signature library adaptively for different users. Meanwhile, the classification of the convolutional neural network model based on a library constructed in actual operation ensures the real-time and accuracy of load monitoring.

Keywords: non-intrusive load monitoring; load identification; convolutional neural network

1. Introduction

The study of demand side management (DSM) is significant for the rational allocation of power resources and the improvement of terminal power efficiency [1]. As an important means to solve the shortage of power supply and support energy development strategies, the long-term and effective implementation of power DSM can reduce environment pollution and enterprise costs, becoming an effective way to promote the sustainable development of society and the economy.

DSM focuses on improving the efficiency of terminal power consumption and adjusting the mode of power consumption to reduce the dependence on power supply. Recently, intelligent DSM has attracted the attention of people [2]. The effective monitoring of load of users can provide data support for intelligent DSM, grasp the power consumption situation and users' behavior in real-time, and guide users to arrange the usage of power consumption reasonably, so as to improve the efficiency of energy utilization [3].

Traditional load monitoring adopts an intrusive method via the installation of an information-collecting device to each user's electrical equipment. Although the collected data are accurate and reliable, the intrusive method has poor achievability due to the high-cost hardware, complicated installation in the early stage and low acceptance of users. In order to overcome this issue, non-intrusive load monitoring (NILM) [4] technology was proposed by Professor Hart in the 1980s. The non-intrusive way monitors the electrical signal only at the power supply point of the user, and analyzes the categories of loads and corresponding operations of the user. The method not only reduces the economic cost of the device, but facilitates the installation for the user. However, monitoring in the non-intrusive way requires high accuracy of load identification.

Lin et al. [5] propose a transient signature extraction scheme under multi-resolution based on the S-transform, and use a 0–1 multi-dimensional knapsack algorithm for load identification. Hong et al. [6] extract seven kinds of load steady-state characteristics and their different combinations to realize load identification. Liang et al. [7,8] use current waveforms, active and reactive power, other steady-state signatures and active transient waveforms for load identification. Adopting two-stage identification, He et al. [9] determine the type of electrical appliances by the steady-state characteristics in the first layer, and then distinguish the type of electrical appliances by the signatures of steady state and transient and user behavior mode in the second layer. This identification structure can divide complex problems into several simple problems and improve the efficiency of the method.

Developing rapidly in recent years, intelligent learning provides a data-driven method to solve problems. Compared with the method of extracting signatures manually, the intelligent learning method can learn complex separable information from multiple dimensions automatically.

Tsai et al. [10] develop a new adaptive NILM system. It uses the k-nearest neighbor rule (k-NNR) and a back propagation artificial neural network (BP-ANN) recognizer to determine the type of each device based on transient signatures. Two low-complexity solutions based on combining k-means and a support vector machine are presented in [11]. Wu et al. [12] adopt the multi-label classification model, and the mixed current signals of multiple loads can be identified without energy decomposition. Chang et al. [13] extract the energy distribution signature of a single event's turn-on/-off transient signal based on Wavelet Multiresolution Analysis (WMRA) technology and Parseval's theorem, and then construct an artificial neural network through a back-propagation classification system. Finally, it classifies the load types according to the detailed energy distribution. Ruzzelli et al. [14] propose that the neural network is trained by signature library data and that online identification is realized by extracting load signatures in real-time. Lin et al. [15] present an IoT-oriented smart Home Energy Management Systems (HEMS) utilizing a novel hybrid Artificial Neural Network-Particle Swarm Optimization (ANN-PSO)-integrated NILM approach. Chang et al. [16–18] train the neural network by the characteristics of active power, reactive power and open transient energy. The method performs well, especially in the identification for loads with similar active and reactive power. Chao et al. [19] propose a semi-supervised NILM system that includes three steps. The transient characteristics of the load are extracted by a constructed one-dimensional convolutional neural network. The load identification can be effectively realized by linear programming boosting and an online process of parameter adjustment.

Existing NILM identification algorithms depend on the signature data, which need to be collected in advance. The monitoring process has a human interference, resulting in the interruption of the monitoring process in the actual operation process. Moreover, a pre-constructed library of load signature templates is fixed and single, meaning it has limitations to be used as a standard library for load signature comparison in different users, affecting the accuracy of identification. Thus, there is an emerging need to achieve automatic monitoring of NILM algorithms and improve the universality of signature libraries to ensure accurate load identification.

Besides, the one-dimensional signatures of loads are more used in load identification algorithms. However, in the actual process of data acquisition, disturbances such as noise and harmonics on the grid side cannot be eliminated, which will affect the one-dimensional signatures of the load. Therefore, it is necessary to consider a more stable signature for identification, which can further improve the accuracy of identification.

In response to the mentioned problems, this paper proposes a new idea to solve the problem of NILM. It mainly includes two stages. Firstly, a dynamic load signature library is constructed adaptively for an independent user after the separation of mixed signals in the actual monitoring process. Then, the load identification is realized by the neural network based on the two-dimensional current data of the load in the library.

In addition, to further focus these problems in regard to NILM research, the contributions of the proposed method are as follows:

- Automatic execution of the monitoring process. After constructing the load signature library for individual users in the early short running phase, the user-specific neural network model is trained based on the data in the library belonging to each user, and then the method of load online identification for each user is realized, which provides a feasible scheme for the automatic implementation of NILM. Moreover, it can solve the problems of weak universality and unsatisfactory identification accuracy of a pre-constructed signature library.
- More stable signatures. The convolutional neural network is used to extract the two-dimensional signatures of a load current, so as to reduce the influence of noise and harmonics on the one-dimensional signature data of a load, strengthen the stability of the extracted signatures, and further improve the accuracy of load identification.

2. Implementation Principle

2.1. Principle and Implementation Structure of Non-Intrusive Load Monitoring

Non-intrusive load monitoring collects a user's power consumption data at the residential power entrance outside the user. All power consumption information of the user is concentrated in the obtained mixed signal data, so the load identification is very important to acquire the detailed power consumption of each load. The signal of each switching load should be separated from the collected mixed signal, and then the load identification is implemented supported by the information in the signature library.

The signature library is the premise of effective identification. In the actual monitoring process, it is difficult to get the independent information data of the load directly, because it is impossible to make the load run independently without disturbing the user. Furthermore, there are various brands of loads and different circuit environments, which may cause a variety of changes on load waveforms, as shown in Figure 1. It is impossible to realize load identification by a priori signature library containing all varied waveforms of different users. Moreover, the classification of separated loads will seriously affect monitoring efficiency. Therefore, the construction method of a library should be universal for the users with different loads. At the same time, complex pre-collection and intervention should be avoided to make the process of NILM automatic and effective.

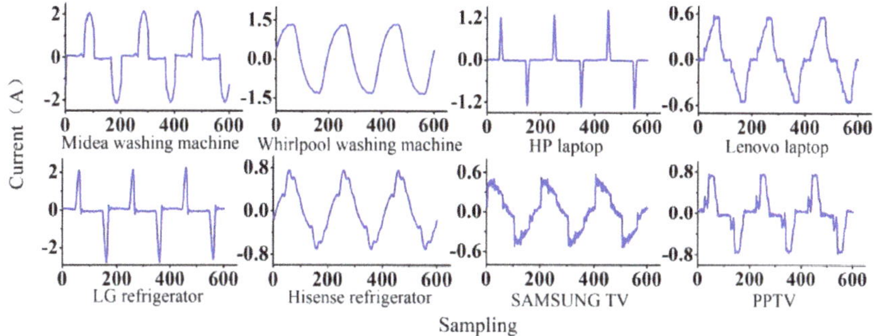

Figure 1. Current waveforms of different categories of electrical appliances with the same load.

The identification accuracy of the unsupervised-based algorithm is unsatisfactory, while the supervised-based depends on the labeled training data. In the existing study of NILM, as the prior knowledge (which is used as the training data) of the load is required in advance, the methods are feasible for a specific user and lack universality for the changes of load data in different users. Thus, a method that can adapt to these changes is required.

Sampling with a high frequency, waveform-based identification method is recommended due to the high accuracy and processing efficiency. Complete waveform and signature information can be obtained in the high-frequency acquisition mode. However, in actual acquisition processes, large amount of data is difficult to store, and the noise and harmonics on the grid side have an impact on the load waveform and signature information, which degenerates the accuracy of identification using one-dimensional data. Considering the limitation of communication capability and economic cost, the identification process is suitable for online execution and, thus, the identification accuracy depends on the real-time processing.

In response to the abovementioned problems in NILM, this paper studies a method that includes two stages: The first step is to analyze the collected mixed signals, classify and determine unknown loads without supervision, and build the load signature library specifically for users adaptively during the short-term process; the second step is to train the convolutional neural network model based on the data of users' libraries constructed in the first step to form the identification model suitable for each independent user, so as to realize load identification. The implementation structure of this paper is shown in Figure 2.

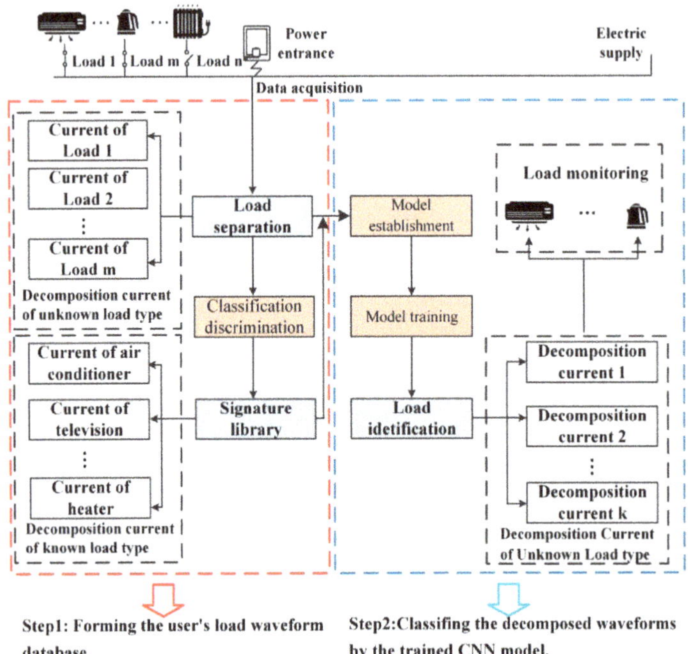

Figure 2. Non-intrusive load identification implementation structure.

2.2. Principle of Electrical Signal Separation

The collected current signal is summed by the current signals on each load branch in operation. It is assumed that the collected current data are denoted by $I(t)$. When M loads operate simultaneously, the mixed current in non-intrusive mode can be shown as Formula (1).

$$I(t) = \sum_{i=1}^{M} I_i(t) + n(t) \tag{1}$$

where $I_i(t)$ is the current signal when the load i operates separately at the time t; $n(t)$ is the noise in the circuit of the user.

The waveform separation is carried out for the electric data which are collected from the power entrance of the independent user. The collected data include the signal of loads of only one user. Two runs of the load switching action cannot be completed simultaneously as there would be a certain time difference between them; thus, the current separation model can be established as the sum of the current signals at two different moments (i.e., before and after the moment of load switching). At the last moment, the load current in the circuit is recorded as $I(t)$. When the load k is switched, the mixed current $I_{new}(t)$ is superimposed by $I(t)$ and the current $I_k(t)$ when the load k runs alone, as shown in Formula (2).

$$I_{new}(t) = I(t) + I_k(t) \tag{2}$$

where $I_k(t)$ represents the current when the latest switching load k runs independently. $I(t)$ can be treated as if there are only two signals in the circuit at this time. They are the circuit signal $I(t)$ before the moment of load switching, and the circuit signal $I_{new}(t)$ after the moment of load switching.

2.3. Construction Principle of Load Signature Library

After load separation, the independent waveforms $I_k(t)$ and $U_k(t)$ can be obtained for load k, but the categories of the waveforms are unknown. The information in the library should include waveform value and the corresponding category label. It is inevitable to judge the load categories of load k, and then a dynamic load signature library can be constructed. There are no prior available data if the library is constructed automatically without interference of users. Moreover, for loads of the same type, the change of model, brand, and even operation environment will cause the variation in load waveform. The load waveform has an infinite variety of forms theoretically, so the problem of load label attachment is attributed to the classification with infinite classes under unsupervised conditions. However, when the label category is infinite, it is very difficult to classify.

Although the waveforms of appliances are variable, the common appliances categories are enumerable. If the loads have the same category, the changing waveforms will share some common signatures. For an independent user, the physical model and waveform of the same load type are fixed, and the operating environment and habits of loads are relatively stable, so the infinite category classification problem is transformed into a limited category problem. It is suitable for each independent user to construct a signature library, which is focused on in this paper. To ensure that the method of library construction is universal for users, the signatures extracted from the unknown load are used as the criteria for load category labeling, so as to realize the classification. The load labeling problem becomes a supervised classification.

The independent load waveforms and signatures of unknown categories in users can be obtained through the proposed method in Section 2.2. The load classifier determines the category of unknown loads by the separated waveform and extracted signatures only. Thus, without prior knowledge, the problem of category classification is transformed into the posteriori knowledge-solving problem, in which the load category under the condition that the load waveform and signatures are known needs to be determined. Here, the Bayesian classification model is a suitable method. On the premise of known sample characteristics, the Bayesian classification model can quantify the probability of samples from each category, and then select the category with the largest posterior probability as the classification result. In addition, due to the variety of independent load waveforms and signatures extracted from different users, the generalization is very important for the classification model. Considering the limited load categories and quantities in one user, the data scale can be limited. With strong generalization, the Bayesian model performs well for limited scale data. Therefore, the Bayesian classification model is established for loads classification in [20]. As for the posteriori knowledge, the signatures are used to calculate the prior probability of the load category. It is assumed that F_k is the signature calculated from the signal U_k and I_k. The probability of load k that belongs to the category ω_n can be obtained by Formula (3).

$$P(\omega_n|F_k) = P(F_k|\omega_n)P(\omega_n)/P(F_k)n = 1,2,\ldots,N \tag{3}$$

where N is the category number of the user. Formula (3) shows that the prior probability $P(\omega_n)$ is converted to the posterior probability $P(\omega_n \mid F_k)$ by the obtained stable signature vector F_k, that is, when the class of load k under the known condition of F_k belongs to the probability of ω_i, the most probable category is the label of load k, as shown in Formula (4).

$$L_k = \text{argmax}P(\omega_n|F_k) \tag{4}$$

where L_k represents the classification result, which is the category label of load k. In this way, unknown category loads separated in succession can be labeled. Then, the waveforms, signatures and categories can be recorded in the library to complete the adaptive library construction of an independent user.

2.4. Convolutional Neural Network Identification Model

After forming the user's library, the independent load waveform of a user is continuously identified based on the data in the library, so as to determine the user's load operation status at the current moment. Stored in the library, the N kind of loads and the corresponding record information are expressed as follows.

$$\begin{cases} \{\hat{U}_1, & \hat{I}_1, & L_1\} \\ \{\hat{U}_2, & \hat{I}_2, & L_2\} \\ \vdots & \vdots & \vdots \\ \{\hat{U}_N, & \hat{I}_N, & L_N\} \end{cases} \tag{5}$$

where the information includes the separated signals of voltage $\hat{U}$, current $\hat{I}$, and label L.

The library has been constructed completely at this time. Real-time identification belongs to a supervised classification problem. The data in the library are suitable for the unique independent users, and the load to be classified is the load in the library. Thus, as the training data, the information in the library enables the classification model for supervised training, which can greatly reduce the invalid sample data for training, build a useful classification model specifically for the independent user, and cut down the impact of over-fitting on the identification results.

Due to the existence of noise and harmonics on the power grid side, extracted from the one-dimensional data of the load waveform, the signature information fluctuates greatly, subsequently affecting the identification process. However, as the load circuit is composed of non-linear components such as diodes, thyristors, transistors, motors and so on, it will also cause the distortion of load current waveform. The harmonic and distortion of the current influenced by the non-linear components in the circuit can also be regarded as the typical signatures of load identification [6–9,20,21]. Direct filtering may destroy the original load signatures useful for identification. Therefore, it is difficult to determine whether the distortion in the current is caused by noise or harmonics accurately and filter it directly. Considering that the waveform of the same independent load is not only relatively stable in a steady state under the high frequency acquisition mode, but is less disturbed by noise and grid side harmonics, the one-dimensional waveform data of a load current stored in the dynamic library are transformed into two-dimensional image data. The image can keep the basic shape and outline of the original current waveform. The amplitude values of current waveform are transformed into the pixel values in the image. The waveform distortion caused by noise or harmonic alters only the position of the pixel points in the original waveform locally and slightly (i.e., changes a few local pixel values of the image), rather than the shape and outline of the current waveform. In the image recognition process, the two-dimensional data can be recognized mainly based on the image features, including contour, shape, contrast and relative position of the marked features. Therefore, the dimensional converting of the waveform for recognition will reduce the influence of noise or harmonics on the recognition results.

Since the one-dimensional current data are transformed into two-dimensional image data, the load identification problem is transformed into the identification problem of the two-dimensional image data. Convolutional neural networks have outstanding performance on two-dimensional image data processing. Images can be input into the network directly to avoid the complexity of data reconstruction in the signature extraction and classification processes. Convolutional neural networks automatically extract multiple image signatures through multiple convolution kernels, approximate complex mapping functions through multi-layer non-linear transformation, and then classify current waveform images to realize real-time load identification. Besides, the distortion of the current waveform caused by noise or harmonic only alters the position of the pixels in the original waveform, resulting in the local translation, rotation and scaling of the waveform position. However, these influences can be weakened by the convolutional neural network with the characteristic of translation invariance. (Invariance means that when the input data are changed locally and slightly, most of the outputs after the pooling function will be not changed. It is extremely significant when we focus on whether a feature appears in two-dimensional data rather than at its location.) As an important layer of the convolutional neural network, the function of the pooling layer is that the output of the network at a certain location in image is replaced by the statistical characteristic output of the pixel value in the surrounding area of that location. Contrastingly, in the constructed convolutional neural network, the average value of the surrounding pixel is extracted by the pooling layer, which weakens the influence of the pixel points affected by the noise and harmonics in the image. Furthermore, multiple pooling layers in the convolutional neural network gradually reduce the influence caused by noise and harmonics. Moreover, in convolution operation, parameter sharing ensures that it is unnecessary to learn a set of parameters for each position in the two-dimensional data, which reduces the computational complexity, training time and storage space of the parameters.

In this paper, the one-dimensional data in a labeled library are converted into the two-dimensional waveform image as a training set, and the convolutional neural network model is trained by supervised learning. The test data consist of the two-dimensional waveform image converted from the one-dimensional data of the separated current signal which is obtained from the mixed signal in real-time. Finally, the load can be identified online by the convolutional neural network, of which the model structure includes convolutional, pooling, a fully connected layer and non-linear function, as shown in Figure 3.

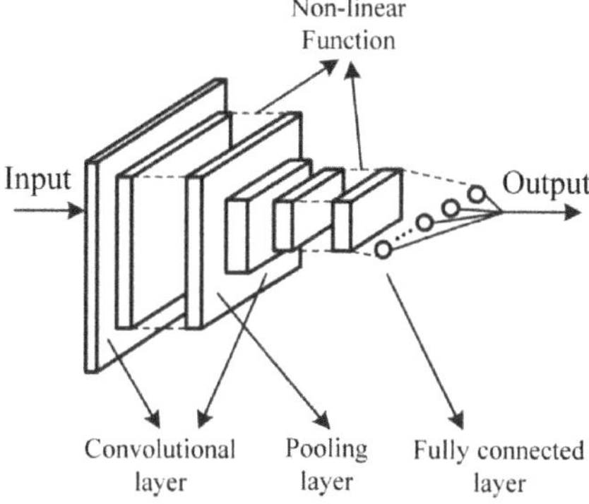

Figure 3. Convolutional neural network model structure.

3. Methodology

In order to achieve the above idea, this paper will introduce the following two stages. In the first stage, switching events are detected and load signals are separated from the collected mixed data. Then the categories of separated loads are judged to form a load signature library. Then, the convolutional neural network is trained by the data in the library. After short-term model training, the classification model suitable for the load signature library is automatically formed, so that the separated load signals separated can be identified in real-time.

3.1. Adaptive Construction of a Load Signature Library

3.1.1. Event Detection and Load Separation

The constructed signature library needs to include the waveform information of each load. However, in the non-intrusive mode, the collected electrical signal of a user is summed by multiple signals of different electrical appliances which are simultaneously in the open state. Thus, the mixed signal requires signal separation. The load switching events can be measured by the current intensity. In [20], if there is an obvious difference in the current intensity of a period compared with that of the previous one, the load switching event can be considered to be increasing.

Then, considering the difference of the load phase when the loads are switched into the user circuit, direct extraction of a switching load signal may result in an error regarding the information of the independent load. The method in [20] extracts the mixed signal on the sample points of the same voltage value in different electrical periods—before and after the switching events. Using the one-dimensional data processing method, signal separation can be realized. The load signal separation method is simple and effective. Therefore, this paper adopts the method in [20] to separate the load electrical signal quickly.

3.1.2. Category Determination of Unknown Load Waveform

After obtaining the separated waveform of load k, the load waveforms require pre-classification. All the unknown loads separated from the mixed signal of independent users are clustered rapidly by load signatures. The process can not only avoid the load waveforms detected and separated repeatedly due to the actual multiple switching by users, but also narrow the range of load categories, so as to reduce the operation burden of subsequent load category determination.

In the initial stage of library construction, the number of load categories and operation modes in each user are unknown. Correspondingly, the number of different categories of waveforms is also unknown. However, for the same user, the waveform of the same load is relatively fixed, and the difference of signatures extracted from the waveform of the same load in each switching is small, while the difference between the waveform of different loads is relatively large. Thus, the unknown load is clustered quickly by the inherent signatures extracted from the separated waveforms, which can greatly reduce the repeatability of the extracted load waveform.

After load signature normalization, the cluster can be achieved by discriminant function, as shown in Formula (6).

$$D_{k,\omega} = ||F_k{}^* - \delta_\omega{}^*|| \tag{6}$$

where $F_k{}^*$ is the normalized signature and $\delta_\omega{}^*$ is the signature of the ω-th load stored in the current time signature library. $D_{k,\omega}$ is the discriminant distance of the signature between load k and category ω. If the minimum value of $D_{k,\omega}$ is less than δ, the type of load k has already been stored in the library. If the minimum value of $D_{k,\omega}$ is greater than δ, the new load is found and its waveform and signatures are recorded in the library.

The method can ensure that the load information in the library has no repeatability. At this time, the category of unknown load waveforms needs to be determined, so as to complete the load information in the signature library. In practice, there are abundant load brands and models among the actual users. Various waveforms cause a signature value fluctuation in different degrees. The method in [20] considers the fluctuated load signatures and calculates the probability of different load signatures. Then, using the Bayesian model and the multiple signatures, the category probability of loads belonging to different load categories is obtained. Eventually, by Formula (4), the most possible load category is selected to solve the problem of category determination for unknown load waveforms.

3.2. Load Identification of the Convolutional Neural Network Based on the Signature Library

After the first stage of the short-term adaptive library building process, the second stage of the sustainable load monitoring begins. Based on the data in the library, this paper transforms a one-dimensional load current signal into two-dimensional image data of periodic current, and the detailed process is divided into the four following steps, where the load current vector $\hat{I}_k$ in one period of steady state is taken as the basic data:

- The sampling point serial number and the current amplitude are taken as abscissa and ordinate, respectively. Connect each data point in turn and draw the binary image of load current waveform;
- A unified range of coordinate axes is selected for binary image, ensuring that the waveform images of different loads are displayed in the same range;
- Hide the axis in the binary image;
- An appropriate image resolution is selected to display the image clearly. The resolution of the binary image is adjusted to adapt to the input of the convolutional neural network.

Then, the input sample of the convolutional neural network can be obtained. In this way, one-dimensional current waveforms can be transformed into two-dimensional image data. Meanwhile, the contour and shape of the waveform can be preserved completely, meaning that they can be directly input into the constructed convolutional neural network for identification.

The training process of the convolution neural network includes forward and backward propagation, in which forward propagation completes signature extraction and sample classification, and backward propagation completes classification error calculation and weight updating.

In forward propagation, signature extraction is realized by convolution and pooling. The upper output is taken as the input of the layer. After calculating in the convolution kernel, the output of the layer is obtained by further calculation with the non-linear activation function, as shown in Formula (7).

$$X_j^l = f(\sum_{i \in M_j} X_j^{l-1} * \kappa_{ij}^l + b_j^l) \tag{7}$$

where X_j^{l-1} represents the j-th signature graph of layer $l-1$, κ_{ij}^l represents the convolutional kernel function of the j-th signature graph mapped from the $(l-1)$-th to the l-th layer, $f()$ is the activation function, b_j^l is the bias parameter, and * represents the convolution. Pooling layer calculation is shown in Formula (8).

$$H_j^l = \mathrm{f}(w_j^l \mathrm{sample}(X_j^{l-1}) + b_j^l) \tag{8}$$

where X_j^l represents the j-th signature graph of layer l. w_j^l and b_j^l are the parameters of weight and bias, sample is pooling function and f is activation function. The pooling layer aims to map the signatures to a smaller range and reduce the dimension of the convolutional signature map. Signature information of load signals with relatively weak power is susceptible to noise and harmonic interference from the power grid. In this case, the maximum pooling will result in only extracting the affected signatures while ignoring the actual signature information of the signal, which in turn will impact the identification effect. Therefore, this effect is weakened by replacing the maximum pooling layer with the average pooling layer.

Softmax layers are usually used as output layers in multi-classification problems, which can output classification results directly in the form of probability vectors. The calculation formula is shown in Formula (9).

$$S_j = e^{a_j} / \sum_{k=1}^{N} e^{a_k} \tag{9}$$

where j represents the result of class j classification and N represents the category of load.

Back propagation depends on the error between the classification result of forward propagation and the given sample label. According to the chain rule, the weight and the error in each layer are updated, as shown in Formulas (10) and (11).

$$\partial E_{total}/\partial w = (\partial E_{total}/\partial out)(\partial out/\partial net)(\partial net/\partial w) \tag{10}$$

$$w^{new} = w - \xi(\partial E_{total}/\partial w) \tag{11}$$

where $\partial E_{total}/\partial w$ represents the partial derivative of the loss function E_{total} to the parameter w, which is updated in each iteration, and ξ represents the learning rate of the convolutional neural network, which determines the magnitude of each adjustment.

The model is trained iteratively with the labeled data in the library, and the connection weights of each layer and the parameter matrices are fully adjusted until the data in the database are exhausted. When the model is trained completely, the load identification can be realized in real-time. The separated waveforms of independent loads will be successively acquired in real-time according to the load separation method in the Section 3.1.1. The waveform data will be converted into two-dimensional images and identified online by the convolution neural network as test data.

The implementation process of the steps in this chapter is shown in Figure 4.

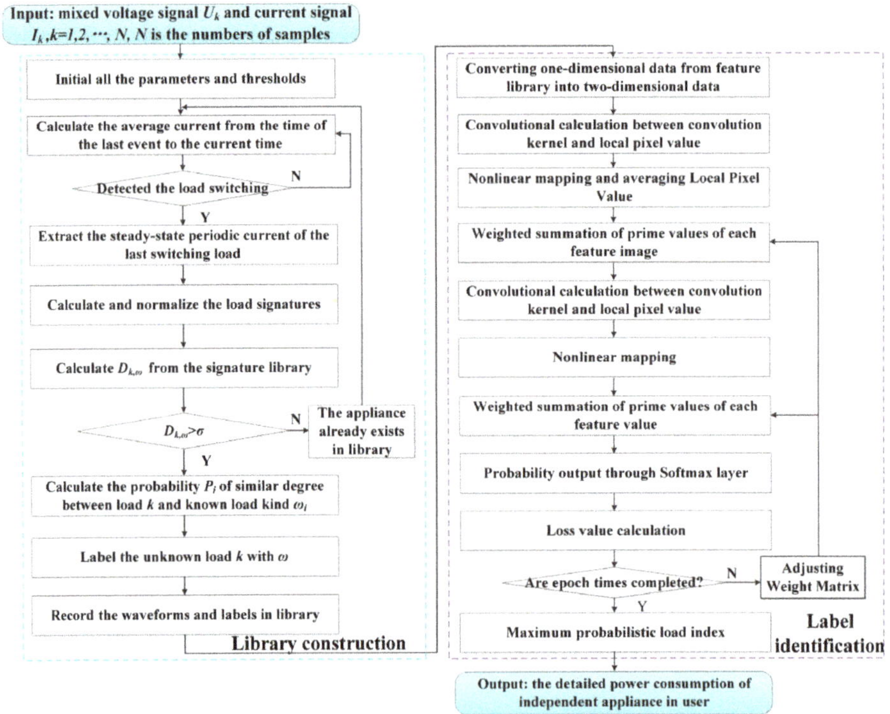

Figure 4. Overall algorithm flow chart.

4. Experiment and Analysis

Actual user data are collected and used for validity verification in this paper. Figure 5 shows the schematic diagram of the experimental system, which is designed for non-intrusive data acquisition from the actual user. By the proposed method in Section 3, the collected data are processed to realize the effective load monitoring.

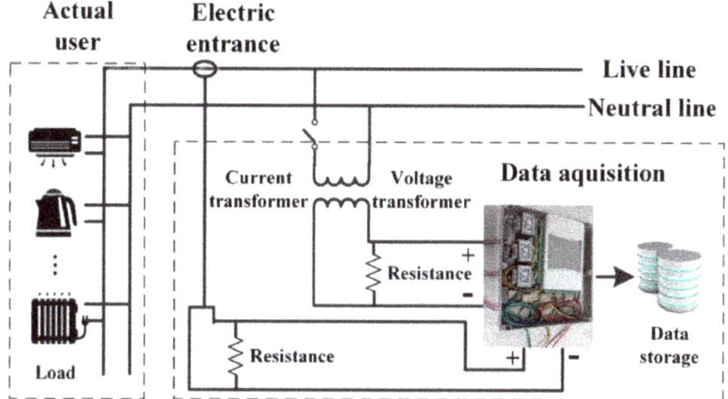

Figure 5. Schematic diagram of the experimental system.

The specific experimental parameters are as follows: the access voltage of the acquisition device is 220 V, and the sampling frequency is 10 kHz. Identification objects include a rice cooker (EC); an electric kettle (EK); a water heater (WH); a water dispenser (WD); a laptop computer (LA); a television (TV); air-conditioning systems A (AC-A), B (AC-B), C (AC-C); a vacuum cleaner (VC); a refrigerator (RE); a microwave oven (MO). Table 1 lists the detailed value of the related threshold involved in the experiment.

Table 1. Threshold parameters.

Experimental Parameters	Value	Experimental Parameters	Value
η	0.1	υ	0.75
σ	0.01	ε	0.05
α	0.1	λ	0.025

4.1. Effectiveness Verification of Library Construction

Figure 6 shows the current separation signal of the load in the experimental environment, and gives the template current for comparison. Blue lines denote the current separation signals which are the periodic current waveforms operating in a stable state separated by the method in Section 3.1.1. Red lines denote the standard currents which are obtained by only switching on a single electrical appliance in the experiment. The high coincidence between the separated signal and the standard current indicates that the load separation has high accuracy. Since the categories of separated load waveforms are unknown before the load labeling, a-l is used to represent the waveforms of loads in this paper.

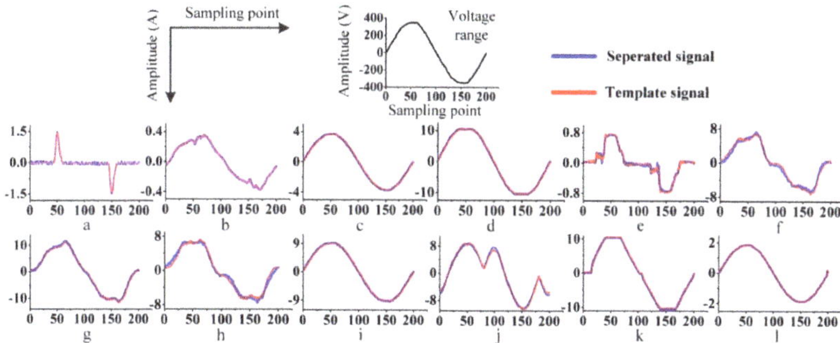

Figure 6. (**a–l**) shows comparison diagram of the separated load current and the template current.

After effective load separation and clustering of independent load waveforms, the categories of loads are judged, and the library of independent users is constructed. Table 2 shows the categories, numbers and actual pre-classification of electrical appliances involved in this paper. The load number is used to represent the category discrimination label of the load.

Table 2. Electrical appliances classification label and pre-classification.

Load	Category Number	Type	Load	Category Number	Type
Electric cooker	1	I	Set top box	12	III
Electric kettle	2	I	Range hood	13	III
Water heater	3	I	Air purifier	14	III
Electric oven	4	I	Air conditioner	15	III
Disinfection cabinet	5	I	Vacuum cleaner	16	III
Water dispenser	6	I	Dehumidifier	17	III
Electromagnetic furnace	7	II	Refrigerator	18	IV
Electric hair dryer	8	III	Microwave oven	19	IV
Laptop	9	III	Washing machine	20	V
Electric fan	10	III	Other loads	21	VI
TV	11	III			

The probability distribution of the several unknown loads separated from load separation parts is shown in Figure 7. It shows the label probability of one load belonging to each category of electrical appliances in the experiment, and quantifies the possibility of the load category by the label probability. The label corresponding to the maximum probability is determined as the category label of the load. The threshold v of unknown class is denoted by the red straight line. If the maximum label probability is still lower than the threshold line, the load will directly be placed in the "unknown load" category.

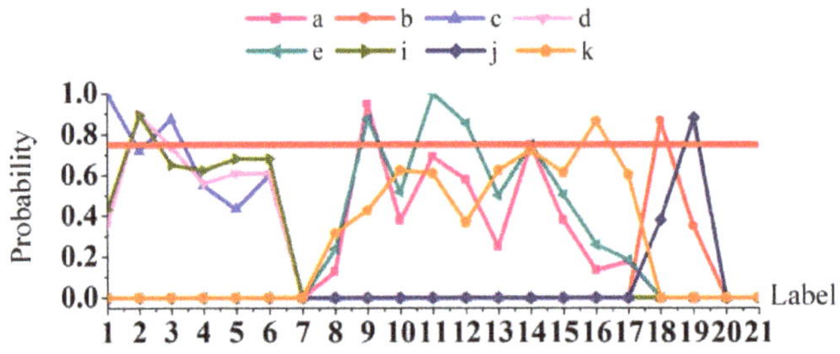

Figure 7. Labeling probability of load.

The probability of possible load labels under their pre-classification is shown in Figure 7. The probability outside their pre-classification is 0. In this paper, the label with the highest probability in the pre-classification of unknown loads is regarded as the result of its category. The waveforms and label probability results are shown in Figure 8. It compares the labels and their probabilities for unknown loads with their real labels. It can be seen that the waveforms are labeled correctly.

4.2. Effectiveness Verification of Load Identification Based on the Convolutional Neural Network

Before the real-time load identification, the convolutional neural network model is trained by the category-labeled data from the established library. After being obtained, the one-dimensional current data (as shown in Section 4.1) are transformed into two-dimensional image data through the data dimension conversion method described in Section 3.2. Because the maximum current of most common household inserts is limited to 10 A, the maximum operating current of most electrical appliances is usually less than or close to 10 A. Therefore, the vertical axis range of the coordinate axis selected in this paper is from −11 A to +11 A, and the horizontal axis range is the sampling point of one current period when the load is in steady-state operation. The maximum current of a few electrical appliances exceeding 10 A applies the same operation, and the identification result is not affected.

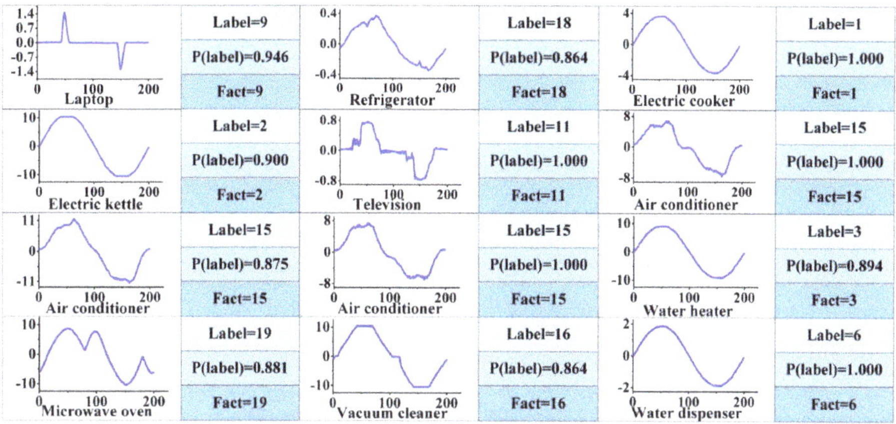

Figure 8. Labeling results of individual appliances.

The labeled appliances in the library are re-numbered to represent the categories of the load in the process of convolutional neural network identification. The load re-numbering of the library established in Section 4.2 is shown in the Table 3.

Table 3. Label number of load image identification.

Load	Identification Number	Load	Identification Number
Laptop	0	Air conditioner 2	6
Refrigerator	1	Air conditioner 3	7
Electric cooker	2	Water heater	8
Electric kettle	3	Microwave oven	9
TV	4	Vacuum cleaner	10
Air conditioner 1	5	Water dispenser	11

The numbers of convolutional layers and pooling layers are extremely important for the classification accuracy of the model. Figure 9 shows the classification accuracy under different numbers of convolutional and pooling layers. When the number of convolutional and pooling layers is less, the parameters are insufficient for the accurate classification of the sample. With an increasing number of layers, the effectiveness of the model's classification process is clearly improved. However, when the layers continue to increase, the increased training parameters raise the difficulty and time of model training. Limited by the current training methods, the increase in layers is more likely to make the classification results fall into the local optimum, leading to over-fitting and other problems. As shown in the figure, when the convolution and the pooling layer are both set as 3, the model is most effective, and the classification accuracy of the test sample can reach 96.73%. Therefore, considering the accuracy and training time of the model, the number of both the convolutional and pooling layers in the convolutional neural network model is determined as 3. In addition, under this optimal layer structure, the kernels 1, 2, and 3 are set as convolutional layers 1, 2, and 3 respectively, as shown in Table 4.

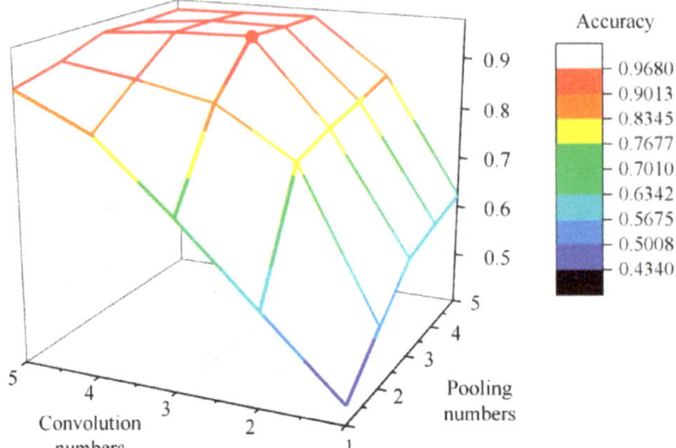

Figure 9. Classification accuracy of the model under different numbers of convolutional and pooling layers.

Table 4. Kernel sizes and numbers in this paper.

Kernel	Sizes	Numbers
1	3 × 3	20
2	5 × 5	20
3	12 × 12	100

In general, 3 × 3 is the popular choice of kernel size in the convolutional neural network, which is determined by the empirical value in the experiment. Specifically, the kernel size of the convolutional kernel is set to be larger than 1 × 1 to enhance the receptive field. A kernel with an even size cannot ensure the same size of the feature map in the input and output. In the case of the same receptive field, the required parameters and computation are increased with the size expansion of the convolutional kernel. Thus, the kernel size of 3 × 3 is used for the first convolutional layer. In order to extract the output image feature of the previous convolutional layer further, the kernel size of the convolutional layer increases gradually, so the second convolutional layer size is set as 5 × 5. In addition, the third layer is the last convolutional layer, followed by the fully connected layer. The input of the fully connected layer needs to be one-dimensional data, which have the same dimension with the output of the third convolutional layer. However, the output dimension of the convolutional layer depends on the input data dimension and kernel size. Considering that the input data dimension of the third layer is 12 × 12, the kernel size of the third convolutional layer is set as 12 × 12. Besides, the size of 12 × 12 is set to reduce the parameter numbers of the fully connected layer significantly. As for kernel numbers, if there are less numbers in the convolutional layers, the extracted image features are not enough for identification, and the model struggles to achieve the desired performance. On the contrary, if the kernel number is set to be oversized, it will incur the problem of model parameters and training speed increasing significantly, as well serious over-fitting problems. Thus, the kernel numbers are empirical values obtained by repeated experiments.

After the determination of the model structure, the model parameters become significant factors in the training process. The parameters of learning rate and epoch are related to the convergence and training speed of the model. The learning rate μ represents the amount of weight updating in each time. If the set value of learning rate is too high, the loss function and model will struggle to converge. On the contrary, if the learning rate is too small, the updating of weights and the change to the model cost will be very small each time, resulting in significantly more epoch times. Epoch times are the training times of all sample data. Figure 10 shows the cost value of the convolutional neural network model training under different learning rates. It can be seen that the identification model tends to converge and the convergence speed is faster at a learning rate of 0.05. When the number of epoch is 500 (i.e., epoch = 500), the loss values in the model are all below 0.003.

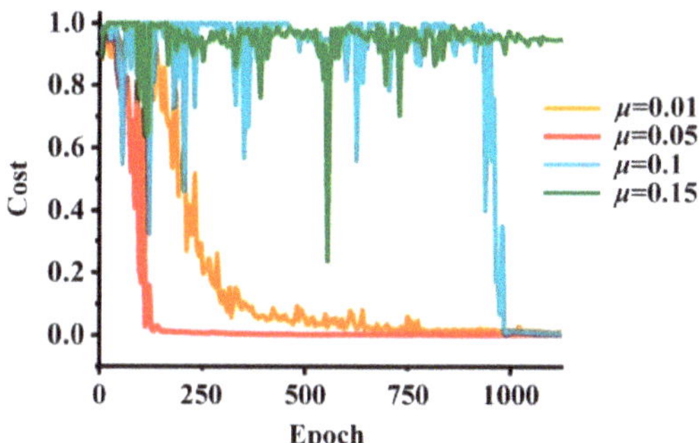

Figure 10. The convergence process of cost under different learning rates.

Figure 11 illustrates the parameters of the convergence process in the model. Six parameters are selected for display. The parameters *kernel_c1* and *bias_c1* are one of the weights and one of the biases in the second convolution layer, respectively. The parameters *kernel_f1* and *bias_f1* are one of the weights and one of the biases in the third convolution layer, respectively. Besides, *weight_f1* is one of the weight parameters between the third convolution layer and the fully connected layer. The parameter of weight output is one of the weights between the fully connected layer and the softmax layer. The detailed values of the above six parameters under different epoch times are shown in Table 5.

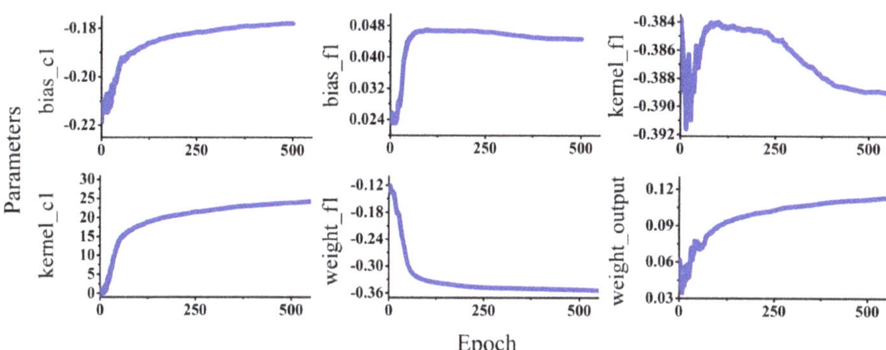

Figure 11. Convergence process of the convolutional neural network model parameter.

Table 5. Parameter values under different epoch times.

Parameter	Epoch Times				
	100	200	300	400	500
bias_c1	−0.188066	−0.182839	−0.180615	−0.178961	−0.178071
bias_f1	0.046807	0.046743	0.045898	0.044920	0.044646
kernel_f1	−0.384234	−0.384602	−0.386321	−0.388277	−0.388930
kernel_c1	17.850862	20.656326	22.131949	23.217981	23.901392
weight_f1	−0.331889	−0.344194	−0.348489	−0.350513	−0.352352
weight output	0.089298	0.099996	0.105751	0.109533	0.111763

It can be seen that the parameters show a gradual increasing trend as the epoch time increases. There is no significant change in the above parameters when the epoch value is greater than 500. It can be considered that the model is trained to converge when the epoch reaches 500. Thus, this paper chooses the number of epoch as 500 to ensure the training efficiency of the model.

In order to display the model identification results, 12 separated current waveforms in Figure 6 are identified by the convolutional neural network model. Figure 12 shows the model input data after dimension conversion by the method proposed in Section 3.2.

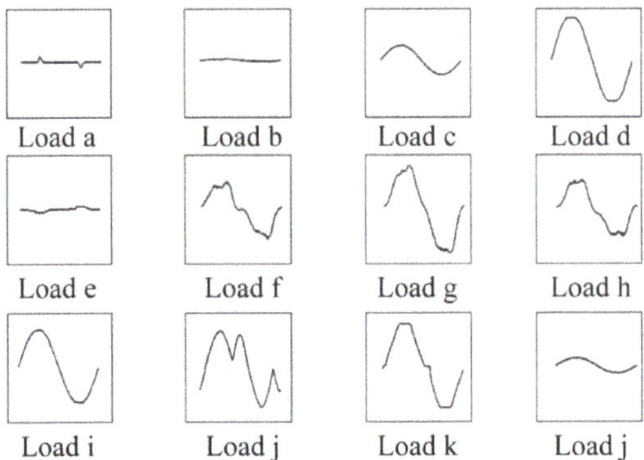

Figure 12. Input image of 12 separated current waveforms.

In the process of identification using the convolutional neural network model, the signature maps extracted from the input data after the first convolutional operation are shown in Figure 13. It can be seen from the figure that the contour edge and other features of each load current waveform in Figure 12 are strengthened and extracted by the kernel in the convolutional layer.

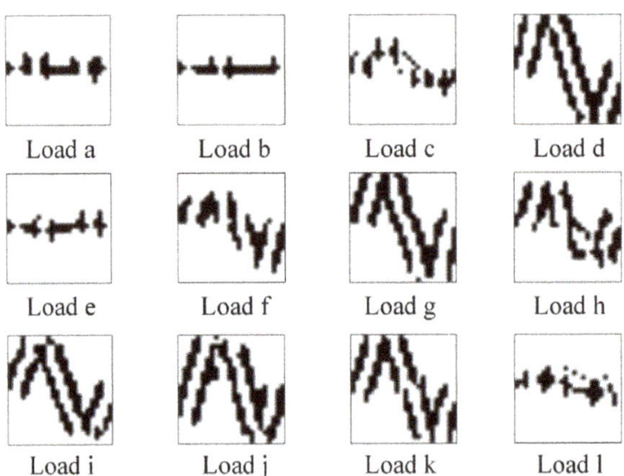

Figure 13. Feature image after the first convolution operation.

After processing of pooling and activation further, the signature maps are shown in Figure 14. It reduces the feature dimensions of the images in Figure 13, and makes non-linear mapping on the feature image, so as to extract the advanced features for identification.

Table 6 gives the classification confusion matrix of the algorithm. The column of the confusion matrix represents the identification label of each load category, the row represents the real label, and the diagonal value represents the accuracy of the correct classification of the load. The identification accuracy increases with the background color deepening.

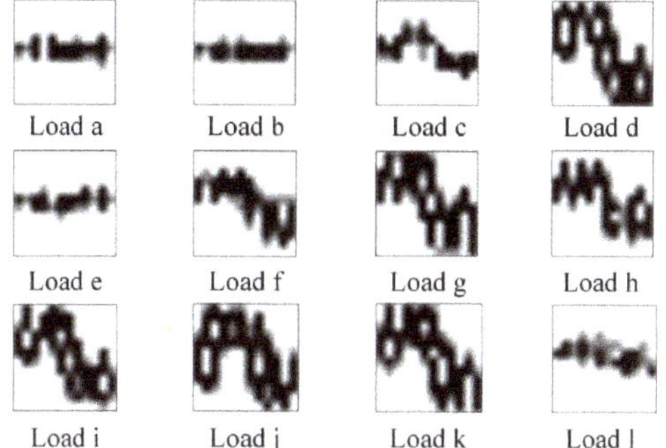

Figure 14. Feature image after pooling and non-linear operation.

Table 6. Load image classification confusion matrix.

	0	1	2	3	4	5	6	7	8	9	10	11
0	0.9985	0.0005	0.0003	0.0000	0.0004	0.0000	0.0000	0.0001	0.0000	0.0000	0.0000	0.0000
1	0.0125	0.7754	0.0743	0.0013	0.1231	0.0002	0.0002	0.0003	0.0018	0.0009	0.0021	0.0079
2	0.0003	0.0017	0.9949	0.0000	0.0009	0.0010	0.0000	0.0002	0.0000	0.0001	0.0000	0.0010
3	0.0002	0.0000	0.0000	0.9914	0.0001	0.0000	0.0023	0.0000	0.0019	0.0000	0.0039	0.0001
4	0.0023	0.1265	0.0059	0.0003	0.8620	0.0001	0.0003	0.0004	0.0007	0.0003	0.0005	0.0008
5	0.0000	0.0000	0.0004	0.0000	0.0000	0.9974	0.0003	0.0010	0.0001	0.0005	0.0000	0.0002
6	0.0000	0.0000	0.0000	0.0003	0.0000	0.0001	0.9977	0.0001	0.0002	0.0000	0.0014	0.0000
7	0.0002	0.0001	0.0001	0.0000	0.0000	0.0005	0.0000	0.9984	0.0002	0.0002	0.0001	0.0001
8	0.0005	0.0001	0.0002	0.0062	0.0003	0.0003	0.0009	0.0005	0.9867	0.0000	0.0039	0.0004
9	0.0000	0.0002	0.0001	0.0005	0.0000	0.0000	0.0012	0.0000	0.0000	0.9977	0.0001	0.0006
10	0.0006	0.0003	0.0007	0.0095	0.0010	0.0005	0.0301	0.0005	0.0071	0.0040	0.9435	0.0023
11	0.0000	0.0001	0.0005	0.0000	0.0020	0.0001	0.0000	0.0000	0.0001	0.0003	0.0001	0.9967

In order to present the accuracy of results of the proposed method, the collected data of day1–day3 are identified by the proposed method, and the power consumption ratio of different loads is presented in Figure 15. As a comparison, smart sockets are installed to the monitoring appliances to obtain the true power consumption, which is shown in the right part of the figure. It can be seen that the total consumption difference between the calculated one and the true one is less than 0.3 kW. The consumption ratio of each load is nearly the same as the true value given by the socket, and the load has the correct label.

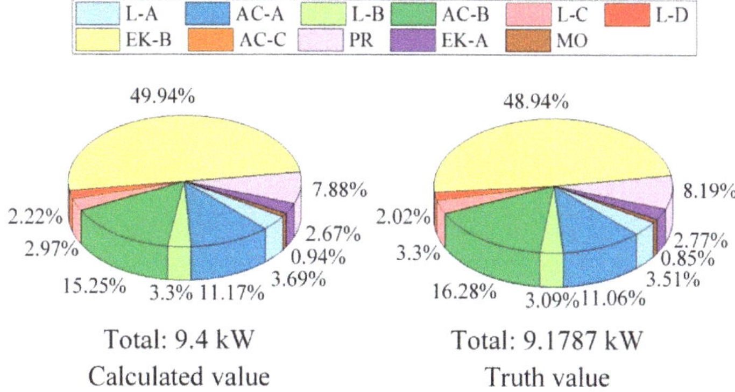

Figure 15. Comparison chart between identification load and actual power consumption results of day1–day3.

In addition, the algorithms proposed by Chao et al. [19], Srinivasan et al. [21], Ahmadi et al. [22] and the genetic algorithm are selected to compare against the proposed method. The above typical algorithm shows good performance on NILM and the feasibility of the proposed method is proved by a comparison of the algorithms. In reference [21], the neural network is also applied to NILM in [21] as proposed by Srinivasan. A typical neural network is used to verify the effectiveness of load identification through the model training of neural networks. Different from our work, the neural networks are trained to extract harmonic signatures from the current for load identification. Then, Ahmadi et al. [22] propose a graph signal processing (GSP) approach for NILM. The graph is also formed by steady-state signatures of loads. It poses the load disaggregation problem as a single-channel blind source separation problem to perform low-complexity classification for load identification. It proves that the load can be identified by processing the signal of a load graph, but it is different from the transformation method of the graph signal in the proposed method. Similarly, with our method, the convolutional neural network is also applied to NILM to form a three-step non-intrusive load monitoring system (TNILM) in Chao' work [19]. Due to the purpose, dimension, structure, input and output data of the convolutional neural network, the proposed algorithm outperforms that in Chao' work. In addition, the traditional intelligent algorithm is widely used in non-intrusive load identification. Genetic algorithm optimization is a conventional intelligence algorithm. Thus, as a supplement, genetic algorithm optimization is used as another method of load identification after the library construction to replace the convolutional neural network in this paper for comparison.

Figure 16 shows the performance comparison curves with the above mentioned method. The comparison of the algorithms' accuracy of load identification is shown in Figure 16a. The increasing load categories have less influence on the algorithm in this paper. The operational efficiency curves are shown in Figure 16b. In the actual stage of load identification, the proposed method has higher operational efficiency and a stable time of load identification. Represented by the violet line, the TNILM in Chao' work [19] includes the convolutional neural network and a multi-label classifier, so it has a relatively long operation time. Denoted by the blue line, the running time of the algorithm in Srinivasan' work [21] increases rapidly with a rising load number. Represented by the green and orange lines, respectively, the GSP algorithm and the genetic algorithm optimization have more stable operational efficiency, but are overall slower than the algorithm in this paper.

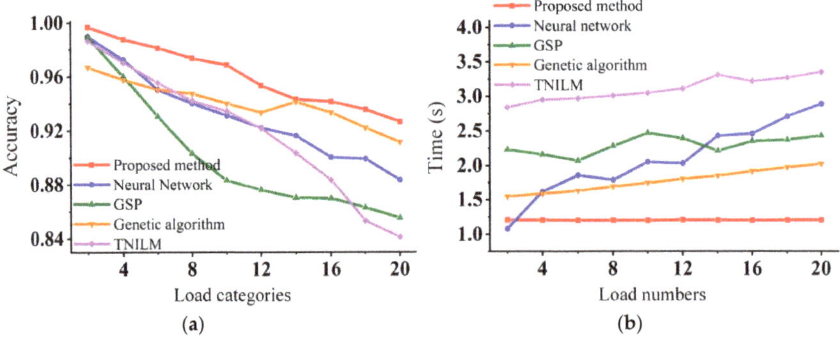

Figure 16. Performance comparison of algorithms. (**a**) Load identification accuracy comparison of different algorithm under different category numbers; (**b**) Operation efficiency comparison of different algorithm under different load numbers.

5. Conclusions

Considering the accuracy and real-time of NILM during actual operation, this paper studies an effective identification method based on the convolutional neural network. Under the high frequency data acquisition mode, this paper adopts the load separation model to obtain the current and voltage waveforms of independent loads and records the corresponding label information using the

Bayesian classification model. Then, the convolutional neural network model is briefly trained by the two-dimensional load data in the library to form a classification model suitable for each signature library, realizing long-term load identification in real-time.

In this paper, the two-dimensional image load data are used for identification. This type of data can preserve the contour and shape signature of a waveform completely and avoid the complexity of data reconstruction as in signature extraction and classification. Furthermore, the contour and shape signatures extracted by the convolutional neural network reduce the influence of noise or harmonics on the identification results. The measured data are used to verify the algorithm proposed in our work. The method performs better than the other compared algorithms. With the increase in load categories and number of users, the advantages of the proposed algorithm are clear. The overall accuracy is higher than 92% and the operation time is less than 1.25 s. Thus, the proposed method can identify the switching load effectively using the convolutional neural network and the obtained corresponding power consumption of each load can be calculated accurately. The whole process provides a complete implementation idea for NILM, which can be automatically executed without intervention.

In future work, the dynamic loads with various transients should be considered. Additionally, the influence of load phase information requires further research, and required the resolution and accuracy for measurement may be another research point.

Author Contributions: X.W. and D.J. conceived and designed the experiments; D.J. and Y.D. contributed experimental tools and analyzed the results; X.W. and D.J. wrote and revised the paper. All authors have read and agreed to the published version of the manuscript.

Funding: This research was funded by [the Fundamental Research Funds for the Central Universities] grant number [2020MS002].

Conflicts of Interest: The authors declare no conflict of interest.

References

1. Sheha, M.; Powell, K. Using real-time electricity prices to leverage electrical energy storage and flexible loads in a smart grid environment utilizing machine learning techniques. *Processes* **2019**, *7*, 870. [CrossRef]
2. Sharifi, R.; Anvari-Moghaddam, A.; Fathi, S.H.; Vahidinasab, V. A flexible responsive load economic model for industrial demands. *Processes* **2019**, *7*, 147. [CrossRef]
3. Cominola, A.; Giuliani, M.; Piga, D.; Castelletti, A.; Rizzoli, A.E. A hybrid signature-based iterative disaggregation algorithm for non-intrusive load monitoring. *Appl. Energy* **2017**, *185*, 331–344. [CrossRef]
4. Hart, G.W. Nonintrusive appliance load monitoring. *Proc. IEEE* **1992**, *80*, 1870–1891. [CrossRef]
5. Lin, Y.-H.; Tsai, M.-S. Development of an improved time-frequency analysis-based nonintrusive load monitor for load demand identification. *IEEE Trans. Instrum. Meas.* **2014**, *63*, 1470–1483. [CrossRef]
6. Hong, Y.; Chou, J. Nonintrusive energy monitoring for microgrids using hybrid self-organizing feature-mapping networks. *Energies* **2012**, *5*, 2578–2593. [CrossRef]
7. Liang, J.; Ng, S.K.; Kendall, G.; Cheng, J.W. Load signature study—Part I: Basic concept, structure, and methodology. *IEEE Trans. Power Deliv.* **2010**, *25*, 551–560. [CrossRef]
8. Liang, J.; Ng, S.K.; Kendall, G.; Cheng, J.W. Load signature study—Part II: Disaggregation framework, simulation, and applications. *IEEE Trans. Power Deliv.* **2010**, *25*, 561–569. [CrossRef]
9. He, D.; Du, L.; Yang, Y.; Harley, R.; Habetler, T. Front-end electronic circuit topology analysis for model-driven classification and monitoring of appliance loads in smart buildings. *IEEE Trans. Smart Grid* **2012**, *3*, 2286–2293. [CrossRef]
10. Tsai, M.-S.; Lin, Y.-H. Modern development of an Adaptive Non-Intrusive Appliance Load Monitoring system in electricity energy conservation. *Appl. Energy* **2012**, *96*, 55–73. [CrossRef]
11. Altrabalsi, H.; Stankovic, V.; Liao, J.; Stankovic, L. Low-complexity energy disaggregation using appliance load modeling. *AIMS Energy* **2016**, *157*, 1–21. [CrossRef]
12. Wu, X.; Gao, Y.; Jiao, D. Multi-label classification based on random forest algorithm for non-intrusive load monitoring system. *Processes* **2019**, *7*, 337. [CrossRef]
13. Chang, H.H.; Lian, K.L.; Su, Y.C.; Lee, W.J. Power-spectrum-based wavelet transform for nonintrusive demand monitoring and load identification. *IEEE Trans. Ind. Appl.* **2014**, *50*, 2081–2089. [CrossRef]

14. Ruzzelli, A.G.; Nicolas, C.; Schoofs, A.; O'Hare, G.M. Real-time recognition and profiling of appliances through a single electricity sensor. In Proceedings of the 7th Annual IEEE Communications Society Conference on Sensor, Mesh and Ad Hoc Communications and Networks (SECON), Boston, MA, USA, 21–25 June 2010; pp. 1–9.
15. Lin, Y.-H.; Hu, Y.-C. Electrical energy management based on a hybrid artificial neural network-particle swarm optimization-integrated two-stage non-intrusive load monitoring process in smart homes. *Processes* **2018**, *6*, 236. [CrossRef]
16. Chang, H.H.; Chen, K.L.; Tsai, Y.P.; Lee, W.J. A new measurement method for power signatures of nonintrusive demand monitoring and load identification. *IEEE Trans. Ind. Appl.* **2012**, *48*, 764–771. [CrossRef]
17. Chang, H.H.; Yang, H.T. Applying a non-intrusive energy-management system to economic dispatch for a cogeneration system and power utility. *Appl. Energy* **2009**, *86*, 2335–2343. [CrossRef]
18. Chang, H.H. Genetic algorithms and non-intrusive energy management system based economic dispatch for cogeneration units. *Energy* **2011**, *36*, 181–190. [CrossRef]
19. Chao, M.; Guo, W.; Zhao, Y.; Xiao, L.; Binrui, L. Non-intrusive load monitoring system based on convolution neural network and adaptive linear programming boosting. *Energies* **2019**, *12*, 2882.
20. Wu, X.; Jiao, D.; You, L. Nonintrusive on-site load-monitoring method with self-adaption. *Int. J. Electr. Power Energy Syst.* **2020**, *119*, 105934. [CrossRef]
21. Srinivasan, D.; Ng, W.S.; Liew, A.C. Neural-network-based signature recognition for harmonic source identification. *IEEE Trans. Power Deliv.* **2005**, *21*, 398–405. [CrossRef]
22. Ahmadi, H.; Marti, J.R. Non-intrusive load disaggregation using graph signal processing. *IEEE Trans. Power Syst.* **2015**, *30*, 3425–3436. [CrossRef]

Article

Multivariate Six Sigma: A Case Study in Industry 4.0

Daniel Palací-López [1,†], Joan Borràs-Ferrís [2,*,†], Larissa Thaise da Silva de Oliveria [2] and Alberto Ferrer [2]

1 International Flavors & Fragrances Inc., IFF (Benicarló), 12580 Benicarló, Spain; daniel.palaci@iff.com
2 Multivariate Statistical Engineering Group, Department of Applied Statistics, Operations Research and Quality, Universitat Politècnica de València, 46022 Valencia, Spain; ladasil@etsii.upv.es (L.T.d.S.d.O.); aferrer@eio.upv.es (A.F.)
* Correspondence: joaborfe@eio.upv.es
† These authors have equal contributions.

Received: 31 July 2020; Accepted: 7 September 2020; Published: 9 September 2020

Abstract: The complex data characteristics collected in Industry 4.0 cannot be efficiently handled by classical Six Sigma statistical toolkit based mainly in least squares techniques. This may refrain people from using Six Sigma in these contexts. The incorporation of latent variables-based multivariate statistical techniques such as principal component analysis and partial least squares into the Six Sigma statistical toolkit can help to overcome this problem yielding the Multivariate Six Sigma: a powerful process improvement methodology for Industry 4.0. A multivariate Six Sigma case study based on the batch production of one of the star products at a chemical plant is presented.

Keywords: Six Sigma; Industry 4.0; multivariate data analysis; latent variables models; PCA; PLS

1. Introduction

Six Sigma is a strategy for process improvement widely used in various sectors such as manufacturing, finance, healthcare, and so on. It is defined by Linderman et al. [1] as "an organized and systematic method for strategic process improvement and new product and service development that relies on statistical methods and the scientific method to make dramatic reductions in customer defined defect rates". Besides, Six Sigma, as a quality tool, has fostered a never-ending improvement culture based on a strong and professionalized organization for improvement, a clear and well thought methodology (DMAIC), and also powerful tools and statistical techniques to carry out the improvement projects within the DMAIC framework that has proved highly effective in a large variety of situations [2].

The DMAIC methodology in Six Sigma is a five-step improvement cycle, i.e., Define, Measure, Analyze, Improve, and Control. Reliable data and objective measurements are critical at each step of the method and, hence the statistical techniques are incorporated into the structured method as needed [1]. Traditionally, classical statistical techniques (e.g., multiple linear regression (MLR)) have been used within the DMAIC framework in a data-scarce context mainly from experimental designs. However, with the emergence of Industry 4.0 and the Big Data movement gaining momentum, data abounds now more than ever, and the speed at which they accumulate is accelerating [3]. Besides, due to the increasing availability of sensors and data acquisition systems collecting information from process units and streams, univariate or low-dimensional data matrices evolve to high-dimensional data matrices. In addition, these data often exhibit high correlation, rank deficiency, low signal-to-noise ratio, and missing values [4].

For all this, the Six Sigma statistical toolkit traditionally focused in classical statistical techniques must incorporate new approaches being able to handle complex data characteristics from this current

Industry 4.0 context. In such context, latent variable-based multivariate statistical techniques are widely recommended.

In the literature there are some examples of this integration of multivariate statistical tools into the Six Sigma toolkit. For example, Peruchi et al. [5] integrated principal component analysis (PCA) into a Six Sigma DMAIC project for assessing the measurement system, analyzing process stability and capability, as well as modeling and optimizing multivariate manufacturing processes in a hardened steel turning case involving two critical-to-quality (CTQ) characteristics. In [6], discriminant analysis and PCA were integrated into the DMAIC Six Sigma framework in order to improve the quality of oil type classification from oil spills chromatographic data.

In addition to that, latent variable regression models (LVRM) have also very attractive features, not only for their ability to build models when good predictions, process monitoring, fault detection, and diagnosis are desired (passive use), but also for being able to use this kind of Industry 4.0 data for process understanding, trouble-shooting, and optimization (active use) [4,7]. Note that for an active use causal models are required and, in contrast to machine learning (ML) and MLR models that can only yield causal models if data come from experimental designs, latent variable regression models (such as partial least squares (PLS) [8,9]) do provide causality in the reduced dimensional space of the latent variables even when using historical data corresponding to the daily production of the processes (happenstance data) [10].

This paper reinforces conclusions from previous works in the literature on how Six Sigma's DMAIC methodology can be used to achieve competitive advantages, efficient decision-making, and problem-solving capabilities within the Industry 4.0 context, by incorporating latent variable-based techniques such as PCA into the statistical toolkit leading to the Multivariate Six Sigma. An important contribution of this paper to past literature is that we advocate the use of more advanced techniques via LVRM such as PLS, and illustrate their successful integration into the DMAIC problem solving strategy of a batch production process, one of the most iconic Industry 4.0 scenarios. This type of process, although it shares many of the characteristics represented by the four V's (volume, variety, velocity, and veracity), may not really be Big Data in comparison to other sectors such as social networks, sales, marketing, and finance. However, the complexity of the questions we are trying to answer is really high, and the information that we wish to extract from them is often subtle. This info needs to be analyzed and presented in a way that is easily interpreted and that is useful to process engineers. Not only do we want to find and interpret patterns in the data and use them for predictive purposes, but we also want to extract meaningful relationships that can be used to improve and optimize a process [11] (García-Muñoz and MacGregor 2016), thus making latent variable-based techniques especially relevant, as they permit making proper use of all the data available. More specifically, this paper addresses a case study based on the batch production of one of the star products at a chemical plant.

2. Methods and Materials

2.1. Six Sigma's DMAIC Methodology

In studying Six Sigma's DMAIC methodology, De Mast and Lokkerbol [12] already commented that there are essentially two options: to study the method as it is prescribed in courses and textbooks (prescriptive accounts), or to study it as it is factually applied by practitioners in improvement Six Sigma projects (descriptive accounts). Here, this work is focused on the second option. However, it is crucial to prescribe initially the main functions of each step. Thus, a rational reconstruction of the DMAIC methodology is shown below [12]:

- Define: problem selection and benefit analysis.
- Measure: translation of the problem into a measurable form, and measurement of the current situation; refined definition of objectives.
- Analyze: identification of influence factors and causes that determine the critical to quality characteristics' (CQCs) behavior.

- Improve: design and implementation of adjustments to the process to improve the performance of the CQCs.
- Control: empirical verification of the project's results and adjustment of the process management and control system in order that improvements are sustainable.

As commented above, due to the Industry 4.0 context of the case study, we propose to incorporate latent variable-based techniques within DMAIC usual framework. To aid the reader's understanding, such techniques are described below.

2.2. Latent Variable Models

Latent variable models (LVMs) are statistical models specifically designed to analyze massive amounts of correlated data. The basic idea behind LVMs is that the number of underlying factors acting on a process is much smaller than the number of measurements taken on the system. Indeed, the factors that drive the process leave a similar signature on different measurable variables, which therefore appear correlated. By combining the measured variables, LVMs find new variables (called latent variables (LVs)) that optimally describe the variability in the data and can be useful in the identification of the driving forces acting on the system and responsible for the data variability [13].

2.2.1. Principal Component Analysis

Principal component analysis (PCA) [14–16] is a latent variable-based technique very useful to apply to a data matrix $\boldsymbol{X}$ ($N \times K$), where N is the number of observations and K the variables measured. The goal of PCA is to extract the most important information from $\boldsymbol{X}$ by compressing the K measured variables into new A latent variables that summarize such important information. This allows simplifying the description of the data matrix and easing the analysis of the correlation structure of the observations and the variables. Thus, PCA can be used not only to reduce the dimension of the original space but also to identify patterns on data, trends, clusters, and outliers. In machine learning terminology PCA is an unsupervised method.

In order to achieve these goals PCA projects $\boldsymbol{X}$ into orthogonal directions obtaining new variables of maximum variance (i.e., principal components (PCs) also called LVs) which are obtained as linear combinations of the original variables. The decomposition carried out by PCA can be expressed as:

$$\boldsymbol{X} = \boldsymbol{T} \cdot \boldsymbol{P}^{\mathrm{T}} + \boldsymbol{E} \tag{1}$$

where $\boldsymbol{P}$ ($K \times A$) is the orthogonal loadings matrix, A being the number of LVs, T ($N \times A$) is the scores matrix composed of the score vectors (columns of T), and $\boldsymbol{E}$ ($N \times K$) is the residuals matrix. Score vectors are orthogonal to each other and explain most of the variance of $\boldsymbol{X}$. Besides, due to the orthogonality in P, the A LVs have independent contributions to the overall explained variation. To calculate the parameters in a sequential manner, the non-linear iterative partial least squares (NIPALS) algorithm can be used [17].

2.2.2. Partial Least Squares Regression

LVMs can be also used to relate data from different datasets: an input data matrix $\boldsymbol{X}$ ($N \times K$) and an output data matrix $\boldsymbol{Y}$ ($N \times L$), where L is the number of output variables measured. This is done by means of latent variable regression models (LVRMs), such as partial least squares (PLS) regression. Thus, LVRMs find the main driving forces acting on the input space that are more related to the output space by projecting the input ($\boldsymbol{X}$) and the output variables ($\boldsymbol{Y}$) onto a common latent space. The number of LVs corresponds to the dimension of the latent space and can be interpreted, from a physical point of view, as the number of driving forces acting on a system [18].

In contrast to classical MLR or ML techniques, PLS regression [8,9] not only model the inner relationships between the matrix of inputs $\boldsymbol{X}$ and the matrix of output variables $\boldsymbol{Y}$, but also provide a model for both. Thus, both $\boldsymbol{X}$ and $\boldsymbol{Y}$ are assumed to be simultaneously modelled by the same LVs providing unique and causal models, which is why PLS yields causal models even with data from daily production (i.e. happenstance data not coming from an experimental design). The PLS regression model structure can be expressed as follows:

$$T = X \cdot W^{*} \tag{2}$$

$$X = T \cdot P^{T} + E \tag{3}$$

$$Y = T \cdot Q^{T} + F \tag{4}$$

where the columns of the matrix $\boldsymbol{T}$ ($N \times A$) are the PLS scores vectors, consisting of the first A LVs from PLS These score vectors explain most of the covariance of $\boldsymbol{X}$ and $\boldsymbol{Y}$, and each one of them is estimated as a linear combination of the original variables with the corresponding "weight" vector (Equation (2)). These weights vectors are the columns of the weighting matrix $\boldsymbol{W}^{*}$ ($K \times A$). Besides, the PLS scores vectors are, at the same time, good "summaries" of $\boldsymbol{X}$ according to the $\boldsymbol{X}$-loadings ($\boldsymbol{P}$) (Equation (3)) and good predictors of $\boldsymbol{Y}$ according to $\boldsymbol{Y}$-loadings ($\boldsymbol{Q}$) (Equation (4)), where $\boldsymbol{F}$ and $\boldsymbol{E}$ are residual matrices. In a predictive use, the sum of squares of $\boldsymbol{F}$ is the indicator of how good the predictive model is, and the sum of squares of $\boldsymbol{E}$ is an indicator of how well the model explains the $\boldsymbol{X}$-space.

To evaluate the model performance when projecting the n-th observation $\boldsymbol{x}_n$ onto it, the Hotelling-T^2 in the latent space, T_n^2, and the Squared Prediction Error (SPE), SPE_{x_n}, are calculated [19]:

$$\tau_n = \mathbf{W}^{*\mathrm{T}} \cdot \boldsymbol{x}_n \tag{5}$$

$$T_n^2 = \tau_n^{\mathrm{T}} \cdot \Lambda^{-1} \cdot \tau_n \tag{6}$$

$$SPE_{x_n} = (\boldsymbol{x}_n - \boldsymbol{P} \cdot \tau_n)^{T} \cdot (\boldsymbol{x}_n - \boldsymbol{P} \cdot \tau_n) = \boldsymbol{e}_n^{\mathrm{T}} \boldsymbol{e}_n \tag{7}$$

where $\boldsymbol{e}_n$ is the residual vector associated to the n-th observation, Λ^{-1} the ($A \times A$) diagonal matrix containing the inverse of the A variances of the scores associated to the LVs, and τ_n the vector of scores corresponding to the projection of the n-th observation $\boldsymbol{x}_n$ onto the latent subspace of the PLS model. T_n^2 is the estimated squared Mahalanobis distance from the center of the latent subspace to the projection of the n-th observation onto this subspace, and the SPE_{x_n} statistic gives a measure of how close (in an Euclidean way) such observation is from the A-dimensional latent space.

PLS model can also be expressed as a function of the input variables (as in a classical regression model) by substituting Equation (2) into Equation (4):

$$Y = X \cdot W^{*} \cdot Q^{T} + F = X \cdot B + F \tag{8}$$

where $\boldsymbol{B}$ ($K \times L$) is the PLS regression coefficient matrix. To calculate the parameters of the model in a sequential manner, NIPALS algorithm can be used [20]. In both PCA and PLS regression, NIPALS algorithm has two main advantages: it easily handles missing data and calculates the LVs sequentially (an important property from a practical point of view).

Although PLS was not inherently designed for classification or discrimination, it can be used for both purposes in the form of PLS discriminant analysis (PLS-DA) [21]. Thus, by means of PLS-DA one can explain differences between overall class properties and classify new observations. In machine learning terminology PLS and PLS-DA are supervised methods.

2.3. LVMs in Batch Processes

Batch processes operate for a finite period with a defined starting and ending point, and a time-varying behavior over the operating period. Thus, the data available on batch processes fall into three categories: summary variables ($\boldsymbol{X}^{SV}$) characteristics of each batch such as initial conditions, charge of ingredients, shift, operator, or features from the trajectories of the process variables throughout batch evolution; time-varying process variables throughout the batch evolution (trajectory variables—X^{TV}; and the CQCs of final product (Y)). The nature of these data is represented in Figure 1.

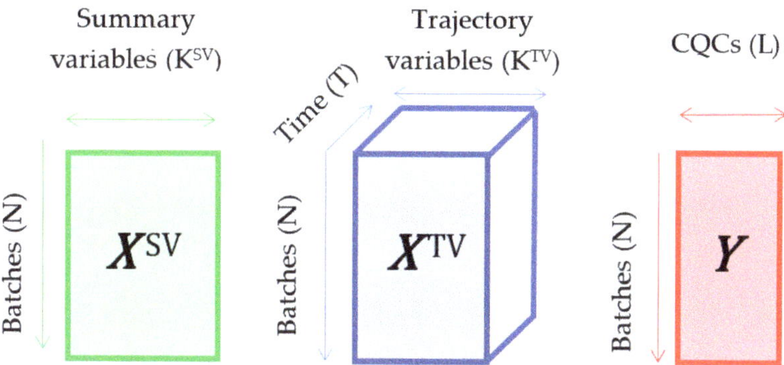

Figure 1. Nature of the batch data set.

The presence of very large arrays, missing data, and the fact that there is temporal and contemporaneous correlation among all variables, provide a strong motivation for using latent variable-based techniques in such data. However, these techniques are based on bilinear modeling approaches and, therefore, a priori they are not able to handle that three-dimensional (3D) X^{TV} arrays. For that, several approaches for analyzing batch data have been proposed where the differences mainly revolve around how to treat X^{TV} [22]. For example, the simplest approach is to extract meaningful landmark features from the trajectories and recording the values of such features for each batch run in a 2D data matrix.

More sophisticated approaches are based on unfolding the 3D X^{TV} in a 2D matrix. One approach is the batchwise unfolding (BWU) [23], that is, to unfold X^{TV} such that all the information for each batch is contained in one row. The data are then mean centered and scaled for each column. Mean centering removes the mean trajectories of each variable, eliminating the main nonlinearity due to the dynamic behavior of the process, and focusing on the variation of all the variables about their mean trajectories. This approach allows exploiting the complex auto and lagged cross correlation structure (i.e., dynamics) of the batch process. Performing any subsequent PCA or PLS analysis on this matrix then summarizes the major sources of variation among the different batches and allows efficient batch-to-batch comparison. This approach also allows for incorporating summary variables (X^{SV}) and final product quality variables (Y) associated with each batch when performing either a PCA or a PLS analysis [19].

Another approach is to unfold the data observation-wise (OWU) [24] with each row corresponding to an observation at some time in each batch, and each column corresponding to the variables measured. Mean centering by column (variable) then simply centers the origin of each variable about zero, but does not remove the average trajectories. The variation remaining is the total trajectory variation for each variable. Performing PCA or PLS (using local batch time or a maturity variable as $\mathbf{y}$) on these OWU data finds a smaller number of components that summarize the major behavior of the complete trajectories of the original variables. It does not initially focus directly on the differences among batches as BWU does, but on summarizing the variable trajectories [22]. The latter can be overcome by carrying out a second model, being similar to the single model BWU, except that it is based on the unfolded score matrix T of the OWU (i.e., OWU-TBWU). The main motivation of the OWU-TBWU approach is dimensionality reduction before BWU stage, which is required when the number of trajectory variables is very large (e.g., when online analytical sensors such as Mass or NIR spectrometers are used). However, the modeling of the OWU data by a single PLS model works well if there is no important dynamics in the process and the instantaneous correlation structure (i.e., the correlation structure or the variables at the same time) remains stable throughout the whole batch evolution. Nevertheless, such assumption does not seem to be realistic at the chemical batch processes analyzed in this case study and, hence, BWU approach is used in this work.

Moreover, batch trajectories are dependent on time (more specifically on the pace the batch is run) and they are rarely synchronized (i.e., the key process events do not overlap at the same time in all batches) [25]. To compare these batch histories and apply statistical analysis one needs to reconcile the timing differences and the key process events among these trajectories [26]. This can be achieved using the dynamic time warping (DTW) method with only a minimal amount of process knowledge [27]. This method nonlinearly warps all batch trajectories to match as closely as possible that of a reference batch. Figure 2 illustrates the BWU approach followed by DTW synchronization.

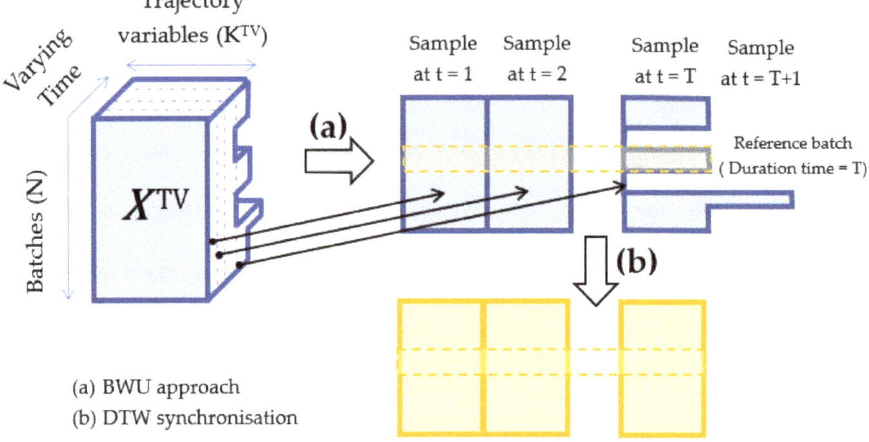

Figure 2. (**a**) Batchwise unfolding (BWU) approach and (**b**) dynamic time warping (DTW) synchronization of the 3D $\boldsymbol{X}^{TV}$.

Note that, once $\boldsymbol{X}^{TV}$ is unfolded and synchronized as seen in Figure 2, it is concatenated with $\boldsymbol{X}^{SV}$, resulting in the matrix $\boldsymbol{X}$ to be used in the PCA or PLS model.

2.4. Software

The software packages used are as follows:

- Multivariate Exploratory Data Analysis (MEDA) Toolbox (for Matlab) [28] for variable and batch screening, and imputation of missing data within a batch.
- MVBatch Toolbox (for Matlab) [29] for batch synchronization.
- Aspen ProMV for calibration by using synchronized batch data, and data analysis.
- Minitab for control chart plotting.

3. Results

In this section, the results from applying each of the DMAIC steps (Define, Measure, Analyze, Improve, and Control) to the process are shown, each in their own subsection. Note that latent variable-based methods such as PCA or PLS are used, instead of more classical ones such as MLR or even ML. Therefore, the tools implemented in some of the steps of the DMAIC cycle differ from more traditional approaches, but the original purpose of each stage remains.

3.1. Define

The aim of this stage is to identify opportunities for improvement that lead to e.g., an increase in benefits, reduced costs or losses, a mitigation of the environmental impact, etc. This requires pinpointing observed problems, framing them within the context of the corresponding processes, evaluating the costs and benefits of addressing them, and locating the most appropriate people to do it given the existing constraints on time and resources.

In this Six Sigma project, the focus was set on the purity (before separation) and volume of production (after purification) of one of the star products at the chemical plant where the continuous improvement program was implemented. This came as a result of an observed increase in the variability of this product's purity, due mostly to significantly lower values (compared to previous operations) being obtained once every four batches, approximately. This also meant a decrease in its average value of around 1%, starting on September of 2014, with an estimated monetary loss of more than 100,000 €/year with respect to previous years.

Figure 3a illustrates the 'Suppliers, Inputs, Process, Outputs, Customers' (SIPOC) diagram identifying the supplier (reaction 1) and inputs (one of the outputs from reaction 1) for the specific process under study (reaction 2), as well as its outputs (with primary focus on the composition of the so-called subproduct 2) and the customer (reaction 3). Figure 3b corresponds to a simplified process block diagram for reaction 2, in which four numbers have been included to indicate the points of the process where the process variables and critical to quality characteristics (CQC) relevant to this work are routinely measured.

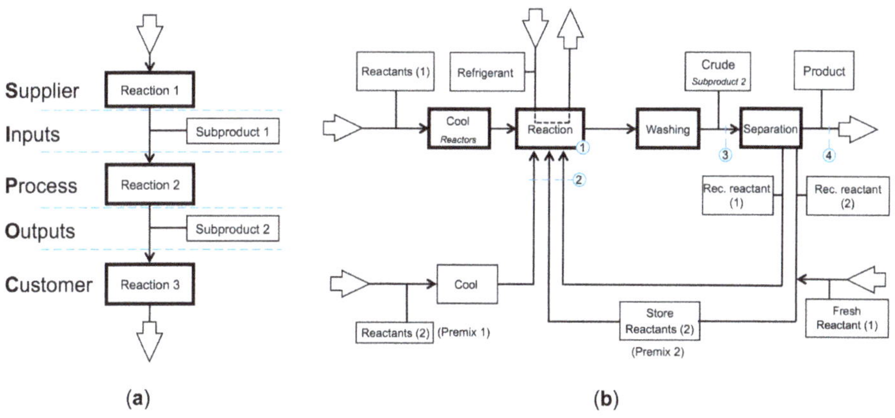

Figure 3. (**a**) 'Suppliers, Inputs, Process, Outputs, Customers' (SIPOC) diagram; (**b**) simplified process block diagram for reaction 2.

Due to confidentiality reasons, the name of the reactants, products, and subproducts cannot be disclosed. However, it must be noted that both 'Reactants (1)' and 'Fresh Reactant (1)' come from reaction 1. Likewise, Premix 1 and Premix 2 are blends constituted by the same chemical species, these being freshly produced raw materials for Premix 1, and one of them (Rec. reactant (2)) having been recovered after the washing and separation following reaction 2. For each batch, each reactor was fed Premix 1 or Premix 2, but never a blend of both. A quick look at already available data revealed that the main concern that motivated this Six Sigma project was related to a loss of purity (in terms of the desired chemical species) of the Crude, (i.e., unrefined product) and therefore a lower amount of Product arriving to reaction 3 per batch of reaction 2.

The project team was constituted by the authors of this paper (Six Sigma Black Belts), championed by two high-profile members from the company and supported by three experts in the process from the technical team. Regarding the planning, the project was estimated to require over six months to be completed. Experimenting in a laboratory or at pilot-plant level was not recommended, since no proper scaling could be done. Additionally, altering or interrupting the production of this process was not allowed to any extent, and therefore experimenting on the plant itself was not an option either. Due to this, only historical data from past production could be used.

3.2. Measure

During this stage, as much available data as possible was collected, and their validity assessed. With these data, an evaluation of the initial situation was done, and potential causes were looked at for the issue that motivated the project.

3.2.1. Available Data

Data from a total of 17,147 batches produced in two different reactors during a nine years period was available, containing information about:

- the averaged values for three process variables (x_1 to x_3) for each batch, measured at point (1) in Figure 3 (i.e., the corresponding reactor);
- amounts (x_4 to x_7) and proportions (x_8 to x_{11}) of some of the most relevant reactants involved in the reaction, measured at point (2) in Figure 3 (i.e., before being introduced into the reactor);
- four categorical variables indicating whether a batch was produced in the first or second reactor (x_{12}) and the use or not of an auxiliary piece of equipment (x_{13}), registered at point (1) in Figure 3; and whether an excess of accumulated reactant had been recovered or not (x_{14}), and whether Premix 1 or Premix 2 had been fed to the reactor for the corresponding batch (x_{15}), registered at point (2) in Figure 3;
- the evolution along the complete duration of each batch for 11 process variables (x_{16} to x_{26}), measured at point (1) in Figure 3, and;
- information on 10 CQC (y_1 to y_{10}), including the purity of the product of interest (y_8), measured at point (4) in Figure 3; and the measure of the total amount of crude coming out of the process (y_4), and its estimation through mass balance (y_6), measured/registered at point (3) in Figure 3.

The first three groups of variables will be referred to as 'summary variables', since they provide summarized information of each batch (e.g., average observed values or setpoint values), disregarding their variation during the evolution of the chemical reaction until completion and discharge of the reactor. The fourth group, on the contrary, is comprises 'trajectory variables' that may show differences in the evolution of the corresponding process conditions among batches even when their average or target values coincide.

Although further experimentation may be suggested during this step to enrich the database used in following stages, such experimentation is not possible in this case, as previously stated, and therefore no such approach will be addressed in this section.

3.2.2. Validation of the Data

In order to detect potential outliers, a PCA model with two latent variables (LVs) was fitted using all available data for the 'summary variables' as provided, resulting in a model that explained 17% of the variability of these data. Adding any more LVs provides no additional information that is useful at this stage, and instead results in PCA models with lower explanatory and predictive capabilities, and less ability for the detection of outliers. A representation of the SPE of all observations in the database resulted in Figure 4.

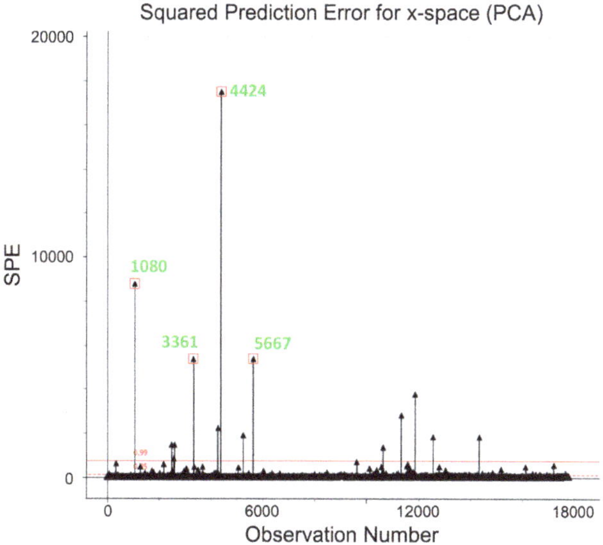

Figure 4. SPE for all observations in the dataset with the summary variables and critical to quality characteristics (CQCs) for a principal component analysis (PCA) model fitted with two LVs [$R^2(X) = 17\%$], SPE 95% (dotted red line) and 99% (continuous red line) confidence limits, and the four observations with highest SPE values highlighted.

This plot allows quickly detecting observations that do not abide by the correlation structure found by the PCA model in the dataset for the 'summary variables'. A contribution plot, such as the one in Figure 5, provides additional information regarding which variables are responsible for the high SPE value for the corresponding observation. In this case, variable x_4 presents an abnormally high value for observation 1080, not following the correlation structure found in the data by the PCA model.

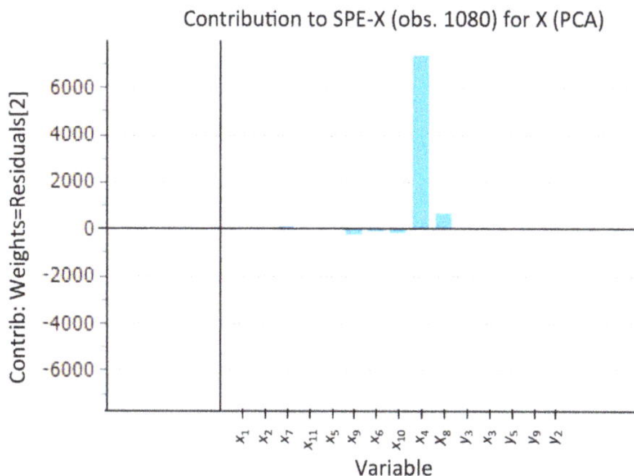

Figure 5. Contribution plot for observation 1080, seen in Figure 4 to have a SPE value significantly above the 99% confidence limit for a PCA model fitted with two LVs [$R^2(X) = 17\%$]. Variable x_4 is seen to be the biggest contributor to the SPE value.

An in-depth analysis of the factors that contribute to the high SPE values of all observed outliers allowed curating the database to either correct wrongly registered data or to eliminate outliers before continuing with the analysis. Consequently, the dataset was reduced from 17,147 original batches to 16,813.

The same procedure was followed for the dataset containing the 'trajectory variables', but no outliers were found among these data other than the ones already identified with the 'summary variables'. As a consequence, these observations were discarded before continuing.

On the other hand, process variable x_5 was found to present almost no variability in the dataset and was also discarded before going on. Additionally, variables 'day', 'month', and 'year' were included only as labels with which observations could be colored, in order to identify possible patterns, stationary effects or changes with time without artificially biasing the model to account for these variables. However, no clustering or displacement of the observations was detected this way, and the presence of outliers was not found to be correlated to these time-related variables.

3.2.3. Quantified Initial Situation and Potential Causes of the Observed Problem

Once outliers were eliminated from the dataset, the starting point of the project was determined according to the remaining information. Figure 6 shows the evolution of the purity of the product of interest (y_8) with time for both reactors. The superimposed dashed blue lines mark the separation between batches produced before and after September 2014.

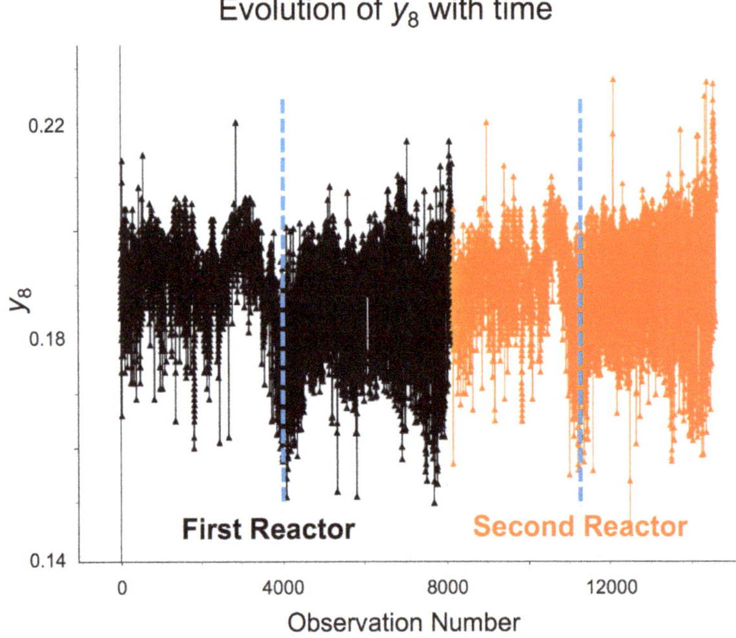

Figure 6. Evolution of the purity of the product of interest (y_8) with time for the first (black) and second (orange) reactors; the dashed blue lines separate batches produced before (left side) and after (right side) September 2014.

The average value for y_8 after September 2014, compared to before, was around 0.1% lower, while its standard deviation had increased to 1.008% (1.47 times that of past batches). Both changes (in average value and variability) were found to be statistically significant (p-value < 0.05), which corroborated, at least partially, the concerns expressed by the technicians at the start of the project. When asked, they mentioned that several changes had taken place at some point during 2014, such as the addition of an auxiliary refrigerating system to one of the reactors, the way the reactants were fed or the recovery of some amount of unreacted raw materials after each reaction.

3.3. Analyze

The main goal of this stage was to identify which process parameters have a significant effect on the product's purity, evaluate the nature of their effect (antagonistic or synergic), and how they relate to each other. In order to achieve it, a PCA model was fitted, and presented in Section 3.3.1, with all summary variables and CQC to explore the correlation structure among them in the database, and to detect clusters of batches that operated in similar way in the past. Afterwards, a PLS-regression model permits predicting the CQCs from the summary variables, and was used to determine which of these factors have a significant effect on the purity of the product of interest (y_8), as seen in Section 3.3.2. Once these variables are identified, a PLS-DA was performed considering the trajectory variables, to assess which of them are responsible for the observed differences between batches with higher and lower performance, as illustrated in Section 3.3.3.

3.3.1. Principal Component Analysis of the Summary Variables and CQCs

This first exploratory analysis was aimed at providing relevant information regarding the existing correlation structure among summary variables and CQCs, and detecting clusters of batches operating in similar conditions and/or providing similar results. Given that outliers were already eliminated from the dataset, a PCA model with five LVs [$R^2(X) = 76\%$] could be directly fit. Additional LVs beyond the fifth corresponded to either variation of individual variables independently of others, or variations not related to the CQCs, and were therefore not considered relevant to the goal of the project.

The Hotelling-T^2 values for the observations used to fit this model can be seen in Figure 7a. Here, batches were colored by variable x_{13} (black: 0; orange: 1). In Figure 7b, the scores plot of LV2 (explaining 20% of the variability of the data) versus LV1 (explaining 22% of the variability) is shown, such that the left red cluster corresponds to the observations in orange in Figure 7a, and the rest correspond to observations in black in Figure 7a. Figure 7c, where the loadings for the variables in the two first latent variables are represented, allows the interpretation of this clustering. In it, variables x_6 and x_{10}, and $x_{13} = 1$, can be seen on the left side, with values close to zero in the second component, while variables x_4 and x_8, and $x_{13} = 0$, are found in the opposite side. This provides two valuable pieces of information (which Figure 7d illustrates, too):

- Variables x_4 and x_8 present higher values for the batches in the blue cluster in Figure 7b, and lower values for the batches in both red clusters, while the opposite is true for variables x_6 and x_{10}. Variable x_{13} takes the value 0 for batches in the blue cluster and in the rightmost red cluster, and 1 for the leftmost red cluster.
- Variables x_4 and x_8 are positively correlated, as are variables x_6 and x_{10}, and the former couple is negatively correlated with the later.

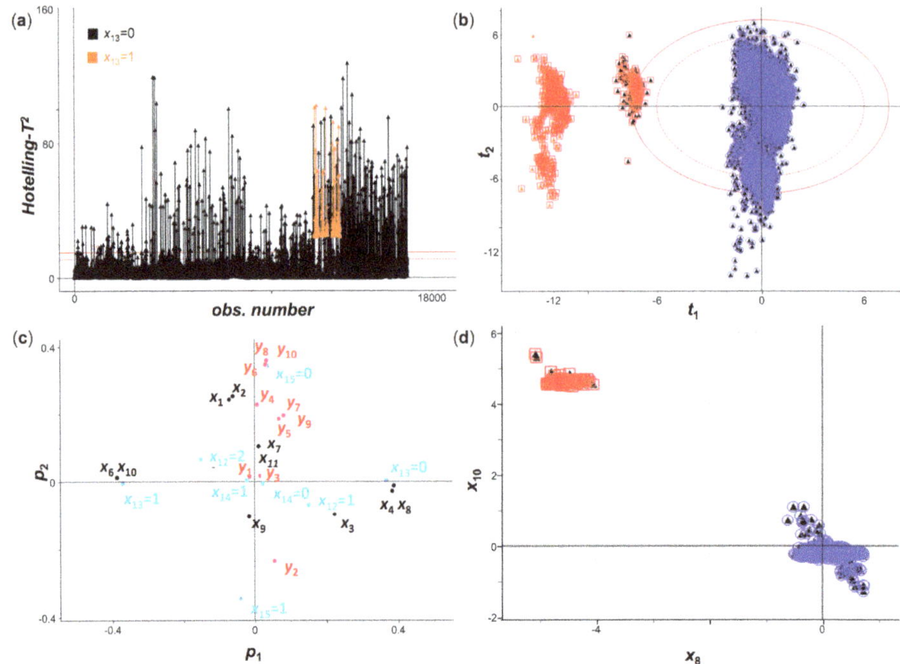

Figure 7. (**a**) Hotelling-T^2 plot of the observations in the dataset with the summary variables and CQCs for a PCA model fitted with five LVs [$R^2(X) = 76\%$], T^2 95% (dotted red line) and 99% (continuous red line) confidence limits, colored by x_{13} (black: 0; orange: 1); (**b**) scores plot for the two first LVs (t_2 vs. t_1) showing three clusters of observations: red circled orange dots associated to $x_{13} = 1$, above average values for x_6 and x_{10} and below average values for x_4 and x_8; red circled black dots associated to $x_{13} = 0$, above average values for x_6 and x_{10} and below average values for x_4 and x_8, and; blue circled black dots associated to $x_{13} = 0$, below average values for x_6 and x_{10} and above average values for x_4 and x_8; (**c**) loadings plot for the two first LVs (p_2 vs. p_1) with CQCs in red, continuous process variables in black, and binary process variables in cyan; (**d**) scatterplot for x_{10} vs. x_8, using the same color code as in Figure 7b.

Note that, more generally, the relationships among all process variables and CQCs in the dataset used to fit the PCA model can also be assessed by looking at the loading plots. In this plot, if the corresponding LVs explain a relevant percentage of the model variability, variables lying close to each other (and far away from the center) will tend to show positive correlation; while if they lay at the opposite site in the plot they will tend to show negative correlation. Figure 7d can be resorted to for carrying out such analysis (latent variables three to five do not, in this case, alter this interpretation). This way, in addition to the aforementioned correlations, positive correlations were found between variables y_4, y_6, y_8, and y_{10}, and variables x_1, x_2, and $x_{15} = 0$, as well as between x_3 and variables y_5, y_7, and y_9. On the other hand, negative correlations were found between y_2, and all other CQCs except for y_1 and y_3, as well as between x_3 and variables x_1 and x_2, and between x_9 and variables x_7 and x_{11}. More importantly, no clear correlation was found between y_8 and variables x_8 to x_{11}. Bivariate dispersion plots for each pair of variables were used to visualize each of these relationships (or lack thereof), and also allowed detecting that not only was $x_{15} = 0$ positively correlated with y_8, but that the intensity of the positive/negative correlations between y_8 and other process variables and CQCs varied when $x_{15} = 1$ (Premix 2 fed to the reactor) with respect to $x_{15} = 0$ (Premix 1 fed to the reactor). As an example, the positive correlation between x_2 and y_8, as well as the relationship between x_{15} and y_8, and the interaction effect between x_2 and x_{15}, are shown in Figure 8.

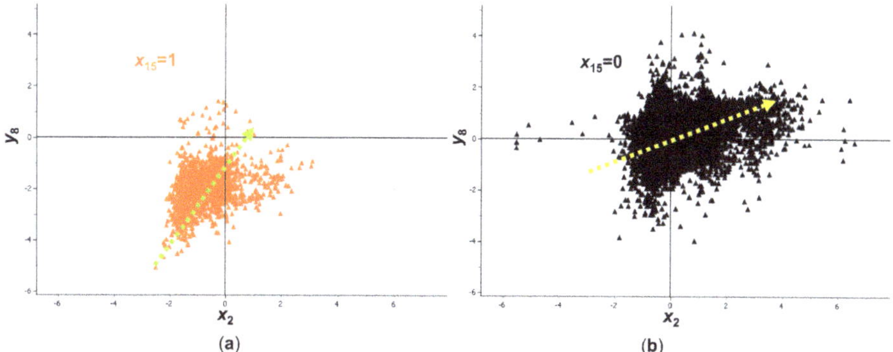

Figure 8. Scatter plot for y_8 vs. x_2, for (**a**) $x_{15} = 1$, and the approximate direction of maximum variability indicated by a green arrow, and (**b**) $x_{15} = 0$, and the direction of maximum variability indicated by a yellow arrow.

From Figure 8a,b, the positive correlation between x_2 and y_8 can be immediately confirmed. Furthermore, the cluster of batches for which $x_{15} = 0$ presents higher values (on average) than those for which $x_{15} = 1$. This is coherent with the conclusions extracted from Figure 7c Additionally, however, the slopes of the green arrow in Figure 8a and the yellow one in Figure 8b differ, pointing to a stronger, more positive correlation between x_2 and y_8 when $x_{15} = 1$, compared to their weaker, but still positive, relationship when $x_{15} = 0$. Therefore, it can be suspected that an interaction exists between x_2 and x_{15}.

3.3.2. Partial Least Squares Regression to Predict the CQCs from the Summary Variables

This analysis was performed in order to identify the sources of variability of the process most related to the product's purity (i.e., variables y_4, y_6, and y_8). This required confirming previous results and quantifying the relationship between the summary variables and the CQCs. For the sake of brevity, only the results regarding the established predictive model for y_8 will be shown in this section, as those to predict y_4 and y_6 provide the same overall conclusions. The potential effects of time related variables ('month' and 'year') and interaction effects between the categorical variables x_{12} to x_{15} and other summary variables were also considered initially. However, no statistically significant differences in the CQCs were found between reactors, and the effect of variables 'month' and 'year' was not statistically significant either. Furthermore, all interaction effects were discarded for the same reason, except for the interactions of both levels of x_{15} with variables x_1 and x_2. Figure 9 presents the coefficients of the resulting PLS-regression model fitted with two LVs [$R^2(Y) = 25.86\%$; $Q^2(Y) = 25.81\%$] to predict y_8.

One important consideration in this analysis is that variable x_3, which is known to be critical in the process, did not appear to have a significant effect on the most relevant CQCs. This is partially because this is a very strictly controlled variable. On the other hand, variables x_1 and x_2, with which x_3 is negatively correlated, are found to apparently have a significant positive effect on y_8. From this, and according to the technicians' knowledge of the process, one may conclude that it is x_3 that has a statistically significant, negative effect on y_8, but one should be cautious given that x_1, x_2, and x_3 do not vary independently. This is illustrated in Figure 10, where batches with higher values for x_1 also present higher values of y_8, on average (i.e., for lower values of x_1, batches with similar and smaller values of y_8 are observed), which points to the positive correlation between x_1 and y_8. On the other hand, batches with the highest values of x_1 only operated at values of x_3 close to its historical minimum, which also illustrates the negative correlation between x_1 and x_3. This could also be seen in the loadings plot of the PCA in Figure 7c. To disentangle this potential aliasing some experimentation should be run in the future, when/if possible.

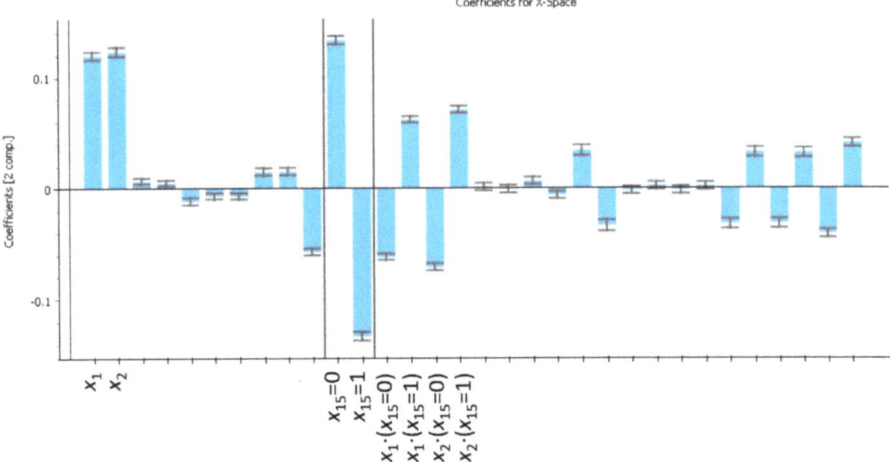

Figure 9. Regression coefficients of the partial least squares (PLS)-regression model fitted with two LVs [$R^2(Y) = 25.86\%$; $Q^2(Y) = 25.81\%$] to predict y_8 from the summary variables and their interactions with x_{15}.

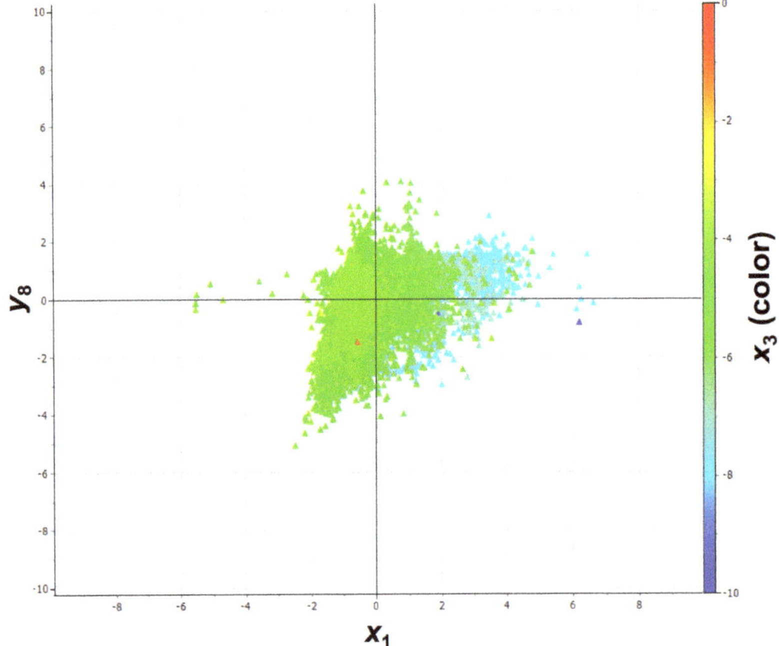

Figure 10. Scatter plot for y_8 vs. x_1, with the observations colored according x_3.

Nevertheless, the negative relationship between $x_{15} = 1$ (using Premix 2) and y_8, already seen in Figure 8, was confirmed once more, as seen in Figure 9, and so finding what is being done differently in such case becomes relevant.

3.3.3. PLS-Discriminant Analysis to Identify Differences in Batches Using Premix 1 and Premix 2

Since x_{15} seems to be one of most important variables affecting the purity of the product, y_8, conclusions obtained by previous analysis were confirmed by means of a PLS-DA model, which was resorted to for finding which variables are responsible for the differences in how batches operated when Premix 2 ($x_{15} = 1$) was fed into the reactor, compared to when Premix 1 ($x_{15} = 0$) was used. This analysis was carried out considering both the summary and trajectory variables, but only the results with the trajectory variables (x_{16} to x_{26}) are illustrated here, for the sake of both brevity and clarity. Figure 11 shows the separation between both clusters of batches in the latent space, while Figure 12 presents the model coefficients associated to each process variable, included the 'warping profile' that results from aligning the trajectories, for a PLS-DA regression model to predict $x_{15} = 1$. This model was fitted with eight LVs, as this number of LVs provided the model with the most discriminant power [$R^2(Y) = 79.80\%$; $Q^2(Y) = 71.90$].

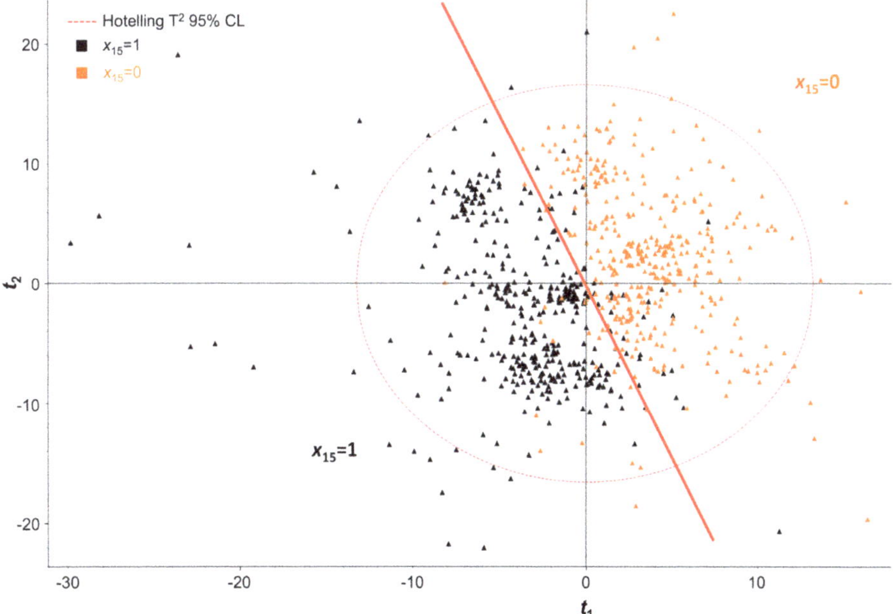

Figure 11. Scores plot for the two first LVs (t_2 vs. t_1) for the PLS-discriminant analysis (DA) regression model fitted with eight LVs [$R^2(Y) = 79.80\%$; $Q^2(Y) = 71.90\%$], showing the separation between batches with $x_{15} = 0$ (orange; right side of the red straight line) and $x_{15} = 1$ (black; left side of the red straight line).

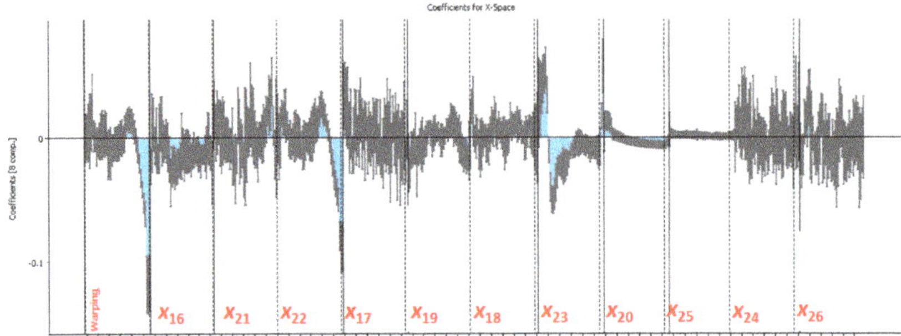

Figure 12. Model coefficients (blue bars) and the confidence interval for a 95% confidence level (grey intervals) associated to each variable for the PLS-DA regression model to predict $x_{15} = 1$ from variables 'warping profile' and x_{16} to x_{26}, fitted with eight LVs [$R^2(Y) = 79.80\%$; $Q^2(Y) = 71.90$]. Positive values indicate that higher values for the corresponding variable at that point in the batch are expected, on average, for batches with $x_{15} = 1$ (Premix 2 used).

From Figure 12 it is concluded that batches where Premix 2 was fed to the reactor ($x_{15} = 1$):

- Proceeded, at the latest stages of the batch, faster than those where Premix 1 was fed instead ($x_{15} = 0$), as seen by the negative values for variable 'warping' near the end of the batch duration.
- Presented lower (and decreasing) values for variable x_{22} (ingredient flow) at the end of the batch.
- Operated at higher values for variable x_{23} (ingredient temperature) at the start of the batch, but lower during the middle part of the batch.

It is worth noting that, although the technicians at the plant were not surprised by the discrepancy in the values observed for variable x_{23} at the start of the batch for the two clusters, the values during the middle part was, according to them, opposite to their expectations.

3.4. Improve

As a result of the analyses performed and summarized in the previous section, the team responsible for this process was able to pinpoint a specific behavior in the process potentially related to the loss of purity/production volume of the desired product (y_4, y_6, and y_8) when Premix 2 was fed into the reactor. When compared to one another, the average value for y_8 for $x_{15} = 1$ was found to be 1.9332% lower than for $x_{15} = 0$, and the standard deviation for y_8 for $x_{15} = 1$ was also 1.12 times that for $x_{15} = 0$, with both differences being statistically significant (p-values < 0.05). A preliminary confirmation of this was provided by the historical data itself, according to which Premix 2 was fed to the reactor for the first time on September 2014, matching the increase in variability and drop in average value for y_8. While the technicians at the plant knew about this operational change (which they themselves implemented), they initially argued that such modification did not justify the change in the process outputs, but the unexpected behavior of variable x_{23} (see Figure 11) pointed to either a contamination issue with Premix 2, or to other, not registered process variables as potential causes. Consequently, they decided to further investigate them and, once detected, to correct the actual issue (the details on how this was done is not disclosed for confidentiality issues) and standardize the treatment of both Premix 1 and Premix 2 to make them equal. The solution was validated during the next two months, and remains a success to this day, with estimated benefits/savings above 140,000 €/year (40,000€ higher that the initial estimation in the 'Define' step).

3.5. Control

Once the causes of the loss in productivity were detected and addressed, an instrumental monitoring scheme was implemented in the plant to detect possible deviations in the process variables, and specially with regards to Premix 1 and Premix 2. This monitoring system was designed in a way that avoids expensive and time-consuming tests (details not disclosed to the project team), the lack of which allowed the actual cause of the issue to go unnoticed before the Six Sigma project was carried out. Currently, the use of multivariate statistical analysis techniques for monitoring purposes supports this instrumental monitoring scheme by quickly signaling out-of-specification batches and delimiting the range of potential causes for such events.

As an example of the achieved improvement, as well as an appropriate Individual-Moving Range (I-MR) control chart to be used for monitoring y_8 after the project goal was achieved, Figure 13 illustrates the evolution of variable y_8 before and after improvement, for batches where Premix 2 was fed to the reactor.

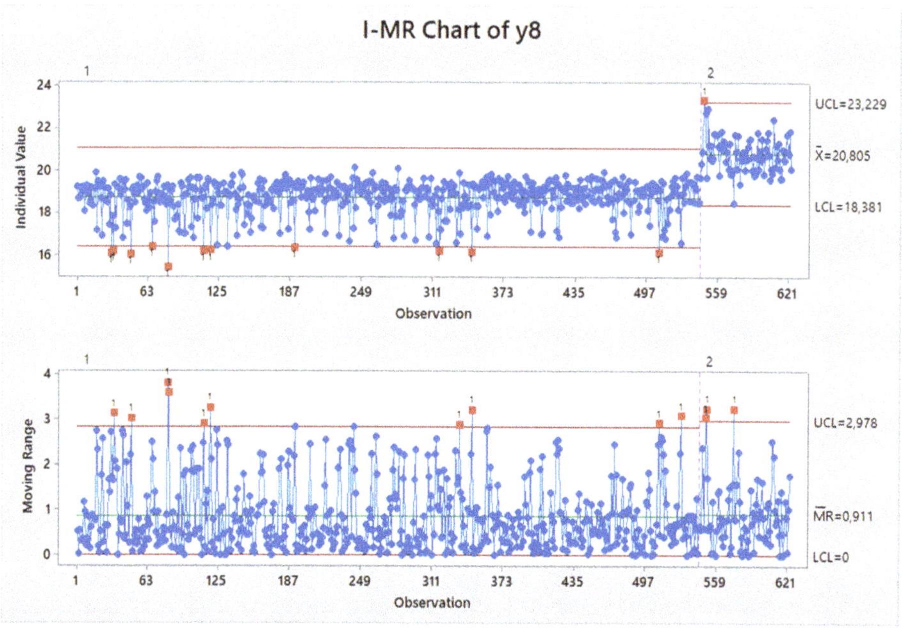

Figure 13. Individual-Moving Range (I-MR) control chart for variable y_8, for batches where Premix 2 was fed into the reactor, before (1) and after (2) improvement in accordance to the conclusions of the Six Sigma project.

4. Discussion

While the success of the Six Sigma methodology has already been documented in the past in numerous industrial case studies, the tools used (mainly based on linear regression and simple graphical displays) are usually those suitable for scenarios where not much information is available yet, a relatively limited number of factors are involved or relevant, and/or experimentation can be carried out to a minimum yet significant extent. In the problem addressed in this project, related to a batch production process, due to the nature of the data registered, typical of Industry 4.0, none of these apply and, therefore, alternative methods had to be resorted to.

In particular, latent variable-based methods such as PCA, PLS, and PLS-DA, applied to historical (i.e., not from DOE) data of both summary variables and trajectory variables (usually just referred to as batch data) were able to extract valuable information to pinpoint the actual causes of the loss of productivity in a real case study. These tools were also implemented for troubleshooting purposes in the future.

In contrast with these latent variable-based methods, data from DOE would have been required to use tools such as linear regression or machine learning tools to infer causality, which is needed for process understanding and optimization purposes. However, as it is typical in Industry 4.0 no data from DOE were available. As an example, when linear regression was applied to the available historical data, due to the highly correlated regressors (process variables), different models using different regressors and having different weights or coefficients on them gave nearly identical predictions and similar to PLS model, but failed to properly identify the relationships between x_1, x_2, x_3, and y_8, as well as the existing interaction effect between the use of Premix 1 ($x_{15} = 0$)or Premix 2 ($x_{15} = 1$) and other process variables, such as the ones shown in Figure 8. Had any of such linear regression models been used in the 'Improve' step of the DMAIC for process improvement, a different set of process operating conditions would have been advised with a high probability of not being feasible in practice, as a result of e.g., the actual relationships between x_1, x_2, and x_3 going unnoticed. Furthermore, a lesser degree of improvement would have been achieved, presumably, if the interaction between e.g., x_2 and x_{15} had not been discovered. Therefore, this constitutes a clear example of the dangers of resorting to more basic linear regression (and also machine learning) techniques for process optimization in scenarios they are not suitable for (i.e., where causality cannot be inferred directly from the raw data), as in this case study, analyzing historical data.

In summary, the use of latent variable-based methods allowed the efficient use of the Six Sigma methodology in a batch production process where this could not have been done using a traditional Six Sigma toolkit, which lead to significant short- and long-term savings, in addition to the implementation of a more robust monitoring system.

5. Conclusions

Traditional Six Sigma statistical toolkit, mainly focused on classical statistical techniques (such as scatterplots, correlation coefficients, and linear regression models from experimental designs), is seriously handicapped for problem solving using process data coming from Industry 4.0. In this context, abundant historical process data involving hundreds/thousands of variables highly correlated with missing values are registered from daily production.

PCA can be used in this context as an exploratory tool not only to reduce the dimension of the original space and visualize the complex variables relationship but also to deal with missing data, identify patterns on data, trends, clusters, and outliers.

As data do not come from a DOE, input-output correlation does not mean necessarily causation, and classical predictive models (such as MLR and ML), proven to be very powerful in passive applications (i.e., predictions, process monitoring, fault detection, and diagnosis), cannot be used for extracting interpretable or causal models from historical data for process understanding, trouble-shooting, and optimization (active use), key goals of any Six Sigma project. This is the essence of the Box et al. (2005) warning [30]: predictive models based on correlated inputs must not be used for process optimization if they are built from observational data (i.e., data not coming from a DOE).

In contrast to classical MLR or ML techniques, PLS regression provides unique and causal models in the latent space even if data come from daily production process. These properties make PLS suitable for process optimization no matter where the data come from.

Therefore, Six Sigma's DMAIC methodology can achieve competitive advantages, efficient decision-making and problem-solving capabilities within the Industry 4.0 context by incorporating latent variable-based techniques, such as principal component analysis and partial least squares regression, into the statistical toolkit leading to Multivariate Six Sigma.

Author Contributions: Conceptualization, A.F.; methodology, A.F., J.B.-F. and D.P.-L.; software, J.B.-F. and D.P.-L.; validation, J.B.-F. and L.T.d.S.d.O.; formal analysis, J.B.-F., L.T.d.S.d.O. and D.P.-L.; data curation, L.T.d.S.d.O. and D.P.-L.; writing—original draft preparation, J.B.-F. and D.P.-L.; writing—review and editing, A.F.; visualization, J.B.-F., L.T.d.S.d.O. and D.P.-L.; supervision, A.F.; project administration, A.F.; funding acquisition, A.F. All authors have read and agreed to the published version of the manuscript.

Funding: This research received no external public funding.

Acknowledgments: The authors would like to acknowledge the invaluable help received from José Antonio Rojas, Antonio José Ruiz, Tim Ridgway, Ignacio Martin, and Steve Dickens, as well as the technicians at the plant, who helped greatly in the acquisition of the data, investigation at the plant, and essential understanding of the process. Without their knowledge and experience, the success of this project would have not been possible.

Conflicts of Interest: The authors declare no conflict of interest.

References

1. Linderman, K.; Schroeder, R.G.; Zaheer, S.; Choo, A.S. Six Sigma: A goal-theoretic perspective. *J. Oper. Manag.* **2003**, *21*, 193–203. [CrossRef]
2. Grima, P.; Marco-Almagro, L.; Santiago, S.; Tort-Martorell, X. Six Sigma: Hints from practice to overcome difficulties. *Total Qual. Manag. Bus. Excell.* **2014**, *25*, 198–208. [CrossRef]
3. Reis, M.S.; Gins, G. Industrial process monitoring in the big data/industry 4.0 era: From detection, to diagnosis, to prognosis. *Processes* **2017**, *5*, 35. [CrossRef]
4. Ferrer, A. Multivariate Statistical Process Control Based on Principal Component Analysis (MSPC-PCA): Some Reflections and a Case Study in an Autobody Assembly Process. *Qual. Eng.* **2007**, *19*, 311–325. [CrossRef]
5. Peruchi, R.S.; Rotela Junior, P.; Brito, T.G.; Paiva, A.P.; Balestrassi, P.P.; Mendes Araujo, L.M. Integrating Multivariate Statistical Analysis Into Six Sigma DMAIC Projects: A Case Study on AISI 52100 Hardened Steel Turning. *IEEE Access* **2020**, *8*, 34246–34255. [CrossRef]
6. Ismail, A.; Mohamed, S.B.; Juahir, H.; Toriman, M.E.; Kassim, A. DMAIC Six Sigma Methodology in Petroleum Hydrocarbon Oil Classification. *Int. J. Eng. Technol.* **2018**, *7*, 98–106. [CrossRef]
7. Jaeckle, C.M.; Macgregor, J.F. Product Design through Multivariate Statistical Analysis of Process Data. *AIChE J.* **1998**, *44*, 1105–1118. [CrossRef]
8. Höskuldsson, A. PLS regression methods. *J. Chemom.* **1988**, *2*, 581–591. [CrossRef]
9. Wold, S.; Sjostrom, M.; Eriksson, L. PLS-Regression—A basic tool of chemometrics. *Chemom. Intell. Lab. Syst.* **2001**, *58*, 109–130. [CrossRef]
10. MacGregor, J.F. Empirical Models for Analyzing "BIG" Data—What's the Difference? In Proceedings of the 2018 Spring Meeting and 14th Global Congress on Process Safety, Orlando, FL, USA, 22–26 April 2018; AIChE: New York, NY, USA, 2018.
11. García Muñoz, S.; MacGregor, J.F. Big Data: Success Stories in the Process Industries. *Chem. Eng. Prog.* **2016**, *112*, 36–40.
12. De Mast, J.; Lokkerbol, J. An analysis of the Six Sigma DMAIC method from the perspective of problem solving. *Int. J. Prod. Econ.* **2012**, *139*, 604–614. [CrossRef]
13. Tomba, E.; Facco, P.; Bezzo, F.; Barolo, M. Latent variable modeling to assist the implementation of Quality-by-Design paradigms in pharmaceutical development and manufacturing: A review. *Int. J. Pharm.* **2013**, *457*, 283–297. [CrossRef] [PubMed]
14. Wold, S.; Esbensen, K.; Geladi, P. Principal component analysis. *Chemom. Intell. Lab. Syst.* **1987**, *2*, 37–52. [CrossRef]
15. Abdi, H.; Williams, L.J. Principal component analysis. *Wiley Interdiscip. Rev. Comput. Stat.* **2010**, *2*, 433–459. [CrossRef]
16. Bro, R.; Smilde, A.K. Principal component analysis. *Anal. Methods* **2014**, *6*, 2812–2831. [CrossRef]
17. Wold, H. Estimation of principal components and related models by iterative least squares. In *Multivariate Analysis*; Krishnaiah, P.R., Ed.; Academic Press: New York, NY, USA, 1966; pp. 391–420.
18. Tomba, E.; Barolo, M.; García-Muñoz, S. General Framework for Latent Variable Model Inversion for the Design and Manufacturing of New Products. *Ind. Eng. Chem. Res.* **2012**, *51*, 12886–12900. [CrossRef]
19. Kourti, T.; Macgregor, J.F. Multivariate SPC Methods for Process and Product Monitoring. *J. Qual. Technol.* **1996**, *28*, 409–428. [CrossRef]
20. Geladi, P.; Kowalski, B.R. Partial least-squares regression: A tutorial. *Anal. Chim. Acta* **1986**, *185*, 1–17. [CrossRef]
21. Barker, M.; Rayens, W. Partial least squares for discrimination. *J. Chemom.* **2003**, *17*, 166–173. [CrossRef]

22. Wold, S.; Kettaneh-Wold, N.; MacGregor, J.F.; Dunn, K.G. Batch Process Modeling and MSPC. *Compr. Chemom.* **2009**, *2*, 163–197.
23. MacGregor, J.F.; Nomikos, P. Multivariate SPC Charts for Batch Monitoring Processes. *Technometrics* **1995**, *37*, 41–59.
24. Wold, S.; Kettaneh, N.; Fridén, H.; Holmberg, A. Modelling and diagnostics of batch processes and analogous kinetic experiments. *Chemom. Intell. Lab. Syst.* **1998**, *44*, 331–340. [CrossRef]
25. Kourti, T. Abnormal situation detection, three-way data and projection methods; robust data archiving and modeling for industrial applications. *Annu. Rev. Control* **2003**, *27 II*, 131–139. [CrossRef]
26. González-Martínez, J.M.; De Noord, O.E.; Ferrer, A. Multisynchro: A novel approach for batch synchronization in scenarios of multiple asynchronisms. *J. Chemom.* **2014**, *28*, 462–475.
27. Kassidas, A.; MacGregor, J.F.; Taylor, P.A. Synchronization of Batch Trajectories Using Dynamic Time Warping. *AIChE J.* **1998**, *44*, 864–875. [CrossRef]
28. Camacho, J.; Pérez-Villegas, A.; Rodríguez-Gómez, R.A.; Jiménez-Mañas, E. Multivariate Exploratory Data Analysis (MEDA) Toolbox for Matlab. *Chemom. Intell. Lab. Syst.* **2015**, *143*, 49–57. [CrossRef]
29. González-Martínez, J.M.; Camacho, J.; Ferrer, A. MVBatch: A matlab toolbox for batch process modeling and monitoring. *Chemom. Intell. Lab. Syst.* **2018**, *183*, 122–133. [CrossRef]
30. Box, G.E.P.; Hunter, W.G.; Hunter, J.S. *Statistics for Experimenters: Design, Discovery and Innovation*, 2nd ed.; John Wiley and Sons: Hoboken, NJ, USA, 2005.

Article

Quality-Analysis-Based Process Monitoring for Multi-Phase Multi-Mode Batch Processes

Luping Zhao *, Xin Huang and Hao Yu

College of Information Science and Engineering, Northeastern University, Shenyang 110819, China; hx1970617@163.com (X.H.); yuhaoneu@163.com (H.Y.)
* Correspondence: zhaolp@ise.neu.edu.cn; Tel.: +86-188-4255-2385

Abstract: In batch processing, not only the characteristics of different phases are different, but also there may be different characteristics between batches. These characteristics of different phases and batches will have different effects on the final product quality. In order to enhance the safety of batch processes, it is necessary to establish an appropriate monitoring system to monitor the production process based on quality-related information. In this work, based on multi-phase and multi-mode quality prediction, a new quality-analysis-based process-monitoring strategy is developed for batch processes. Firstly, the time-slice models are established to determine the critical-to-quality phases. Secondly, a multi-phase residual recursive model is established using each quality residual of the phase mean models. Subsequently, a new process-monitoring strategy based on quality analysis is proposed for a single mode. After that, multi-mode quality analysis is carried out to judge the relevance between the historical modes and the new mode. Further, online quality prediction is achieved applying the selected model based on multi-mode quality analysis, and an according process-monitoring strategy is developed. The simulation results show the availability of this method for multi-phase multi-mode batch processes.

Keywords: multi-phase residual recursive model; multi-mode model; quality prediction; process monitoring

Citation: Zhao, L.; Huang, X.; Yu, H. Quality-Analysis-Based Process Monitoring for Multi-Phase Multi-Mode Batch Processes. *Processes* **2021**, *9*, 1321. https://doi.org/10.3390/pr9081321

Academic Editors: Marco S. Reis and Furong Gao

Received: 31 May 2021
Accepted: 27 July 2021
Published: 29 July 2021

1. Introduction

Batch process products play an increasingly important role in modern human life. In order to meet the ever-changing market demand of modern society, the safe and reliable operation of batch processes and continuous and stable product quality have gradually become the focus of attention in the processing industry [1,2]. The characteristics of batch operation processes are more complex than that of continuous industrial processes and have more abundant statistical characteristics. In order to enhance the safety of the batch production process and its control system, it is urgent to establish a suitable process-monitoring system to monitor the production process.

Currently, data-driven methods [3–5] of extracting information from process data and modeling monitoring have become a hotspot in process-monitoring research. With the advancement of sensor technology, almost all industrial objects are equipped with different types of sensing devices. This results in a large amount of data being obtained in an industrial process. The data-driven methods extract information hidden in data by analyzing and mining collected industrial data, which may help reveal the operation mode of the industrial process and trace the fault reasons. In recent years, data-driven methods are continuously developed and perfected, batch process monitoring and fault diagnosis technologies based on data-driven methods have increasingly become research hotspots of people, and theories of the batch process monitoring and fault diagnosis technologies are continuously and deeply developed.

Multivariate statistical analysis methods do not require the acquisition of process mechanism knowledge; they only require the use of historical data to build models. These methods can effectively extract key information in data, eliminate redundancy and remarkably reduce data dimensionality so that the process running state can be directly displayed in a two-dimensional statistical monitoring graph. Before the 1990s, researchers have generally simply treated batch processes as special continuous processes of limited duration, with no theoretical system of research specifically directed to batch process monitoring. Due to the essential difference of the characteristics of the batch process and the continuous process, a satisfactory effect is difficult to obtain in the batch process.

Aiming at the three-dimensional data characteristics of the batch process, a trilinear decomposition model can be established to directly investigate the three-dimensional data structure [6]. The data are stored and analyzed by using a trilinear decomposition model, and structural information of the data can be retained. In summary, there are six different two-dimensional matrix unfolding modes [7], which mainly reflect the different arrangement modes inside the data.

Nomikos [8–10] proposed multi-way principal components analysis (MPCA) and multi-way partial least squares (MPLS) methods, innovatively extending the successful application of multivariate statistical analysis methods to batch processes. Different internationally academic institutions and teams, including Wold professor [11] of Umea University, English Martin professor [12] of Newcastle University, proposed their own methodology, which facilitates the study of batch process monitoring. Corresponding models for monitoring have been established based on the model under normal conditions. When influenced by abnormal disturbances, the process variable correlations are changed, thereby deviating from the laws and characteristics under normal conditions. Corresponding multivariate statistics are calculated and compared with the monitoring control limits defined in advance, and the occurrence of abnormal working conditions can be detected.

In the batch process, the multi-phase nature is another important nature. In recent years, many scholars have conducted considerable research into process monitoring and quality analysis of batch processes [13–15]. Most researches were carried out by establishing different models to obtain different characteristics and dividing a cycle of a batch into phases, due to the cognition that the correlation of variables in the same phase is similar, and the correlation of variables in different phases is very different. Some scholars study the characteristics of the phases, e.g., the problem of transitions between adjacent phases [16] and the problem of non-uniform durations [17]. In addition, the scholars suggested that phases contribute to the final quality together, and individual phase models should be connected in some way during the modeling process. Therefore, a recursive quality regression method aiming at the multi-phase characteristics of the batch process was proposed [18], where the regression on the process variables in the current phase and the residual quality obtained in the previous phase was carried out to extract important quality information between the phase.

In addition, due to the influence of various factors, there are multi-mode characteristics in the batch process. In the whole operation process, process changes in batch direction lead to different process states and different process characteristics. In this way, monitoring and quality prediction for only one process state may lead to inaccurate analysis and monitoring results. In order to solve this problem, some scholars proposed to build an integrated model that can include both the common model and the specific model [19]. However, these methods barely evaluate the changes along the batch direction, in which models are in general updated arbitrarily, decreasing the efficiency of the monitoring system, as well as increasing the chance of introducing disturbances into the process model. Some scholars have proposed a specific modeling method for a specific process state [20]. However, the process variation along the batch direction may be too slow to be divided into several states. In addition, in the batch production process, when a new mode is generated, the corresponding model is built in the mode library and saved in the mode library. However, the relationship between these modes is not analyzed and judged. As new modes are generated one after another, all new modes must be saved, which makes the mode library larger and larger. Therefore, a quality prediction method based on the relationship between modes is proposed to extract information from historical modes [21].

In recent years, monitoring of multiple characteristics of batch processes has also been the direction of many scholars. A process-monitoring method based on multi-mode Fisher discriminant analysis to solve the problem of multi-mode monitoring of batch process was proposed [22], which overcomes the limitation of the single operation mode assumption. Taking the whole batch trajectory as the research object, based on the dynamic time warping method, the obtained data are automatically classified from the perspective of data distribution to reflect the differences in batch direction. For the batch process with multi-phase characteristics, a two-phase PLS regression model based on phase analysis and different statistical analyses was proposed [23]. At the first level, multiple PLS models are used to monitor a single point in time. At the second level, the final quality is predicted. Through these two different levels of models, real-time monitoring and accurate quality prediction are organically combined. Due to the calibration and modeling problems caused by operation switching (or moving to different phases), a new evolutionary PLS method is proposed, which can be used to predict intermediate quality measurement and to detect process faults avoiding false positives [24].

In this work, both multi-phase quality analysis and multi-mode quality analysis are conducted at the same time to develop a comprehensive process-monitoring strategy based on the quality prediction of batch processes. The multi-phase and multi-mode batch process concerned here involves variety in two directions. One is the multi-phase direction, the other is the multi-mode direction, and the processing methods of the two directions are different due to different process characteristics. In the multi-phase direction, the phase residual recursive model is unitized to connect the contributions of the successive phases on the final quality together, while in the multi-mode direction, the relationship between the current mode and the historical mode is analyzed and extracted to obtain more quality-related information for quality prediction. Firstly, the time-slice modeling method and the goodness-of-fit index are used to analyze the influence of different phases on the final quality and identify the critical-to-quality phases. Then, the phase mean model is introduced to analyze the phase characteristics and monitor the phase based on quality information. After that, single modes are analyzed, where the residual regression model of each phase is established with the quality variables of the current phase and the quality residual of the previous phase, and the current mode is predicted and monitored. In addition, for the quality prediction and monitoring of multiple modes, it is emphasized to extract the relationship between historical mode and new mode by between-mode modeling. This model contains more modal quality-related information and can better predict and monitor multiple modes. Finally, the strategy is applied to an injection molding process to illustrate the effectiveness of the strategy.

The rest of this paper includes four parts: the proposed method is presented in Section 2, including critical-to-quality phase identification, phase mean model, multi-phase residual recursive modeling for a single mode, between-mode modeling for multiple modes and model comparison and selection. In Section 3, the injection molding process is briefly introduced, and the method used is illustrated through an example to obtain the results and make a comparative analysis. At last, the conclusion is drawn.

2. Methodology

2.1. Critical-to-Quality Phase Identification Based on Time-Slice Model

In the batch process, there will be different process requirements in the whole operation process, causing obvious phase characteristics. The batch process can be divided into several phases according to process variable relevance. Due to the phase characteristic, there is no significant change in the correlation between process variables and quality variables at different sampling times in the same phase; that is to say, the effect of process operation behavior on quality is similar in the same phase. However, in different phases, the influences of process variables on quality are different, and they show different statistical relationships. Because of the above characteristics of batch processes, a phase that has a significant contribution to the final quality is defined as the critical-to-quality phase. There may be several critical-to-quality phases in the batch process. If production has multiple quality variables, the critical-to-quality phases may be different or the same for different quality variables, depending on the characteristics of the process. Therefore, it is important to find out the critical-to-quality phases that contribute the most to the quality change.

Batch process data are generally represented by $\mathbf{X}(I \times J_x \times K)$, where I is the number of batches, J_x is the number of process variables, and K is the sample times. The quality data is generally represented by $\mathbf{Y}(I \times J_y)$, where J_y is the number of measurement values. The measurement values of all J_x variables at the sampling interval k (k = 1, … ,K) are stored in $\mathbf{X}_k(I \times J_x)$, which is called the kth time slice of $\mathbf{X}_k$. The relationship between process variables and quality variables at time interval k can be collected from matrices $\mathbf{X}_k$ and $\mathbf{Y}$. By applying PLS, the kth time-slice PLS model is realized.

$$\begin{aligned} \mathbf{X}_k &= \mathbf{T}_k\mathbf{P}_k^{\mathrm{T}} + \mathbf{E}_k \\ \mathbf{Y} &= \mathbf{U}_k\mathbf{Q}_k^{\mathrm{T}} + \mathbf{F}_k \end{aligned} \tag{1}$$

The previous model can be expressed by the regression model as:

$$\hat{\mathbf{Y}}_k = \mathbf{X}_k\mathbf{B}_k \tag{2}$$

Where $\mathbf{T}_k$ and $\mathbf{U}_k$ are the score matrices, $\mathbf{P}_k$ and $\mathbf{Q}_k$ are the loading matrices, $\mathbf{E}_k$ and $\mathbf{F}_k$ are the residual matrices, $\mathbf{B}_k$ is the regression parameter matrix, k = 1,2, … ,K, and $\hat{\mathbf{Y}}_k$ is the predicted quality. When considering a single quality variable $\mathbf{y}(I \times 1)$, the regression model can be simply expressed as:

$$\hat{\mathbf{y}}_k = \mathbf{X}_k\boldsymbol{\beta}_k \tag{3}$$

$\boldsymbol{\beta}_k$ is the regression parameter and $\hat{\mathbf{y}}_k$ is the predicted quality at the current time. In the regression model, the number of latent variables needs to be determined, and the four-fold cross-validation method is used in this work [25,26].

In this paper, the index R^2, which is used to describe the goodness of fit of the regression model in the field of multivariable linear regression, is used to measure the influence of each time slice on the final quality. Those time slices with high R^2 are identified to be critical to quality, and the phases with these time slices are identified as critical-to-quality phases. The kth sampling time is defined. The prediction accuracy R_k^2 of the quality prediction model for the quality index $\mathbf{y}$ is as follows:

$$R_k^2 = \frac{\sum_{i=1}^{I}(\hat{\mathbf{y}}_{i,k} - \overline{\mathbf{y}})^2}{\sum_{i=1}^{I}(\mathbf{y}_i - \overline{\mathbf{y}})^2} \tag{4}$$

where $\mathbf{y}_i$ is the quality variable measurement value of the ith batch operation in the test batch, $\hat{\mathbf{y}}_{i,k}$ is the model prediction value of the ith batch operation quality variable of the predicted kth time slice, and $\overline{\mathbf{y}}$ is the average value of the quality variable measurement value of the test batch. The value range of R_k^2 is 0–1. When R_k^2 approaches 1, it indicates that the accuracy of the quality prediction model is high, which indicates that the bigger impact on the quality variables is in this phase. On the contrary, the smaller the R_k^2 is, the smaller the impact on quality variables is. Therefore, by observing the R_k^2 size of different phases, the critical-to-quality phases in the batch process can be determined.

2.2. Phase Mean Model

According to the characteristics of batch processes, the whole process can be divided into several phases. There are obvious differences in the process variables in different phases, and the same phase can be almost considered to have similar process variables. In this work, for the multi-phase and multi-mode quality analysis, it is supposed that the characteristics along the time direction in each phase are constant. It is considered to establish such a model that can represent the process variable relationships of the entire phase. The phase mean model is achieved as follows.

First, the average variable matrix is calculated in phase c,

$$\overline{\mathbf{X}}_c = \sum_{k=1}^{K_c} \mathbf{X}_{k,c}/K_c \tag{5}$$

where K_c is the data length of phase c. $\mathbf{X}_{k,c}$ is the data matrix of the process variables at the k moment in phase c. Thus, $\overline{\mathbf{X}}_c$ is the average variable matrix of phase c.

Within phase c, phase regression models can be built using the PLS method,

$$\begin{aligned} \overline{\mathbf{X}}_c &= \mathbf{T}_c\mathbf{P}_c^{\mathrm{T}} + \mathbf{E}_c \\ \mathbf{Y} &= \mathbf{U}_c\mathbf{Q}_c^{\mathrm{T}} + \mathbf{F}_c \end{aligned} \tag{6}$$

The previous model can be expressed by the regression form as:

$$\hat{\mathbf{Y}}_c = \overline{\mathbf{X}}_c\mathbf{B}_c \tag{7}$$

where the concepts of $\mathbf{T}_c$, $\mathbf{U}_c$, $\mathbf{P}_c$, $\mathbf{E}_c$, $\mathbf{F}_c$, and $\mathbf{B}_c$ are the same as those of the time-slice model, except that each matrix is with the meaning of the phase mean. $\hat{\mathbf{Y}}_c$ is the predicted quality of the cth phase mean model. When a single quality variable $\mathbf{y}(I \times 1)$ is considered, the regression model can be simply expressed as:

$$\hat{\mathbf{y}}_c = \overline{\mathbf{X}}_c\boldsymbol{\beta}_c \tag{8}$$

where $\boldsymbol{\beta}_c$ is the regression parameter, and at present $\mathbf{T}_c$ is a matrix of the dimension $I \times H$, $\mathbf{U}_c$ is a matrix of the dimension $I \times 1$.

For process monitoring, Hotelling-T^2 and SPE statistics are calculated in systematic and residual subspaces, respectively [8,27].

$$T_c^{\,2} = \mathbf{x}_c^{\mathrm{T}}\mathbf{R}_c\left(\frac{\mathbf{T}_c^{\mathrm{T}}\mathbf{T}_c}{I-1}\right)^{-1}\mathrm{R}_c^{\mathrm{T}}\mathbf{x}_c \tag{9}$$

$$SPE_c = \|\tilde{\mathbf{x}}_c\|^2 = \|(\mathbf{I}_{J_x} - \mathbf{P}_c\mathbf{R}_c^{\mathrm{T}})\mathbf{x}_c\|^2 \tag{10}$$

where $\tilde{\mathbf{x}}_c$ is the residual vector; $\mathbf{R}_c = \mathbf{W}_c(\mathbf{P}_c^{\mathrm{T}}\mathbf{W}_c)^{-1}$; $\mathbf{W}_c$ is the weight matrix; $\delta_{T_c^2}(\alpha)$ is the control limit with α confidence of $T_c^{\,2}$; and $SPE_c(\alpha)$ is the α confidence limit of SPE. The detailed properties and calculations can be found in reference [28].

The corresponding control limits are:

$$\delta_{T_c^2}(\alpha) = \frac{H(I^2-1)}{I(I-H)}F_{c\alpha}(H, I-H) \tag{11}$$

$$SPE_c(\alpha) = g_c\chi^2_{h,\alpha} \tag{12}$$

where $F_{c\alpha}(H, I-H)$ is the F distribution with α confidence and H and $I-H$ degrees of freedom, and H is the number of retained latent variables; $g_c\chi^2_{h,\alpha}$ is the χ^2 distribution with the same confidence level of α and the proportional coefficient of $g_c = s/2\mu$; $h = 2\mu^2/s$; and μ is the mean value of SPE; s is the variance of SPE.

2.3. Multi-Phase Residual Recursive Modeling for Single Mode

In the phase-based PLS method, a phase regression model is established between the process variables and the final quality variables in each phase. It is assumed that in each phase, the model can capture the relationship between process variables and final quality variables. However, these individual models are not related to each other, and each phase seems to contribute to the final quality individually. This is in contradiction with the nature of the multi-phase batch process; that is, multi-phase acts on the final quality together in sequence. In addition, it should be noted that in the multi-phase batch process, the former phase may affect the later phase and the final process quality. In the current phase of quality regression modeling, the influence of the previous phases should be considered. Therefore, a recursive quality regression method for the multi-phase batch process is proposed, which uses the quality residuals of the previous phase model to establish the current phase regression model. All phases that are critical to quality are correlated by phase-based recursive regression residuals so that they together contribute to the final quality.

The establishment of a multi-phase residual recursive model is shown in Figure 1.

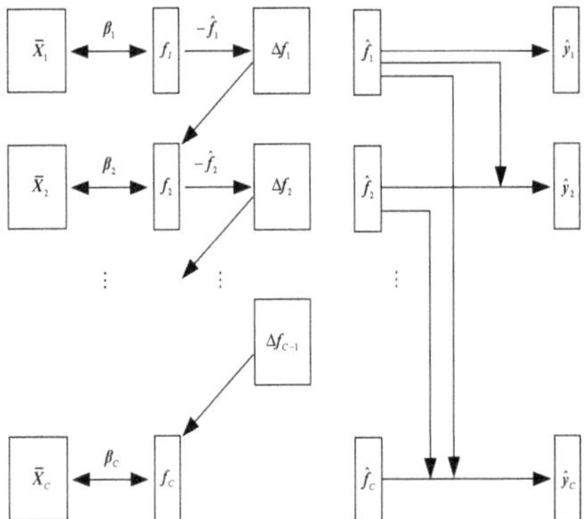

Figure 1. Illustration of multi-phase residual recursive modeling.

For a single phase, each phase is modeled by the regression between the average variable matrix $\overline{\mathbf{X}}_c$ and the current quality residual $\mathbf{f}_c$, then the regression parameter $\boldsymbol{\beta}_c$ and the residual prediction quantity $\hat{\mathbf{f}}_c$ are obtained,

$$\begin{aligned} \mathbf{f}_c &= \overline{\mathbf{X}}_c \boldsymbol{\beta}_c + \mathbf{f}^* \\ \hat{\mathbf{f}}_c &= \overline{\mathbf{X}}_c \boldsymbol{\beta}_c \end{aligned} \tag{13}$$

The quality residual in the first phase $\mathbf{f}_1$ is the quality measurement itself. The residual of the second phase is the deviation between the prediction quality of the first phase and the residual of the first phase, and so on.

The current phase quality prediction results are the sum of the completed phase and the current phase quality residual prediction,

$$\hat{\mathbf{y}}_c = \sum_{i=1}^{c} \hat{\mathbf{f}}_i, c = 1, 2, \ldots, C \tag{14}$$

The final online quality prediction results are as follows:

$$\hat{\mathbf{y}}_k = \begin{cases} \mathbf{X}_{k,c_1}\boldsymbol{\beta}_1 & k \in c_1 \\ \hat{\mathbf{f}}_1 + \mathbf{X}_{k,c_2}\boldsymbol{\beta}_2 & k \in c_2 \\ \hat{\mathbf{f}}_1 + \hat{\mathbf{f}}_2 + \mathbf{X}_{k,c_3}\boldsymbol{\beta}_3 & k \in c_3 \\ \hat{\mathbf{f}}_1 + \hat{\mathbf{f}}_2 + \hat{\mathbf{f}}_3 + \mathbf{X}_{k,c_4}\boldsymbol{\beta}_4 & k \in c_4 \end{cases} \tag{15}$$

where $c_1, \ldots, c_4$ are four phases, respectively, and $\mathbf{X}_{k,c_1}, \ldots, \mathbf{X}_{k,c_4}$ are the phase mean variable matrices.

The Hotelling-$\boldsymbol{T}^2$ and $\boldsymbol{SPE}$ statistics for the current time k are:

$$\boldsymbol{T}_k{}^2 = \mathbf{x}_k{}^{\mathrm{T}}\mathbf{R}_k\left(\frac{\mathbf{T}_k{}^{\mathrm{T}}\mathbf{T}_k}{I-1}\right)^{-1}\mathbf{R}_k{}^{\mathrm{T}}\mathbf{x}_k \tag{16}$$

$$\boldsymbol{SPE}_k = \|\widetilde{\mathbf{x}}_k\|^2 = \|(\mathbf{I}_{J_x} - \mathbf{P}_k\mathbf{R}_k{}^{\mathrm{T}})\mathbf{x}_k\|^2 \tag{17}$$

where $\widetilde{\mathbf{x}}_k$ is the residual vector at the current time.

The corresponding control limits are:

$$\delta_{\boldsymbol{T}_k{}^2}(\alpha) = \frac{H(I^2-1)}{I(I-H)}\boldsymbol{F}_{k\alpha}(H, I-H) \tag{18}$$

$$\boldsymbol{SPE}_k(\alpha) = g_k\chi^2_{h,\alpha} \tag{19}$$

where $\boldsymbol{F}_{k\alpha}(H, I-H)$ is the $\boldsymbol{F}$ distribution with α confidence and H and $I-H$ degrees of freedom, and H is the number of retained latent variables; $g_k\chi^2_{h,\alpha}$ is the χ^2 distribution with the same confidence level of α and the proportional coefficient of $g_k = s/2\mu$; $h = 2\mu^2/s$; and μ is the mean value of $\boldsymbol{SPE}$; s is the variance of $\boldsymbol{SPE}$.

2.4. Between-Mode Modeling for Multiple Modes

The multi-phase problem has been addressed in the previous part; thus, in this part, the multi-mode problem is the key interesting issue. While it does not mean the multi-phase problem is not considered any longer and without a special statement, the methodology below is proposed based on the above multi-phase analysis.

To solve the multi-mode problem, the main idea is to extract the relationship between the historical modes and the new mode. This proposed model contains more modal information and can better predict and monitor multiple modes. The framework of this section is shown in Figure 2. The model is established not only based on the new mode but also on the historical modes in the modal library. Firstly, the process variables and quality variables in historical modes are regressed and analyzed using the single-mode model. Secondly, the new mode process variables are applied to those single-mode models of historical modes, and the assumed predicted qualities of the new mode are obtained. Then, the regression analysis is carried out on the assumed predicted qualities and the final actual quality, obtaining the between-mode model. Finally, by applying the between-mode model, the final prediction quality is obtained. The details of between-mode quality regression modeling are introduced as follows.

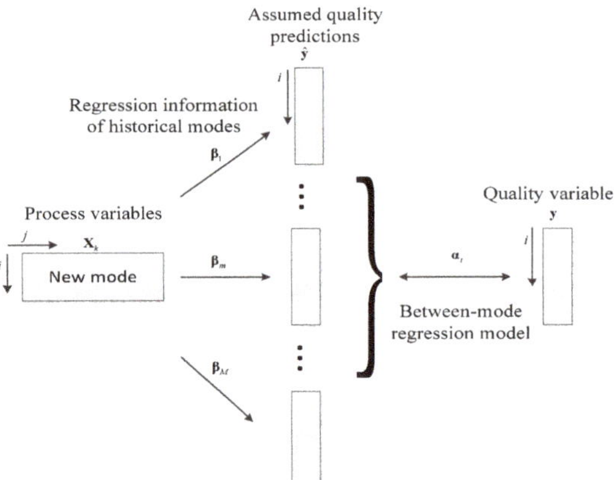

Figure 2. Illustration of between-mode quality regression modeling.

Within phase c, for the new mode with the normalized time-slice process variables, $\mathbf{X}_{t,k}(I_t \times J_x)$, and the quality variable $\mathbf{y}_t$, process variables are first applied to the regression models obtained from the historical modes to obtain the assumed quality predictions,

$$\hat{\mathbf{y}}_{t,m,k}=\mathbf{X}_{t,k}\boldsymbol{\beta}_{m,c} \tag{20}$$

where m is the number of the historical modes, m = 1,2, ... ,M, and t represents the new mode. $\hat{\mathbf{y}}_{t,m,k}$ are the assumed prediction quality. $\boldsymbol{\beta}_{m,c}$ are the regression parameters of mode m of phase c for the historical modes. By obtaining the assumed quality predictions, the quality information of historical modes is shared by the new mode. Further, the quality information of historical modes will be judged and extracted by the next regression.

Then, the assumed quality predictions will be regressed with the quality data of the new mode. All these assumed predictions of the historical modes can comprise a new matrix $\mathbf{Z}_{t,k}(I_t \times M)$, $\mathbf{Z}_{t,k} = \left[\hat{\mathbf{y}}_{t,1,k}, \ldots, \hat{\mathbf{y}}_{t,m,k}, \ldots, \hat{\mathbf{y}}_{t,M,k}\right]$. Then, the kth time-slice PLS regression model is built between $\mathbf{Z}_{t,k}$ and $\mathbf{y}_t$ as follows [29]:

$$\begin{aligned} \mathbf{Z}_{t,k} &= \mathbf{T}_{t,k}\mathbf{P}_{t,k}^{\mathrm{T}} + \mathbf{E}_{t,k} \\ \mathbf{y}_t &= \mathbf{U}_{t,k}\mathbf{Q}_{t,k}^{\mathrm{T}} + \mathbf{F}_{t,k} \end{aligned} \tag{21}$$

where $\mathbf{T}_{t,k}$ and $\mathbf{U}_{t,k}$ are the score matrices of the new mode, $\mathbf{P}_{t,k}$ and $\mathbf{Q}_{t,k}$ are the loading matrices of the new mode, and $\mathbf{E}_{t,k}$ and $\mathbf{F}_{t,k}$ are the residual matrices of the new mode. Novel predictions are obtained,

$$\hat{\mathbf{y}}^{*}_{t,k}=\mathbf{Z}_{t,k}\boldsymbol{\alpha}_{t,k} \tag{22}$$

where $\hat{\mathbf{y}}^{*}_{t,k}$ shows this new regression relationship of the between-mode relationship analysis, k = 1,2, ... ,K_c, and $\boldsymbol{\alpha}_{t,k}$ is the regression parameter of the kth time-slice model.

The regression parameters of phase c can be obtained from the regression parameters of the time-slice model,

$$\boldsymbol{\alpha}_{t,c} = \frac{1}{K_c}\sum_{k=1}^{K_c} \boldsymbol{\alpha}_{t,k} \tag{23}$$

where K_c is the number of the time intervals within phase c. Then the predictions, $\hat{\mathbf{y}}^{*}_{t,c,k}$, based on the regression parameter of the whole phase, $\boldsymbol{\alpha}_{t,c}$, are obtained,

$$\hat{\mathbf{y}}^{*}_{t,c,k}=\mathbf{Z}_{t,k}\boldsymbol{\alpha}_{t,c} \tag{24}$$

Then for phase c, corresponding coefficients can be obtained:

$$\mathbf{Z}_{t,c} = \frac{1}{K_c}\sum_{k=1}^{K_c}\mathbf{Z}_{t,k} \tag{25}$$

$$\mathbf{T}_{t,c} = \frac{1}{K_c}\sum_{k=1}^{K_c}\mathbf{T}_{t,k} \tag{26}$$

$$\mathbf{P}_{t,c} = \frac{1}{K_c}\sum_{k=1}^{K_c}\mathbf{P}_{t,k} \tag{27}$$

$$\mathbf{W}_{t,c} = \frac{1}{K_c}\sum_{k=1}^{K_c}\mathbf{W}_{t,k} \tag{28}$$

$$\mathbf{R}_{t,c} = \mathbf{W}_{t,c}(\mathbf{P}_{t,c}\mathbf{W}_{t,c})^{-1} \tag{29}$$

where K_c is the number of time intervals in phase c, k = 1,2, … ,K_c.

In online monitoring, the score matrix $\mathbf{T}_{t,c}$, the load matrix $\mathbf{P}_{t,c}$, and the weight matrix $\mathbf{W}_{t,c}$ are obtained according to the offline model. The online T^2 statistics and online $\boldsymbol{SPE}$ statistics are calculated:

Online $\boldsymbol{T^2}$ statistics:

$$T_k^{\ 2} = \mathbf{z}_k^{\ \mathrm{T}}\mathbf{R}_{t,c}\left(\frac{\mathbf{T}_{t,c}^{\ \mathrm{T}}\mathbf{T}_{t,c}}{I-1}\right)^{-1}\mathbf{R}_{t,c}^{\ \mathrm{T}}\mathbf{z}_k \tag{30}$$

Online $\boldsymbol{SPE}$ statistics:

$$\boldsymbol{SPE}_k = \|\widetilde{\mathbf{z}}_k\|^2 = \|(\mathbf{I}_M - \mathbf{P}_{t,c}\mathbf{R}_{t,c}^{\ \mathrm{T}})\mathbf{z}_k\|^2 \tag{31}$$

where $T_k^{\ 2}$ and $\boldsymbol{SPE}_k$ are the T^2 and $\boldsymbol{SPE}$ statistics calculated at the kth time interval, respectively, and $\widetilde{\mathbf{z}}_k$ is the residual vector of the kth time interval.

The corresponding control limits are:

$$\delta_{T_k^2}(\alpha) = \frac{H(I^2-1)}{I(I-H)}F_{k\alpha}(H, I-H) \tag{32}$$

$$\boldsymbol{SPE}_k(\alpha) = g_k\chi^2_{h,\alpha} \tag{33}$$

where $F_{k\alpha}(H, I-H)$ is the F distribution with α confidence and H and $I-H$ degrees of freedom, and H is the number of retained latent variables; $g_k\chi^2_{h,\alpha}$ is the χ^2 distribution with the same confidence level of α and the proportional coefficient of $g_k = s/2\mu; h = 2\mu^2/s;\mu$ is the mean value of $\boldsymbol{SPE}$; and s is the variance of $\boldsymbol{SPE}$.

2.5. Model Comparison and Selection

In this section, two models are compared, which are the single-mode model and the between-mode model. To be clear, the single-mode model is introduced in Section 2.3. This model only considers one mode, and the quality is forecasted and monitored in its own mode on the basis of the critical-to-quality phase residual recursive analysis. The other model is developed in Section 2.4, and the between-mode model, which is established based on the historical modes and the new mode to obtain the assumed quality predictions and involve the quality information of the historical modes in the regression model for the new mode. It should be noticed that in both models, the multi-phase issue is addressed in the same way, by the residual recursive modeling, for the fair comparison as well as strategy consistency.

First, for the new batches $\mathbf{X}_{new}(I_{new} \times J_x)$, the single-mode quality predictions $\hat{\mathbf{y}}_{new,t,m,k}$ are gained at kth time. The multi-mode quality predictions $\hat{\mathbf{y}}^{*}_{new,t,c,k}$ are gained at kth time.

$$\begin{aligned} \hat{\mathbf{y}}_{new,t,c,k} &= \mathbf{X}_{new}\boldsymbol{\beta}_{t,c} \\ \hat{\mathbf{y}}_{new,t,m,k} &= \mathbf{X}_{new}\boldsymbol{\beta}_{m,c} \end{aligned} \tag{34}$$

$$\begin{aligned} \mathbf{Z}_{new,t,k} &= [\hat{\mathbf{y}}_{new,t,1,k}, \ldots, \hat{\mathbf{y}}_{new,t,m,k}, \ldots, \hat{\mathbf{y}}_{new,t,M,k}] \\ \hat{\mathbf{y}}^{*}_{new,t,c,k} &= \mathbf{Z}_{new,t,k}\boldsymbol{\alpha}_{t,c} \end{aligned} \tag{35}$$

Then, the root-mean-square error (RMSE) values are obtained,

$$\mathbf{RMSE} = \sqrt{\frac{1}{I_{new}}\sum_{i=1}^{I_{new}}(\mathbf{y}_{new,t,i} - \hat{\mathbf{y}}_{new,t,i,k})^2} \tag{36}$$

$$\mathbf{RMSE} = \sqrt{\frac{1}{I_{new}}\sum_{i=1}^{I_{new}}(\mathbf{y}_{new,t,i} - \hat{\mathbf{y}}^{*}_{new,t,i,k})^2} \tag{37}$$

The RMSE values can well reflect the precision of prediction. The smaller the RMSE values, the higher the prediction accuracy.

3. Illustration and Discussions

3.1. Process Description

Injection molding technology is one of the important means of plastic processing, and it is also a typical batch process. In order to accurately predict the quality of products, it is necessary to know enough about the injection molding process. A complete injection molding process is mainly composed of mold closing, injection, packing-holding, plasticizing, cooling, mold opening, part ejection, and other processes. There are four phases that are the most important operation phases to determine the quality of parts: the first one is the injection phase, which injects the molten plastic into the mold; secondly, in the packing-holding phase, the packaging materials are used under a certain pressure; then, in the plasticizing phase, the material is transported forward, plasticized and melted, and then transferred to viscous fluid for storage; the final phase is the cooling phase, where the plastic is cooled in the mold until the part becomes sufficiently rigid for ejection. The process variables that have an important influence on the final quality can be read online by high-precision sensors.

In this work, high-density polyethylene (HDPE) was used as the injection material. The quality index analyzed in this experiment is the weight of injection molded parts. According to the different settings of packing pressure (PP) and barrel temperature (BT), the experimental batches can be divided into five different modes. The experimental conditions are shown in Table 1. The process data of each mode is stored in X ($23 \times 11 \times 525$). The quality data of each mode is stored in y (23×1). The data used in the modeling process are all real data obtained from experiments.

Table 1. Different operation modes.

Modes	PP/Bar	BT/°C
Mode 1	25	180
Mode 2	35	180
Mode 3	25	200
Mode 4	30	200
Mode 5	35	200

3.2. Critical-to-Quality Phase Identification

In the injection molding process, different phases have different effects on the quality of products. For example, in the injection phase, the main variables affecting the final product weight are the injection speed and the barrel temperature. In general, the higher the barrel temperature is, the lower the product weight is. The faster the injection rate increases, the more melt injection and the greater the product weight. In addition, the pressure variables (such as the nozzle pressure, the cylinder pressure), the screw stroke, the injection speed, and the barrel temperature are positively correlated with the sputtering quality of injection products. That is to say, the faster the injection speed, the higher the pressure and the temperature are, and the more likely the sputtering phenomenon will appear in the intermittent operation. In the packing-holding phase, the weight of the injection molded part is mainly determined by the nozzle pressure, the cylinder pressure, and the cavity pressure. Two temperature variables, the cavity temperature and the barrel temperature, also affect the weight of the product. The lower the temperature, the greater the weight.

Taking mode 3 as an example, the critical-to-quality phase analysis is carried out. There are 23 batches in mode 3. A total of 18 train batches are selected as the prediction batches to analyze the phase characteristics. The R_k^2 and the phase mean of R_k^2 are shown in Figure 3. It can be seen from the figure that the R_k^2 values of the injection phase and the packing-holding phase are larger, which means these two phases have greater impacts on the final prediction quality than other phases.

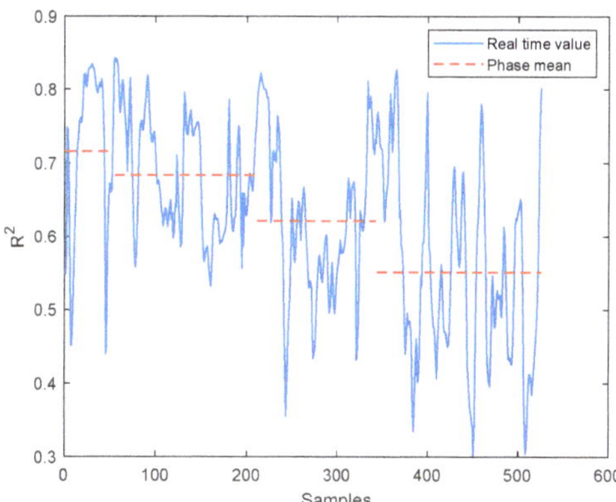

Figure 3. R_k^2 contribution rate of the different batches in mode 3.

The phase mean value of R_k^2 of the four phases under three different modes is shown in Table 2.

Table 2. Phase mean value R_k^2 of the different modes and different phases.

Phases	Mode 1	Mode 2	Mode 3	Mode 4	Mode 5	Mean
Injection phase	0.4567	0.7867	0.7161	0.8219	0.7643	0.7091
Packing-holding phase	0.4013	0.7500	0.6836	0.8234	0.7801	0.6877
Plasticizing phase	0.3893	0.7175	0.6214	0.8058	0.7839	0.6636
Cooling phase	0.3407	0.6415	0.5510	0.8002	0.8075	0.6282

According to the data in the above table, for mode 1, mode 2, mode 3, and mode 4, the phase mean values of R_k^2 of the injection phase and the packing-holding phase are greater than the phase mean values of R_k^2 of the plasticizing phase and the cooling phase. For mode 5, the phase mean value of R_k^2 of the cooling phase is the largest. Based on the mean R_k^2, the injection phase and the packing-holding phase are selected as the critical-to-quality phases for subsequent monitoring and analysis.

3.3. Multi-Phase Monitoring for Single Mode

In this part, the single-mode model is adopted for quality prediction and process monitoring. The first 18 batches of mode 3 are selected for modeling, and the last 5 batches of mode 3 are tested. According to the four-fold cross-validation method, in the modeling of the injection molding phase, the number of reserved latent variables of the traditional method and the proposed method is four. In the packing-holding phase, the number of latent variables of the traditional method is three, while the number of latent variables of the proposed method is two. The confidence level of α is set to 0.99. The simulation result of the predicted quality of one test batch is shown in Figure 4 and compared with the traditional partial multi-phase least squares method [30], in which for each phase, one single model is built for quality prediction. The mean RMSE predicted for the five test batches under different prediction methods are shown in Table 3. It can be seen from Table 3 that the mean RMSE predicted by the traditional method is 0.0702, while the mean RMSE predicted by the proposed method is 0.0632, which indicates that the proposed recursive method of phase residuals shows a more accurate prediction effect. The results of monitoring of the first test batch are shown in Figures 5 and 6. Because the traditional method also divides the batch process into four phases, in each phase, the quality is directly predicted and monitored, and in the first phase the proposed method regards the actual quality as the residual of the first phase, so the prediction and monitoring effects of the first phase, namely the injection phase, of the traditional method and proposed method is the same. The monitoring results of the injection phase are shown in Figure 5. It can be seen that T^2 and SPE are not beyond the control limits. In Figure 6, the monitoring results of the packing-holding phase are shown. It can be seen that T^2 and SPE of both the traditional method and the proposed method are not beyond their respective control limits. This shows that the proposed modeling method based on a single mode can monitor the corresponding test batches.

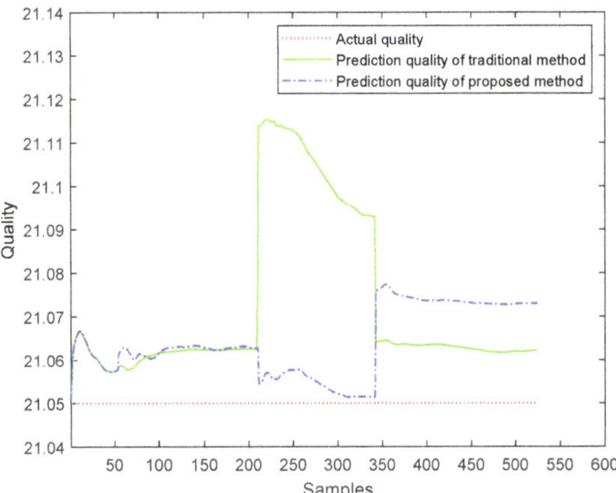

Figure 4. Single-mode online prediction of mode 3.

Table 3. RMSE of the different prediction modes in single-mode prediction.

Prediction Modes	Mode 3	Mode 1
Traditional method	0.0702	0.1398
Proposed method	0.0632	0.1154

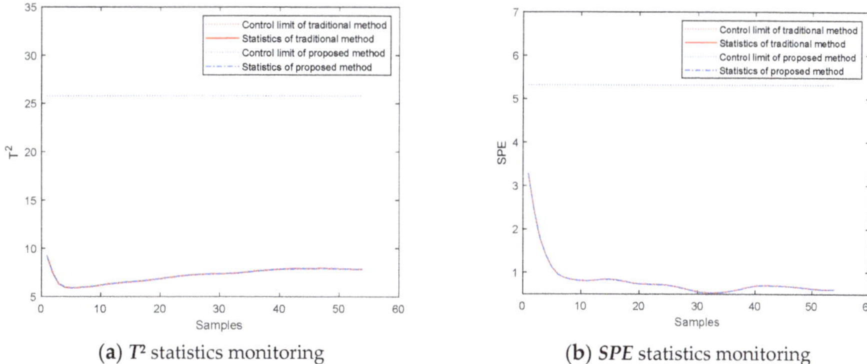

(**a**) T^2 statistics monitoring (**b**) *SPE* statistics monitoring

Figure 5. Single-mode online monitoring of injection phase of mode 3.

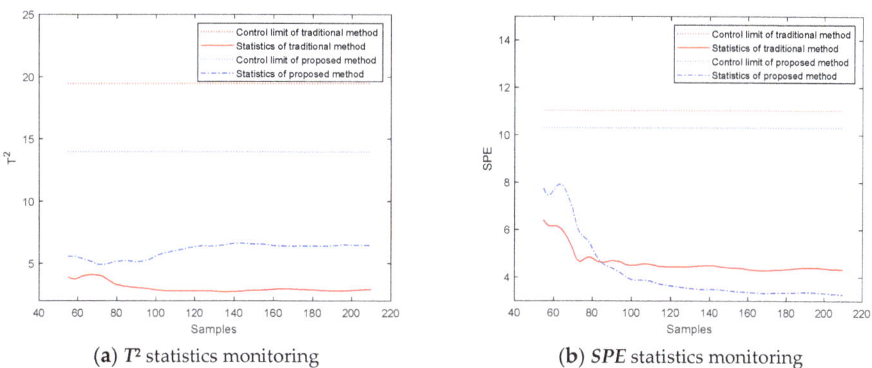

(**a**) T^2 statistics monitoring (**b**) *SPE* statistics monitoring

Figure 6. Single-mode online monitoring of packing-holding phase of mode 3.

In addition, batches from mode 1 are tested using the monitoring model built based on mode 3; that is, the first 18 batches of mode 3 are selected for modeling, and 5 batches of mode 1 are tested. The results of each phase of one batch of five test batches in mode 1 are displayed. The quality prediction result is shown in Figure 7 and compared with that of the traditional partial least squares method. The mean RMSE predicted for the five test batches under different prediction methods are shown in Table 3. The mean RMSE predicted by the traditional method is 0.1398, while the mean RMSE predicted by the proposed method is 0.1154, which indicates that the proposed recursive method of phase residuals shows a more accurate prediction effect. The monitoring results of the injection phase of mode 1 are shown in Figure 8. It can be seen that T^2 statistics do not exceed the control limit, but *SPE* statistics have exceeded the limit. The monitoring results of mode 1 in the packing-holding phase are shown in Figure 9. It can be seen that *SPE* statistics of the proposed method have exceeded the control limits in the beginning part. However, *SPE* statistics of the traditional method do not exceed the control limit. So the proposed method can distinguish this batch of mode 1 and is better than the traditional method. Thus, when a single mode is used for modeling, the other modes can be distinguished by the proposed method.

In order to compare the prediction effect of different modes and different methods under the single-mode modeling, RMSE of prediction results of five test batches of mode 3 and mode 1 are calculated respectively on the basis of the model of mode 3, as shown in Table 3.

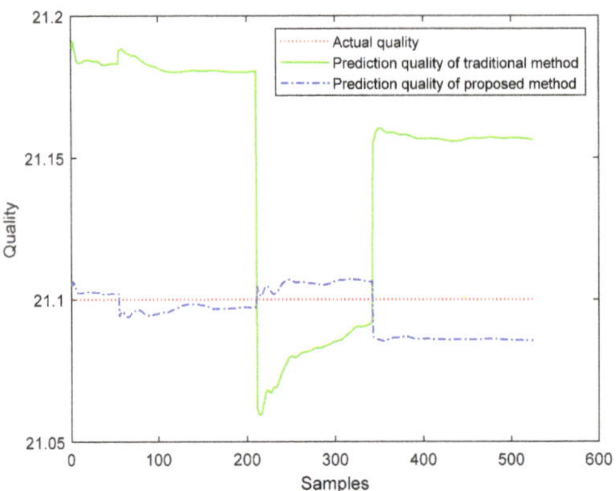

Figure 7. Single-mode online prediction of mode 1.

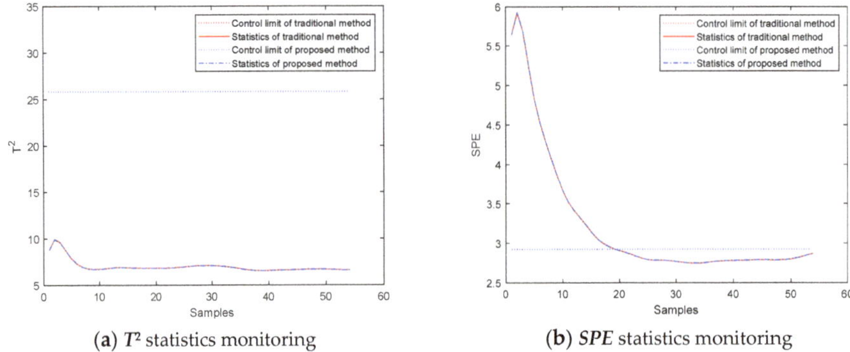

(**a**) T^2 statistics monitoring (**b**) *SPE* statistics monitoring

Figure 8. Single-mode online monitoring of injection phase of mode 1.

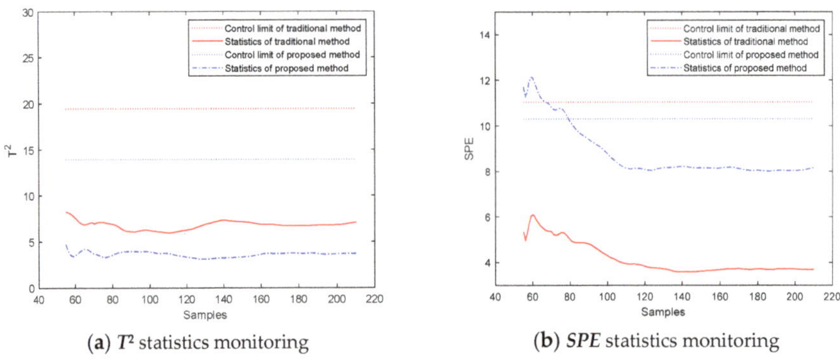

(**a**) T^2 statistics monitoring (**b**) *SPE* statistics monitoring

Figure 9. Single-mode online monitoring of packing-holding phase of mode 1.

It can be seen from the above table that in the single-mode modeling and prediction, the prediction effect of the test mode, which is the same as the modeling mode, is better than that of other test modes. In addition, according to the comparison of different methods, it can be concluded that the prediction effect of the proposed method is more accurate than that of the traditional method.

In the injection molding process, there are two main faults. One is material disturbance. A small amount of polypropylene (PP) is mixed into the original material HDPE. Because the viscosity of PP is higher than that of HDPE, higher heat will be generated in the operation process, resulting in the melt temperature in the nozzle being higher than the normal state. The second is the sensor fault. Due to the sensor fault, no data can be detected, resulting in a fault in the process.

First, a faulty batch caused by material disturbance is selected for monitoring, where the temperature variable is increased by 5 °C at the 60th sampling point. Therefore, according to the actual process situation, a batch is selected in the test batch of mode 3, and the temperature variable is increased by 5 °C at the 60th sampling point. The monitoring effects of the traditional method and the proposed method are shown in Figure 10. Compared with the traditional method, the monitoring effect of the proposed method is better since the statistics will rise rapidly when the fault occurs, especially for the T^2 statistics.

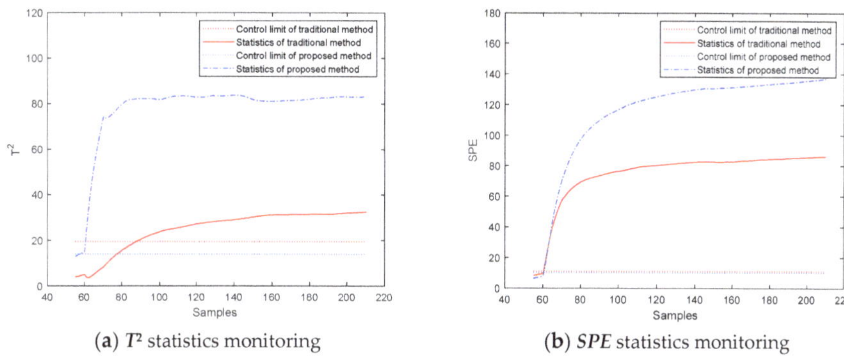

(**a**) T^2 statistics monitoring (**b**) *SPE* statistics monitoring

Figure 10. Single-mode online monitoring of material disturbance fault.

For the sensor fault, a test batch with the pressure variable removed after the 150th sampling point is monitored. The T^2 and ***SPE*** monitoring effects of the traditional method and the proposed method of the single-mode model are shown in Figure 11. Compared with the traditional method, the statistics of the proposed method rise more rapidly, and the amplitudes are relatively large.

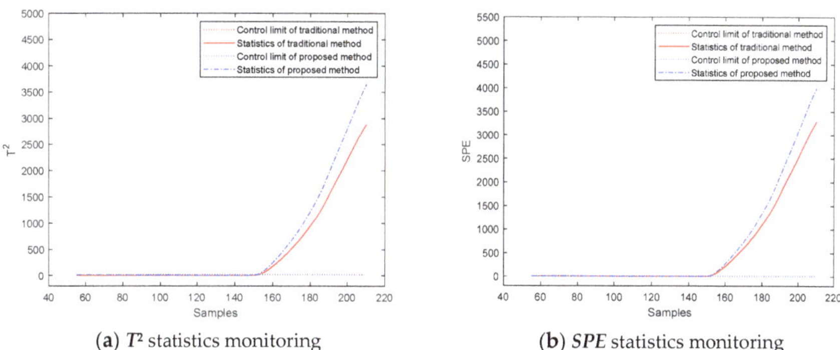

(**a**) T^2 statistics monitoring (**b**) *SPE* statistics monitoring

Figure 11. Single-mode online monitoring of sensor fault.

3.4. Multi-Mode Monitoring

For the between-mode modeling analysis, 18 batches in mode 1, mode 2, mode 4, and mode 5 are selected respectively as historical modes. Mode 3 with 18 batches is used as the new mode for modeling. The test data are constructed by the five test batches of mode 3. According to the four-fold cross-validation method, in each phase of modeling, the number of reserved latent variables of the traditional method and the proposed method is two. The confidence level of α is set to 0.99.

In order to illustrate the advantages of the method proposed in this paper, it is compared with the traditional multi-mode and multi-phase methods, in which individual models are built for a single phase within a single mode. One batch of five test batches in mode 3 is selected to show the results. The simulation results of the prediction are shown in Figure 12. The mean RMSE predicted for the five test batches under different prediction methods are shown in Table 4. The mean RMSE predicted of mode 3 by the traditional method is 0.0496, while the mean RMSE predicted by the proposed method is 0.0458, which indicates that the proposed method shows a more accurate prediction effect. The monitoring results of the injection phase and the packing-holding phase of the first test batch of mode 3 are shown in Figures 13 and 14, respectively. In Figure 13, it can be seen that T^2 and $\boldsymbol{SPE}$ do not exceed the control limits in the injection phase. In Figure 14, in the packing-holding phase, T^2 and $\boldsymbol{SPE}$ do not exceed their respective control limits.

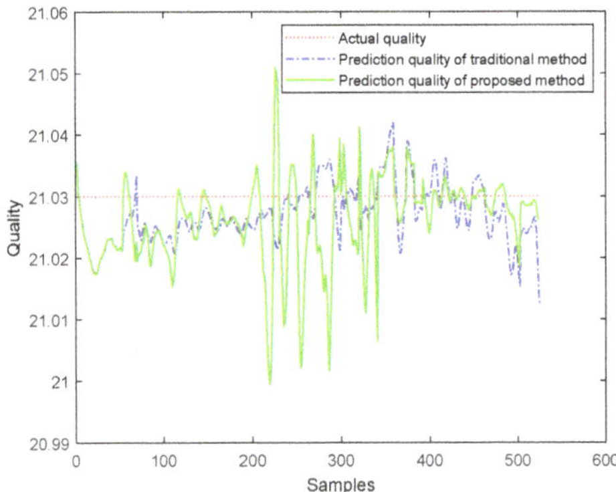

Figure 12. Multi-mode online prediction of mode 3.

Table 4. RMSE of different prediction method.

Prediction Model	Prediction Method	Mode 3	Mode 1
Single-mode model	Traditional method	0.0702	0.1398
	Proposed method	0.0632	0.1154
Between-mode model	Traditional method	0.0496	0.1010
	Proposed method	0.0458	0.0876

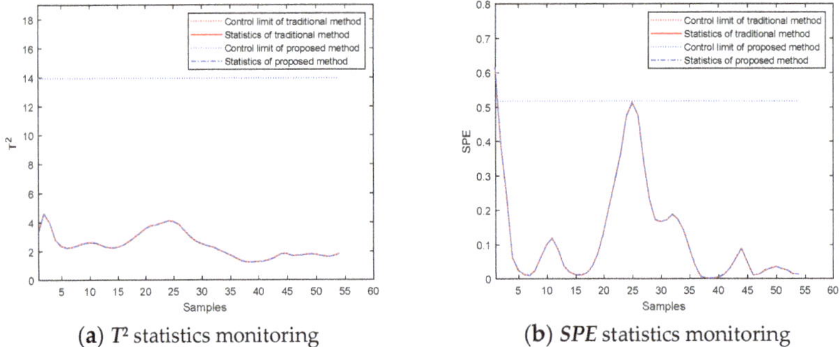

(**a**) T^2 statistics monitoring (**b**) *SPE* statistics monitoring

Figure 13. Multi-mode online monitoring of injection phase of mode 3.

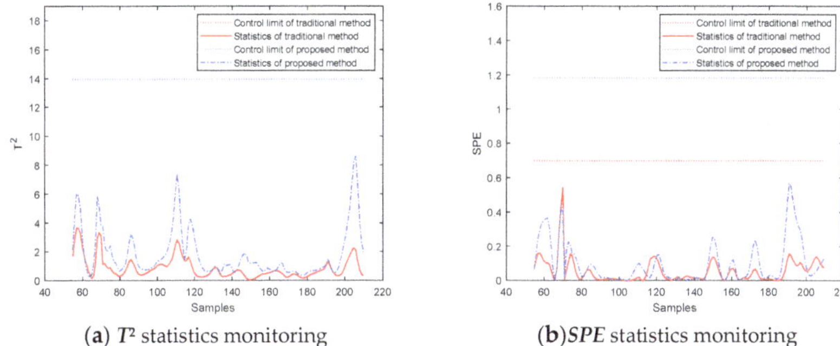

(a) T^2 statistics monitoring (b) *SPE* statistics monitoring

Figure 14. Multi-mode online monitoring of packing-holding phase of mode 3.

In addition, in order to illustrate the monitoring of new test modes by the multi-phase multi-mode model, 18 batches of mode 2, mode 3, mode 4, and mode 5 are selected respectively as the historical modes for each phase, and the historical regression parameters are obtained. Mode 3 with 18 batches is used as the new mode for modeling to predict and monitor the new test batches of mode 1. The results of each phase of one batch of five test batches in mode 1 are displayed. The simulation results of quality prediction are shown in Figure 15. The mean RMSE predicted of mode 1 for the five test batches under different prediction methods are shown in Table 4. The mean RMSE predicted by the traditional method is 0.1010, while the mean RMSE predicted by the proposed method is 0.0876, which indicates that the proposed method shows a more accurate prediction effect. Figures 16 and 17, respectively, show the monitoring results of the injection phase and the packing-holding phase of one test batch of mode 1. Because the historical mode and training data do not contain the information of mode 1, when monitoring, T^2 and ***SPE*** in the injection phase exceed the control limit, which will lead to an alarm.

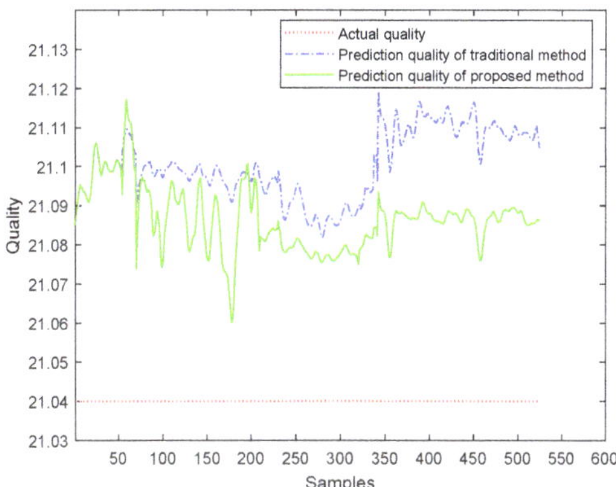

Figure 15. Multi-mode online prediction of mode 1.

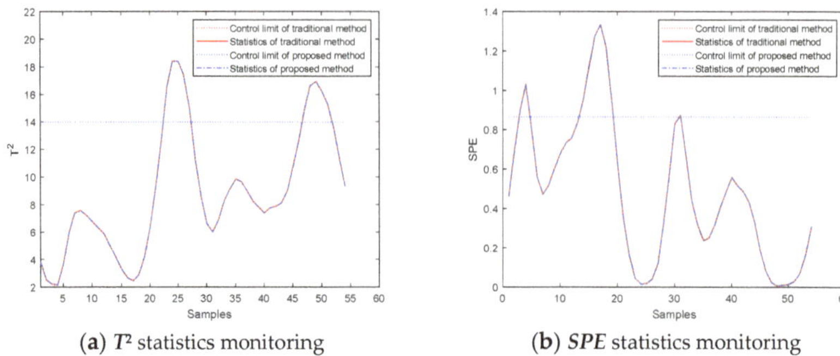

(a) T^2 statistics monitoring **(b)** *SPE* statistics monitoring

Figure 16. Multi-mode online monitoring of injection phase of mode 1.

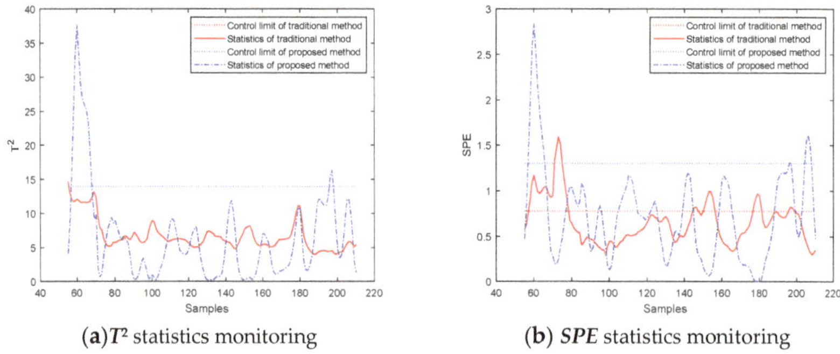

(a) T^2 statistics monitoring **(b)** *SPE* statistics monitoring

Figure 17. Multi-mode online monitoring of packing-holding phase of mode 1.

In order to compare the prediction results of the single-mode model and the between-mode model of different prediction methods, RMSE values of five test batches in mode 3 and mode 1 are used for judgment, as shown in Table 4.

According to the simulation results, it can be concluded that the between-mode model extracts the related information in the historical modes, so it contains more necessary information. It can be seen from Table 4 that the prediction results of the between-mode model are better than those of the single-mode model. Comparing the RMSE of the traditional method and the proposed method, it can be seen that the proposed method is more accurate for quality prediction. From the monitoring figures, it can be seen that if part of the mode information has been included in the modeling process, the statistics do not exceed the control limit, leading to a suitable monitoring effect. In contrast, if the modeling process does not contain the mode information, the statistics will exceed the control limits. To sum up, compared with single-mode modeling, the between-mode modeling contains more historical modal information, leading to better prediction, and can achieve the purpose of information selecting for monitoring. Therefore, for the current modes modeling, the between-mode modeling method can be selected.

For faulty batch monitoring using the between-mode modeling, the faulty batch data is consistent with the single-mode modeling faulty batch data. First, the faulty batch caused by material disturbance is monitored. The monitoring results of the traditional method and the proposed method are shown in Figure 18. Both the proposed method and the traditional method can detect the fault.

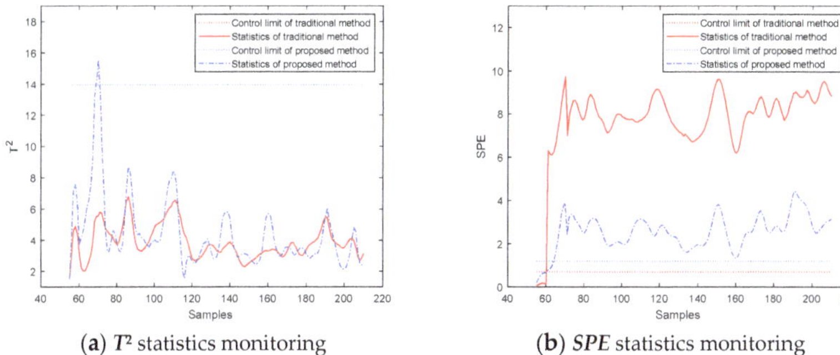

(a) T^2 statistics monitoring (b) *SPE* statistics monitoring

Figure 18. Multi-mode online monitoring of material disturbance fault.

Secondly, the monitoring effects of the traditional method and the proposed method for the sensor fault are shown in Figure 19. Compared with the traditional method, the statistics of the proposed method rise more sharply, and the amplitudes are relatively large.

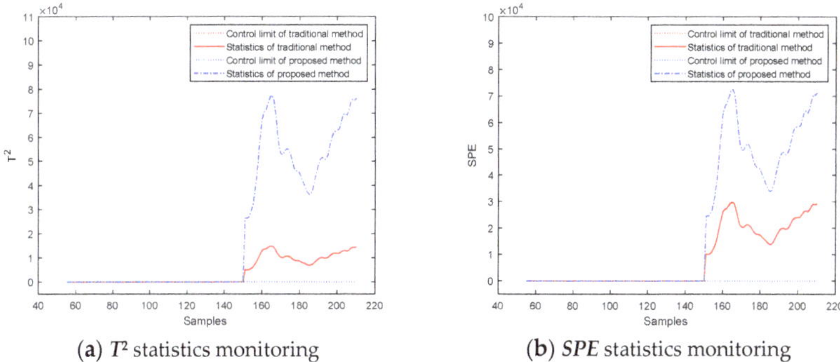

(a) T^2 statistics monitoring (b) *SPE* statistics monitoring

Figure 19. Multi-mode online monitoring of sensor fault.

4. Conclusions

In this work, based on the analysis of multi-phase multi-mode batch processes, a combination of the multi-phase quality residual recursion model for multiple phases and the between-mode model for multiple modes is proposed, and according process-monitoring strategies based on quality analysis are developed. Firstly, the critical-to-quality phases are identified and selected based on the influence of different phases on the final quality of the batch process. Then, the phase mean model is established, and based on the multi-phase quality residual recursive model, the quality predictions of critical-to-quality phases are obtained, and those phases are monitored. On the other hand, the between-mode model is used to analyzes the regression relationship between the process variables and the quality of the new mode through the historical modes, and online monitoring is carried out on this basis. Through the simulation of the experimental data of an injection molding process, it is proved that due to better quality predictions, the proposed strategy can provide better process-monitoring results for multi-phase multi-mode batch processes.

However, the experimental data used in this paper are all processed so that the length of the same phase of different batches is equal, which is often difficult to achieve in the actual industry due to various reasons, such as the influence of climate, the quality difference of raw materials, the data acquisition system based on a non-time coordinate, etc. In order to solve this problem, the effect of this method on the data of the batch process with unequal data lengths should be considered. For this, further research will be conducted in the future.

Author Contributions: Conceptualization, L.Z.; methodology, L.Z., X.H. and H.Y.; software, X.H. and H.Y.; validation, L.Z; formal analysis, L.Z., X.H. and H.Y.; investigation, L.Z.; resources, L.Z.; data curation, X.H.; writing—original draft preparation, X.H. and H.Y.; writing—review and editing, L.Z. and X.H.; visualization, X.H.; supervision, L.Z.; project administration, L.Z.; funding acquisition, L.Z. All authors have read and agreed to the published version of the manuscript.

Funding: This research was funded by the National Natural Science Foundation of China (No. 61503069) and the Fundamental Research Funds for the Central Universities (N150404020).

Institutional Review Board Statement: Not applicable.

Informed Consent Statement: Not applicable.

Data Availability Statement: The data presented in this study are openly available in https://share.weiyun.com/GCsqQcET (accessed on 29 July 2021).

Acknowledgments: This work is supported in part by the National Natural Science Foundation of China (No. 61503069) and the Fundamental Research Funds for the Central Universities (N150404020).

Conflicts of Interest: The authors declare no conflict of interest.

References

1. Martin, E.; Morris, A.; Papazoglou, M.; Kiparissides, C. Batch process monitoring for consistent production. *Comput. Chem. Eng.* **1996**, *20*, 599–604. [CrossRef]
2. Wold, S.; Kettaneh, N.; Fridén, H.; Holmberg, A. Modelling and diagnostics of batch processes and analogous kinetic experiments. *Chemom. Intell. Lab. Syst.* **1998**, *44*, 331–340. [CrossRef]
3. Zhao, C.H.; Yu, W.K.; Gao, F.R. Data Analytics and Condition Monitoring Methods for Nonstationary Batch Processes—Current Status and Future. *Acta Autom. Sin.* **2020**, *46*, 2072–2091.
4. He, Q.P.; Wang, J. Statistical process monitoring as a big data analytics tool for smart manufacturing. *J. Process. Control* **2018**, *67*, 35–43. [CrossRef]
5. Qin, S.J. Process data analytics in the era of big data. *AIChE J.* **2014**, *60*, 3092–3100. [CrossRef]
6. Bro, R. Multiway calibration multilinear PLS. *J. Chemom.* **1996**, *10*, 47–61. [CrossRef]
7. Zhao, C.H. *Online Monitoring and Quality Prediction for Multi-phase Batch Processes*; Northeastern University: Shenyang, China, 2009.
8. Nomikos, P.; MacGregor, J.F. Multivariate SPC charts for monitoring batch processes. *Technometrics* **1995**, *37*, 41–59. [CrossRef]
9. Nomikos, P.; MacGregor, J.F. Monitoring batch processes using multiway principal component analysis. *AIChE J.* **1994**, *40*, 1361–1375. [CrossRef]
10. Nomikos, P.; MacGregor, J.F. Multi-way partial least squares in monitoring batch processes. *Chemom. Intell. Lab. Syst.* **1995**, *30*, 97–108. [CrossRef]
11. Wold, S.; Sjöström, M. Chemometrics, present and future success. *Chemom. Intell. Lab. Syst.* **1998**, *44*, 3–14. [CrossRef]
12. Lane, S.; Martin, E.B.; Kooijmans, R. Performance monitoring of a multi-product semi-batch process. *J. Process. Control* **2001**, *11*, 1–11. [CrossRef]
13. Dong, D.; McAvoy, T.J. Multi-stage batch process monitoring. In Proceedings of the 1995 American Control Conference, Seattle, WA, USA, 21–23 June 1995; pp. 1857–1861.
14. Ündey, C.; Cinar, A. Statistical monitoring of multistage, multi-phase batch processes. *IEEE Control Syst. Mag.* **2002**, *22*, 40–52.
15. Lu, N.Y.; Wang, F.L.; Gao, F.R. Sub-PCA modeling and online monitoring strategy for batch processes. *AIChE J.* **2004**, *50*, 255–259. [CrossRef]
16. Zhao, C.H.; Lu, N.Y.; Wang, F.L.; Jia, M.X. Stage-based soft-transition multiple PCA modeling and on-line monitoring strategy for batch processes. *J. Process. Control* **2007**, *17*, 728–741. [CrossRef]
17. Lu, N.Y.; Gao, F.R.; Yang, Y.; Wang, F.L. PCA-based modeling and on-line monitoring strategy for uneven-length batch processes. *Ind. Eng. Chem. Res.* **2004**, *43*, 3343–3352. [CrossRef]
18. Zhao, L.P.; Wang, F.L.; Chang, Y.Q.; Wang, S.; Gao, F.R. Phase-based recursive regression for quality prediction of multi-phase batch processes. In Proceedings of the 13th IEEE International Conference on Control and Automation, Ohrid, Macedonia, 3–6 July 2017; pp. 283–288.
19. Hwang, D.H.; Han, C.H. Real-time monitoring for a process with multiple operating modes. *Control Eng. Pract.* **1999**, *7*, 891–902. [CrossRef]
20. Chen, J.; Liu, J. Mixture principal component analysis models for process monitoring. *Ind. Eng. Chem. Res.* **1999**, *38*, 1478–1488. [CrossRef]
21. Zhao, L.P.; Zhao, C.H.; Gao, F.R. Between-mode quality analysis based multi-mode batch process quality prediction. *Ind. Eng. Chem. Res.* **2014**, *53*, 15629–15638. [CrossRef]
22. Qin, Y. *Research on Data Driven Batch Process Monitoring and Quality Control*; Zhejiang University: Hangzhou, China, 2018.

23. Zhao, C.; Wang, F.; Mao, Z.; Lu, N.; Jia, M. Improved Batch Process Monitoring and Quality Prediction Based on Multi-phase Statistical Analysis. *Ind. Eng. Chem. Res.* **2008**, *47*, 835–849. [CrossRef]
24. Gunther, J.C.; Conner, J.S.; Seborg, D.E. Process monitoring and quality variable prediction utilizing PLS in industrial fed-batch cell culture. *J. Process. Control* **2009**, *19*, 914–921. [CrossRef]
25. Ni, L.J.; Zhang, L.G. *Basic Chemometrics and Its Applications*; East China University of Science and Technology Press: Shanghai, China, 2011; p. 237.
26. Liu, D.J. *Research of Partial Least Squares Regression Algorithm Based on Optimal Selection of Latent Variables*; Northeastern University: Shenyang, China, 2015.
27. Ündey, C.; Ertunç, S.; Çinar, A. Online Batch/Fed-Batch Process Performance Monitoring, Quality Prediction, and Variable-Contribution Analysis for Diagnosis. *Ind. E. Chem. Res.* **2003**, *42*, 4645–4658. [CrossRef]
28. Zhao, L.P.; Zhao, C.H.; Gao, F.R. Inter-batch-evolution-traced process monitoring based on inter-batch mode division for multi-phase batch processes. *Chemom. Intell. Lab. Syst.* **2014**, *138*, 178–192. [CrossRef]
29. Lu, N.Y.; Gao, F.R. Stage-based process analysis and quality prediction for batch processes. *Ind. Eng. Chem. Res.* **2005**, *44*, 3547–3555. [CrossRef]
30. Yuan, Q.; Zhao, L.; Wang, S.; Chang, Y.; Wang, F. Quality Analysis and Prediction for Multi-phase Multi-mode Injection Molding Processes. In Proceedings of the 30th Chinese Control and Decision Conference, Shenyang, China, 9–11 June 2018; pp. 3591–3596.

Article

Quality 4.0 in Action: Smart Hybrid Fault Diagnosis System in Plaster Production

Javaneh Ramezani * and Javad Jassbi

Faculty of Sciences and Technology and Uninova CTS, NOVA University of Lisbon, Campus de Caparica, 2829-516 Caparica, Portugal; j.jassbi@uninova.pt

* Correspondence: m.ramezani@campus.fct.unl.pt or ramezanijavaneh@gmail.com

Received: 9 April 2020; Accepted: 21 May 2020; Published: 26 May 2020

Abstract: Industry 4.0 (I4.0) represents the Fourth Industrial Revolution in manufacturing, expressing the digital transformation of industrial companies employing emerging technologies. Factories of the future will enjoy hybrid solutions, while quality is the heart of all manufacturing systems regardless of the type of production and products. Quality 4.0 is a branch of I4.0 with the aim of boosting quality by employing smart solutions and intelligent algorithms. There are many conceptual frameworks and models, while the main challenge is to have the experience of Quality 4.0 in action at the workshop level. In this paper, a hybrid model based on a neural network (NN) and expert system (ES) is proposed for dealing with control chart patterns (CCPs). The idea is to have, instead of a passive descriptive model, a smart predictive model to recommend corrective actions. A construction plaster-producing company was used to present and evaluate the advantages of this novel approach, while the result shows the competency and eligibility of Quality 4.0 in action.

Keywords: statistical process control; control chart pattern; disruptions; disruption management; fault diagnosis; Industry 4.0; construction industry; plaster production; neural networks; decision support systems; expert systems; failure mode and effects analysis (FMEA); discriminant analysis

---

1. Introduction, Background, and Problem Statement

1.1. Introduction

In today's globally complex and competitive business environments, quality is one of the crucial issues for ensuring the success of enterprises [1]. In order to produce with the desired quality and meet the customer's expectations, production processes need to be monitored to avoid any defect and deviation [2]. Traditionally, statistical process control (SPC) was used as a powerful approach for monitoring and identifying variations manually [1,2]. Developments in manufacturing and information technology enabled SPC to move from merely statistical control to real-time diagnosis purposes with minimum human intervention [3]. Control charts, invented by Shewhart in the 1920s, are essential tools in SPC to assist in controlling the behavior of the process. These tools are used to decide if the process is behaving as intended or in the presence of some unnatural causes of variations. X-bar and R charts are basic Shewhart control charts for drawing a series of process measured data with control limits [3,4]. Process variation emerges from either common causes (natural variations) or specific causes (assignable reasons). Specific causes are those that cause changes and short-term fluctuations, and, if they occur, they destroy the stability of the process, which ought to be known and eliminated as quickly as time permits. Common causes are because of the inherent characteristics of the process, and, if they exist, deviations (background noise) are in control [5,6]. However, the most crucial ability of control charts is detecting various types of patterns consisting of a series of consecutive points that are observed on these charts, which reflects fluctuations in the process [7]. The control chart patterns (CCPs) are generally divided into natural and unnatural patterns. Natural patterns usually exist in the manufacturing process and indicate that the process is statistically stable. As long as the measured data are inside the control limits or only natural random patterns exist, the process is under control. When some measurements fall out of the control limits or the measured data within the control limits signify a non-random pattern, the process is deemed out of control. Unnatural patterns displayed in control charts can be of various types, and each class can be related to specific causes unfavorably influencing the process stability. For example, "*Shift*" patterns may be related to variations in raw material, supplier, or machine, whereas "*Trend*" patterns may occur due to gauge wear or environmental changes [1,8]. Different common patterns that regularly emerge in control charts can be found in Figure 1. Over time, various further decision rules such as "*zone tests*" and "*run rules*", including "*Western Electric,*" "*Nelson*", etc., were developed to assist quality control engineers and operators in detecting unnatural CCPs and circumstances leading to a change in the process [9]. Table 1 shows the most recommended rules for the Shewhart control charts to identify abnormal patterns and interpret their characteristic signs in the control chart. In general, the use of run rules can result in quickly signaling a shift in the process. However, the application of all these rules, when no particular cause exists, increases the risk of false alarms (Type I errors) to an unacceptable extent. In addition, run rules do not provide valuable pattern-related information because of a lack of sufficient pattern discrimination capability. Furthermore, control charts do not consider prior knowledge or adequate historical data. Therefore, these decision rules are not particularly useful for CCP recognition [10,11]. Since the analysis of control charts is complicated, because it relies on considerable statistical knowledge, skill, and experience of the practitioners (quality control personnel), developing an efficient automated pattern recognition system that can ensure steady and unbiased analysis of CCPs can compensate for this gap [12].

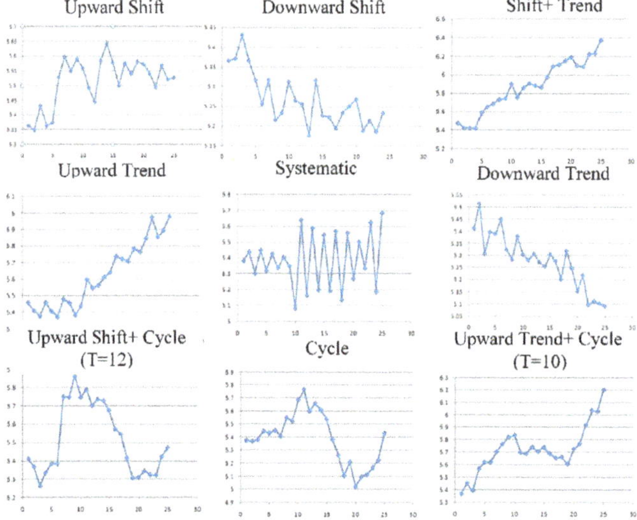

Figure 1. Typical patterns in control charts.

Table 1. Mostly recommended rules for detecting typical unnatural patterns.

No.	Unnatural Pattern	Characteristic Signs in Control Chart
1	Over-control	Single point beyond control limits (above $+3\sigma$ or below -3σ).
2	Shift	Sudden change (series of 9 points) above or below the central line.
3	Trend	Continuous movement (rise or fall) of 6 consecutive points.
4	Systematic	A point-to-point fluctuation (14 consecutive points alternating up and down).
5	Cycle	Periodic peaks and troughs (4 out of 5 points above $+2\sigma$ or below -2σ).
6	Mixtures	A run of consecutive points on both sides of the central line, all far from the central line (8 points in a row more than $+1\sigma$ from centerline).

1.2. Background and Problem Statement

With the development in manufacturing and computing technology, several approaches were proposed using artificial intelligence technologies such as artificial neural networks (ANNs), expert systems (ESs), and fuzzy sets to automatically and intelligently CCP recognition [13]. In the domain of SPC, fast and accurate control, as well as observing the variation of quality characteristics and, consequently, recognition of unnatural patterns, is the primary purpose of each fault detection and diagnosis system. There are numerous studies in this field on CCP recognition that used different machine learning algorithms and other intelligent approaches, namely, K-nearest neighbors (KNN), decision trees (DT), NN-based models, ES-based models, support vector machine (SVM), wavelet-based models, and fuzzy logic [14–16]. These approaches aim at extracting meaningful information from a large amount of data to detect instabilities in the process with minimal time and cost and maximum accuracy [17]. To sum up, the most significant approaches are explained briefly in Table 2 by highlighting their advantages and disadvantages.

Table 2. Related works.

Model	Advantage	Disadvantage	Related Works
KNN	- Very fast training (instance-based learning). - Very easy to implement.	- Weakness in working with large dataset. - Memory limitation. - Sensitive to noisy data.	[15,18]
DT	- Simple to understand, interpret, and generate rules.	- May suffer from overfitting. - Unstable classifier.	[15,16]
NN	- Does not need precise knowledge of interactions between the parameters. - Learns to recognize patterns during the training phase. - Able to handle noisy data. - High performance.	- NN's topology cannot be systematically determined. - Training of the network is prolonged, and processing for large NNs is difficult. - Needs a large amount of useful training samples. - Problem of overfitting.	[3,4,7,10–15,19–21]
ES	- Availability, consistency, extensibility, and testability of the information. - Rules can be updated easily.	- Problems of incorrect recognition for similar statistical properties (features overlapping).	[9,12,14,19,22,23]
SVM	- Easily handles nonlinear, un/semi-structured, and high-dimensional data. - Overfitting problem is not as much as other methods. - With an appropriate kernel function, complex problems can be solved.	- Computationally expensive. - Difficult understanding and interpreting the final model. - Long training time for large datasets.	[2,14,24–26]
Fuzzy	- High precision.	- Low speed and the long run time of the system.	[5,14,21,26]

The literature review shows that ANNs and ESs are the most widely used approaches, being easier to understand and implement and having higher performance in comparison to other CCP recognition approaches mentioned above. NNs are suitable for SPC as they are good at classification and pattern recognition, and they are able to handle the noisy measurements with no requirement for the provision of explicit rules regarding the monitored data [20]. Notably, ESs are useful for quality control applications due to their potential for identifying causes of deviations and recommending preventive and corrective actions [23]. There are two approaches to applying ANNs to CCP recognition: (1) using neural networks (NNs) to detect variation in X-bar and/or R charts, and (2) using NNs to identify unnatural patterns [19]. In this regard, NNs can be classified into two main categories: supervised NNs, involving multilayer perceptron (MLP) and radial basis function (RBF), and unsupervised NNs, including learning vector quantization (LVQ) and adaptive resonance theory (ART) [10]. Among the ANNs, the multilayer perceptron (MLP) was successfully exploited by many researchers in order to address the unnatural CCP recognition problem. Learning vector quantization (LVQ) is a well-applied alternative method to solve the problem of slowness in training the MLP network [4,12,20,27].

The fault diagnosis is an essential issue in SPC, to reduce downtime and disruption cascades that can ensue [24]. In recent years, various diagnostic systems were developed to automate fault diagnosis, but none of them fit our problem in the plaster production process discussed here. Most fault diagnosis approaches in the literature only considered a particular control chart, often X-bar or R (range) chart, to examine the process changes (mean or variance). However, in practice, in many processes, it is required to combine the two charts as multiple assignable causes may occur [28]. On the other hand, identification of unnatural patterns combined with specific knowledge of the process results in a more targeted diagnosis. Unfortunately, none of the CCP recognition models in the literature provide this combination automatically, which can be valuable for diagnostic purposes. Moreover, the performance of the model was not evaluated when developing these approaches in a real case study.

Yet, the common problem reported in these studies is the inability to recognize various single and concurrent CCPs, as well as a high rate of false recognition [4,29]. On the other hand, most applications of NNs and ESs to CCP recognition do not obtain more detailed information about the patterns and their change point (when these patterns are observed on control charts). This information is essential for practical assignable cause analysis and, in turn, accelerates the accomplishment of proper remedial activities [21].

Therefore, in this paper, designing a hybrid fault diagnosis system is proposed using NNs and a rule-based ES to help the quality control personnel in recognizing the roots of deviations, and in taking needed predictive or corrective actions. In the design process, for the structure of the NN, a modular approach comprising an LVQ network and seven multilayer perceptrons (MLPs) is used. Therefore, our work provides a neural expert system in intelligent real-time monitoring and predictive, corrective, and remedial diagnosis of process control in plaster production. To develop the proposed neural expert system (Figure 2), we address the following notable features of the model:

- Ability to detect various natural and unnatural (single and concurrent) CCPs.
- Monitoring and analyzing X-bar and R charts abnormalities simultaneously.
- Capability to estimate nonrandom patterns' corresponding parameters, different directions, and change points (starting point) in control charts.
- Identifying the responsible variables on the occurrence of unnatural patterns.
- Recognizing causes of process instability.
- Recommending predictive and/or corrective actions in a time of crisis.

The idea is to have, instead of a passive descriptive model, a smart predictive model to assist quality control engineers for the fault diagnosis of the process, particularly from a practical perspective regarding a Quality 4.0 era.

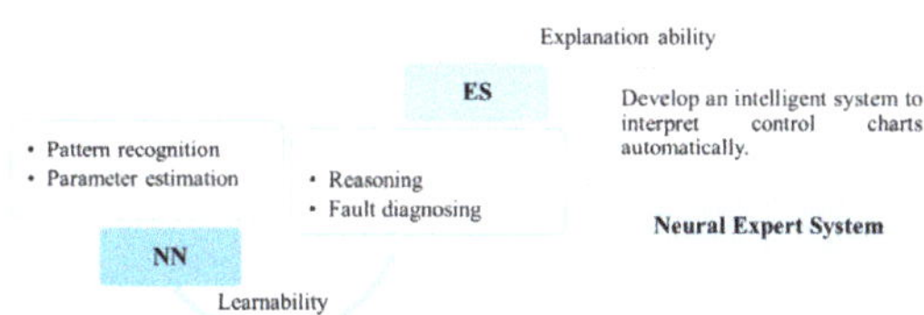

Figure 2. Combination of NN and ES: A neural expert system.

1.3. Contribution to Industry 4.0

Quality "4.0" is a branch of the Industry 4.0 (I4.0) movement associated with the digital transformation process connected with emerging technologies. Quality 4.0 could be defined as the application of Industry 4.0 technologies to quality management methods and tools [30]. According to Reference [31], "Quality 4.0 does not replace traditional quality methods, but rather builds and improves upon them". This concept covers all issues of advanced quality management in the digital era [32]. For quality (technology, processes, and people), Industry 4.0 enables the transformation of existing capabilities (culture, management, collaboration, and competencies) to drive value [31].

The impact of I4.0 on manufacturing is beyond just the physical production of goods, involving targeting all processes and functions to achieve flexibility, smartness, cost-effectiveness, and resilience. Artificial intelligence and machine learning are among the aforementioned technologies that can be utilized to enhance the quality as the heart of smart manufacturing [12,33,34].

On the other hand, construction projects face different sources of disruptions, as they are time-limited, expert-dependent, and highly influenced by process fluctuations caused by weather conditions, material quality, etc., which leads to a high level of complexity and uncertainty in the construction ecosystem [35]. Industry 4.0 challenged the construction industry ecosystem by demonstrating the construction digitalization potential for real-time data collecting, processing, and sharing tools to enhance alignment between demand and supply [35,36].

SPC is an essential tool to monitor process disruptions, safety assurance, and reliability analysis in construction projects [37]. Industry 4.0, with its automation, connectivity, and digital access capacity, is anticipated to be capable of increasing the efficiency and productivity of SPC. This could happen through enabling intelligent monitoring and diagnosis, automatic tracking of equipment and material, and real-time decision-making, especially in situations where the process is becoming more volatile and complex (Figure 3) [12,30].

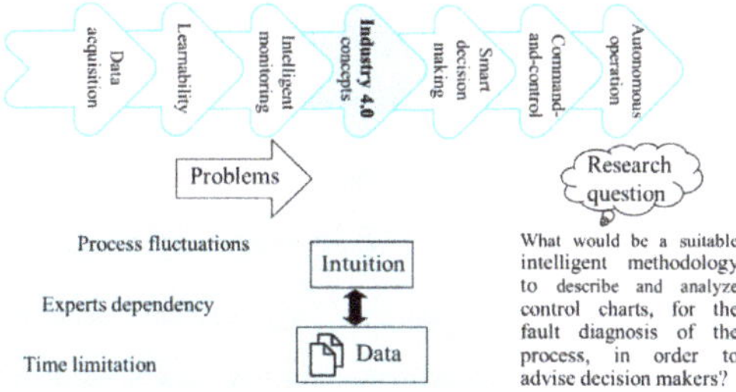

Figure 3. Quality 4.0: integration of traditional statistical process control (SPC) with Industry 4.0.

Digitalization and automation are the two pillars of smart manufacturing [38]. In this work, our effort was to develop a model based on traditional quality systems driven by disruptive technology. In this paper, artificial intelligence in the form of ANN and ES is employed. This proposes new values for the value chain of the manufacturing system on the factory floor level.

In fact, the innovation associated with digitalization, automation, communication, optimization, and customization of Industry 4.0 concepts and trends allows for real-time analysis and interpretation of production, industries, and service processes to improve quality by detecting failures and justifying possible causes while staying competitive in volatile business environments [30,39]. This work provides just a step to move forward and make the dream of "smart manufacturing" happen under the light of Industry 4.0.

The remainder of this paper is organized as follows: Section 2 concisely outlines the methodology of the research; Section 3 presents the proposed model; Section 4 describes the detailed structure of the model; Section 5 presents a comparative analysis and shows some results from a real case study; finally, Section 6 concludes the paper.

2. Materials and Methods

The methodology of this research is descriptive–experimental research, which is a systematic mapping study based on Reference [40] and an implemented case study in a plaster production plant. Figure 4 represents the schematic diagram depicting the proposed procedure. Using the review of the literature and benchmarking on the extraction of intelligent models used in process control, the structure of a hybrid fault diagnosis system using ANN and ES in the process control of plaster production is presented in this research. Mapping is used to present structuring to synthesize the three main research areas that include statistical process control, neural networks, and expert systems in this research. The case study, based on experiences from model implementation and validation in a plaster production plant, is reflected in Section 4. The plaster production process, which was selected as a case study, is a fluctuating process that has many influencing parameters. On the other hand, because the final production, i.e., construction plaster or so-called "plaster of Paris (PoP)", is mixed in a silo, the monitoring and modification processes in a short period can prevent the entire silo storage product from crashing. On the contrary, if the process is not monitored with statistical process control over an extended period, non-compliance of part of the product with the standard can crash the entire stored production. For example, in the case of filling more than 10% of the silo from a mismatched product, the whole product in the silo will crash. In order to improve the process quality, a survey was done of experts using a questionnaire and interviews to identify critical control parameters. The "initial setting time" of plaster was detected as the critical parameter of the production process. The initial setting time is dependent on the "crystal water" of baked plaster and ought to last between 7 and 15 min in the intended case study. The acceptance range in our case study was between low (LSL = 5.0) and upper (USL = 5.08) specification limits. The process was deemed in control with the lower and upper control limits of LCL = 5.26 and UCL = 5.56. Then, based on existing records, causes of process failures and defects in construction plaster, which were connected with the plaster's qualitative characteristics, were examined using a "cause and effect" diagram" [41]. Finally, parameters that could improve customer satisfaction after identifying and prioritizing the foreseeable failure modes were determined and analyzed applying failure mode and effects analysis (FMEA) [42]. The statistical population of this research comprised "PoP", which is baked at a particular time in the "low burn" kiln and moved from baking salon to storage silo. The sampling method was a stratified random sampling method. Because of the characteristics of the plaster production process and consistent with the background studies, 25 subgroups of n = 125 samples were taken from multiple samples from different shifts. In this research, data were analyzed using three approaches of FMEA, ANN, and discriminant analysis (DA) [1]. To perform discriminant analysis, an understandable database for "SAS" software using Excel software was provided, and discriminant analysis was performed using programming ("Proc Discrim") in SAS software. For the case study of the present study, the data related to the critical parameter of the process were firstly collected, and the causes of product failure were identified and prioritized. Then, given that the proposed model is an intelligent hybrid model that can learn the patterns from input data (samples) using the power of learning neural networks, data were detected. Finally, the error of identifying training and test datasets was compared with the statistical method of discriminant analysis. In this research, in order to monitor and troubleshoot the process, a model for combining SPC and artificial intelligence was designed using "MATLAB" software. The program codified in MATLAB is able to produce, present, and quickly encode neural network input data, as well as execute expert system rules. The program itself can also perform traditional SPC operations.

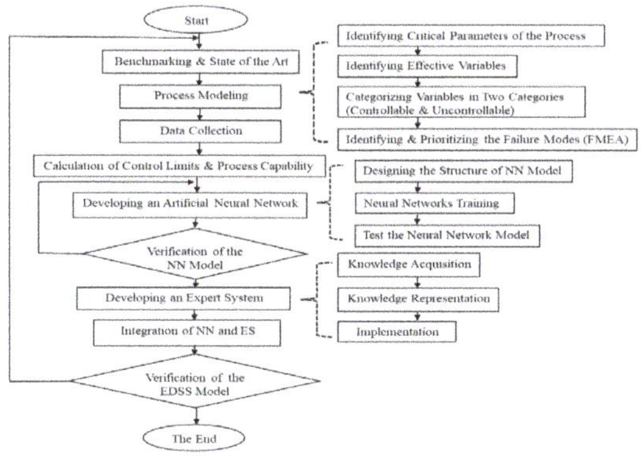

Figure 4. The flow diagram of the study procedure.

3. The Proposed Hybrid Fault Diagnosis Model

Based on what was said earlier, this research is based on the integration of NNs and ESs to provide analysis and interpretation for CCPs. The main focus of this study is to introduce a neural expert system-based pattern identifier, which will allow identifying abnormal patterns in order to correct their assignable causes. The operator will be warned if an abnormal pattern occurs in the process. By replacing human skills with a detection algorithm, human intervention is greatly reduced, and an intelligent manufacturing environment could be achieved. In this study, NNs are used to recognize control chart patterns, and an expert system is also used to interpret the identified pattern and determine the causes of the abnormal pattern. The general model of the research is depicted in Figure 5. As Figure 6 indicates, the proposed system consists of three subsystems:

- The SPC subsystem controls the traditional statistical process and, using statistical formulas, draws mean and variance for the sampled data of the process. It also sets control limits and determines the capability of the process and, in cases where any point on the charts is out of control, alerts "out of control" mode.
- The pattern recognition subsystem is accountable for detecting abnormal CCPs. Here, unnatural patterns in the X-bar chart are detected using neural networks, and abnormal patterns in the R chart are identified using "*Western Electric*" with the rule-based expert system.
- The reasoning subsystem is responsible for interpreting the purpose of process variations and proposing corrective or preventive actions. In this subsystem, using process-specific knowledge provided as if–then rules in the knowledge base, the cause of the abnormal patterns in the X-bar chart is interpreted. On the other hand, the cause of the unnatural patterns in the R chart is interpreted using general process knowledge, presented as if–then rules in the knowledge base (Figure 6).

Overall, the model design structure can be divided into three stages of neural network creation, expert system development, and integration of neural network and expert system, as explained below.

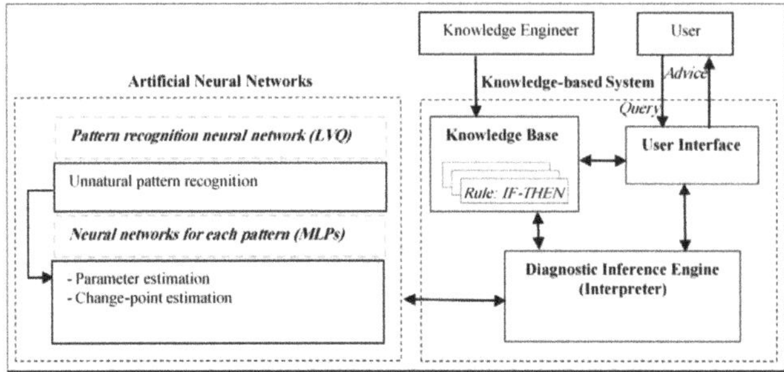

Figure 5. The structure of the neural expert system.

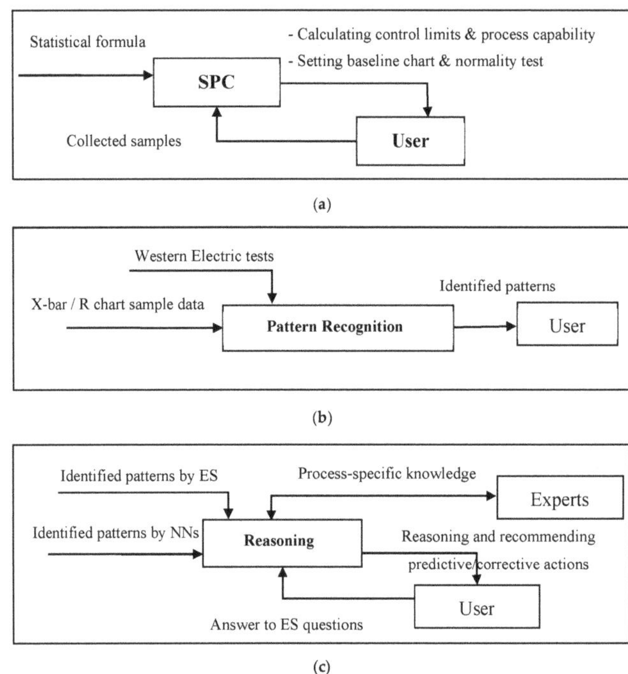

Figure 6. (**a**) SPC subsystem; (**b**) pattern recognition subsystem; (**c**) reasoning subsystem.

4. Experimental Results

This section describes the detailed characteristics of the structure of the proposed model and provides the results of their performances.

4.1. Developing the Neural Network Model

In the subsections below, the procedure for simulating normal and abnormal patterns in this research is firstly described. Then, the steps of creating a neural network, including neural network model structure design, neural network training, and neural network model validation, are presented.

4.1.1. Simulation of Unnatural CCPs and Their Corresponding Parameters

In the present research, because of the lack of a large number of useful samples to investigate abnormal CCPs, simulation of the models for training and testing networks was required; however, it was attempted to simulate the data in line with the underlying process data. In the statistical issues, there is a probability distribution function for every random variable based on which the relevant parameters are also determined. Therefore, any natural deviations could be determined according to the probability distribution function of the corresponding random variable [12]. With these explanations, the parameters and functions employed to simulate control chart patterns are presented in Table 3. In this table, the parameters, on the one hand, represent the number of non-random disturbances. Furthermore, they reflect the process improvement during the implementation of recovery programs. In designing the proposed model, it is intended to identify the X-bar chart patterns using NN. The simulator function for the natural variation of the X-bar chart includes normal distribution: $x(t) = n(t)$, and the parameters of this distribution in our case are $\mu = 5.4$ and $\sigma = 0.1$. Given these values, the corresponding parameters of the other abnormal patterns in this chart were calculated, as shown in Table 3.

Table 3. Simulator functions of CCPs and the range of corresponding parameters' changes.

Pattern Type	Simulator Functions	Parameter Change Range
Natural	$x(t) = n(t)$	-
Shift (Sh.)	$x(t) = n(t) + u \times b$ [1]	$b = [1\sigma\sim3\sigma] \Rightarrow [0.1,0.3]$ $b = [-3\sigma\sim-1\sigma] \Rightarrow [-0.3, -0.1]$
Trend (Tr.)	$x(t) = n(t) + s \times t$ [2]	$s = [0.1\sigma\sim0.3\sigma] \Rightarrow [0.01, 0.03]$ $s = [-0.3\sigma\sim-0.1\sigma] \Rightarrow [-0.03, -0.01]$
Cycles (Cyc.)	$x(t) = n(t) + l$ [3] $\times \sin((2\pi t)/T$ [4]$)$	$l = [1\sigma\sim3\sigma] \Rightarrow [0.1,0.3]$ $T = 8,12, \ldots$
Systematic (Sys.)	$x(t) = n(t) + g$ [5] $\times \cos(\pi t)$	$g = [1\sigma\sim3\sigma] \Rightarrow [0.1,0.3]$
Shift + Trend (Sh. + Tr.)	$x(t) = n(t) + u \times b + s \times t$	$b = [1\sigma\sim3\sigma] \Rightarrow [0.1,0.3]$ $s = [0.1\sigma\sim0.3\sigma] \Rightarrow [0.01, 0.03]$ $b = [-3\sigma\sim-1\sigma] \Rightarrow [-0.3, -0.1]$ $s = [-0.3\sigma\sim-0.1\sigma] \Rightarrow [-0.03, -0.01]$
Shift + Cycle (Sh. + Cyc.)	$x(t) = n(t) + u \times b + l \times \sin((2\pi t)/T)$	$b = [1\sigma\sim3\sigma] \Rightarrow [0.1,0.3]$ $l = [1\sigma\sim3\sigma] \Rightarrow [0.1,0.3]$ $T = 8, 12 \ldots$
Trend + Cycle (Tr. + Cyc.)	$x(t) = n(t) + s \times t + l \times \sin((2\pi t)/T)$	$s = [0.1\sigma\sim0.3\sigma] \Rightarrow [0.01, 0.03]$ $l = [1\sigma\sim3\sigma] \Rightarrow [0.1,0.3]$ $T = 8, 12 \ldots$

[1] Shift magnitude. [2] Trend slope (s). [3] Amplitude. [4] Period (T). [5] Magnitude of variations (g).

4.1.2. Designing the Structure of the Neural Network Model

In designing the general structure of the NN model, a modular approach was used. The overall model structure consisted of two separate sets of Module I and Module II. In the modular approach, the inputs and outputs of each network can be better managed, and the results of each network performance can be traced.

- **Module I**

Module I was developed to diagnose the behavior of the plaster production process. To this end, the classification power of competing algorithms was used, and a learning vector quantization (LVQ) network was designed to classify input patterns.

o Topology of LVQ Network

In the LVQ network (Figure 7), the connection type between layers is semi-connected, and the input vector, according to the process requirements, includes 25 neurons (25 samples taken from the process). The first layer contains 175 neurons, and the second layer includes eight neurons, while there was no need to consider the term bias. Each of the second-layer neurons represents one of the simulated patterns. Accordingly, neurons # 1 of the natural patterns and other neurons detect abnormal patterns of shift, trend, cycle, systematic, shift + trend, shift + cycle, and trend + cycle, respectively. Due to minimizing the number of outputs, there are patterns in this network, such as "upward shift" and "downward shift", which represent a pattern with the same equations but values of different parameters. The main criterion for determining the number of subgroups required per class was reducing the incorrect identification of patterns. On the other hand, we attempted to assign almost identical neurons to patterns with the same number of parameters (Table 4).

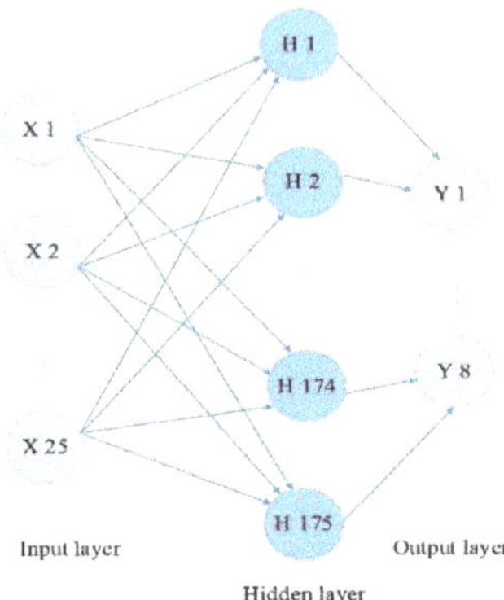

Figure 7. Learning vector quantization (LVQ) network.

Table 4. Number of inputs, hidden, and output layer neurons.

		Neuron
Input		25
Hidden Layer	Natural (Network 1)	1
	Upward Shift (Network 2)	12
	Downward Shift (Network 2)	12
	Upward Trend (Network 3)	18
	Downward Trend (Network 3)	18
	Cycles (Network 4)	20
	Systematic (Network 5)	4
	Upward Shift + Upward Trend (Network 6)	18
	Downward Shift + Downward Trend (Network 6)	18
	Shift + Cycles (Network 7)	27
	Trend + Cycles (Network 8)	27
	Total	175
Output		8

o Learning Algorithm Used in Module I

The LVQ network in module I was trained by "enabling competition to take a place among the "*Kohonen*" neurons. The competition is based on the Euclidean distances (d_i) between the weight vectors (W_i) of these neurons (i) and the input vector (x).

$$d_i = \|Wi - X\| = \sqrt{\left(W_{ij} - X_j\right)^2}. \tag{1}$$

The neuron which has the least distance is the winner in the competition and is allowed to change its connection weights. The weights of the other neurons remain unchanged. The new weights can be obtained from

$$W_{\text{new}} = W_{\text{old}} + \lambda(X - W_{\text{old}}), \tag{2}$$

if the winner neuron is in the correct output category, or

$$W_{\text{new}} = W_{\text{old}} - \lambda(X - W_{\text{old}}), \tag{3}$$

if the winner is in the wrong category [18,27].

In the above equations, λ is the learning rate, which decreases monotonically with the number of iterations. In this research, $\lambda = 0.01$ was considered. Appendix A provides the Matlab code.

• **Module II**

Module II in the proposed model was formulated to estimate the parameters of unnatural patterns of process control diagrams and estimation of the change point of the unnatural patterns.

o Topology of MLP Networks

In Module II, seven multilayer perceptron networks or MLPs do basic (single) and concurrent (mixture) pattern analysis. In this module, each of the networks in this set performs only the interpretation of one of the abnormal patterns. In these networks, the main parameters are estimated based on the definitions set out in Table 3. Moreover, in Module II, conditions for estimating the change point of abnormal behaviors in control charts are provided. In MLP networks (Figure 8), the type of layer connection is fully connected, and the number of inputs to all MLP networks is 26, 25 of which are the number of input neurons and one of which is the network bias which is equivalent to 1 for all simulated data. The number of hidden-layer neurons is optimized such that, for a certain amount of error for MLP on all network networks, the number of iterations required to achieve the error desired was calculated. Then, the number of neurons with the least number of repetitions until the desired error is chosen as the number of optimal neurons (Table 5). In each output layer, the neuron is embedded considering the number of related parameters of every pattern. In order to approximate different orientations of process changes, for example (upward or downward) given the outputs in the interval [−1, 1], a bipolar sigmoid function with the A = 0.1 constant is used. The relationship of the desired transfer function can be seen below.

$$g(x) = \frac{1-e^{-x}}{1+e^{-x}}\ x = \text{A.net}. \tag{4}$$

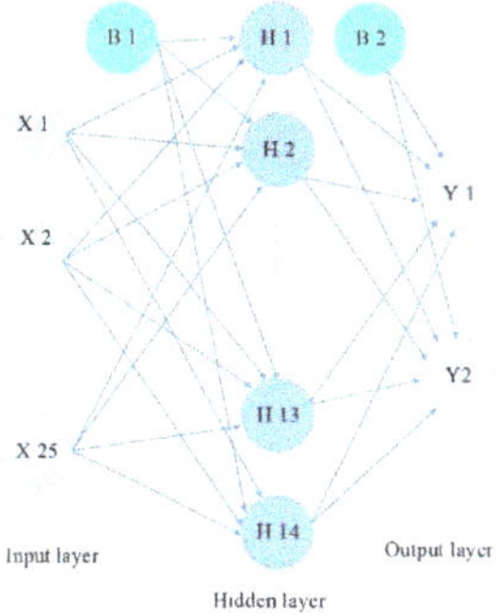

Figure 8. MLP network for shift pattern.

Table 5. Number of input, hidden, and output layer neurons.

Network Name	Input	Hidden Layer	Output
Shift		14	2
Trend		15	2
Cycles		22	3
Systematic	25 + 1 (bias) = 26	17	2
Shift + Trend		25	3
Shift + Cycle		23	4
Trend + Cycle		21	4

o Learning Algorithm Used in Module II

The training method in MLP networks is "backpropagation with an adaptive learning rate, where the weight of each layer, by the output and output derivative, is corrected until the network is fully trained". In this study, the training dataset is applied to corresponding networks in a category form, and errors are calculated at each step until the learning process is performed. Learning rate (λ) also changes according to the following command, so that the $E(t)$ is the network error at the time step t:

$$\lambda = \{0.99\lambda(t-1) \quad if \quad E(t) \leq E(t-1) 0.01\lambda(t-1) \quad if \quad E(t) > E(t-1). \tag{5}$$

There is a condition of training stop on the network error, which tries to minimize the error square between network outputs and the objective function using the gradient descent method. The network error, which is a cumulative error, is defined below in which p stands for pattern number, o represents the output neurons, d_{ij} is desired value for the j output of i pattern, and o_ij is the actual output of the network for the i pattern [43]. Appendix B provides the source code written in Matlab.

$$E = \frac{1}{2}\sum_{i=1}^{p}\sum_{j=1}^{o}\left(d_{ij} - o_{ij}\right)^2. \tag{6}$$

o Change Point of Unnatural Patterns

Estimation of the change point (starting point) and subsequent length of the unnatural pattern sequence expressed when the problem started and how long it lasted can help discover the causes of disorders. Since the neural network gives the change point of the abnormal patterns as a Module II output (MLP network) when generating training data, the change point is randomly generated between one and 10, and it is trained on the network to estimate its value as the network output. It should be noted that, in Module I, a fixed number is assumed to be the starting point for each pattern. In opting for the assumed fixed number as the change point, the conditions for pattern formation are considered with regard to its parameters. For example, since the cycle pattern has a period parameter (T = 8–12), the starting point should be earlier.

4.1.3. Neural Network Training

Training examples were introduced randomly to the NN. Before training, connection weights were generated with small random values. Weights were adjusted by training and presenting each pattern to the network. A maximum of 200,000 repetitions was considered as the stopping criterion for training. During the training phase, a series of vectors are provided to the network. The training vector consists of two sub-vectors: An input pattern and a target pattern. There are a total of 33 values for each training vector. A series of 25 coded observations, called the input pattern, is presented to the input layer, and the target pattern, which has an integer output of its inputs, is presented to the output layer (Table 6). Since each pattern has two orientations of changes (e.g., positive and negative shift), the desired output is set to 1 or −1. An output of 1 corresponds to a positive change, and an output of −1 corresponds to a negative change. For example, the output vectors of the "natural" pattern would be [10000000] and downward shift [0–1000000].

- **Training dataset**

In this study, the simulated data for the neural network model were divided into two subsets of training data and test data. Since there was no prior knowledge of the relative importance of unnatural patterns here, the training set contained approximately an equal number of training data for each type of pattern. In total, there were 11,000 training samples in the study set. The total number of training data in the LVQ network was 4000 samples, which equally considered 500 for each pattern. The total number of training samples for MLP networks was 7000, in which the same amount of 1000 samples was generated and applied for each of the seven MLP networks. In order to produce the training dataset with the specifications mentioned above, a program was codified, which was capable of producing an unlimited number of natural and unnatural patterns with different parameters (for example, with different mean and standard deviations).

Table 6. Scaling range for outputs of MLPs.

	Pattern Type	Shift	Trend	Cycle	Systematic	Shift-Trend	Shift-Cycle	Trend-Cycle
Output 1	min	−0.4	−0.04	0.01	0.01	−0.4	0.01	0.001
	max	0.4	0.04	0.4	0.4	0.4	0.4	0.04
Output 2	min	0	0	6	0	−0.04	0.01	0.01
	max	12	12	14	12	0.04	0.4	0.4
Output 3	min	-	-	0	-	0	6	6
	max	-	-	12	-	12	14	14
Output 4	min	-	-	-	-	-	0	0
	max	-	-	-	-	-	12	12

o Dataset scaling

In order to scale the dataset, the upper and lower boundaries of the input data were firstly specified by matching the maximum and minimum values of the input parameters. Then, the desired data were scaled and expanded into fitting values according to the type of the used transfer functions. Using the scaling method, all of the above operations were performed with the program written in Matlab. In the given formula, *A* is the "original value", *Ascale* is the "normalized value", *Amin* is the "minimum observable value", and *Amax* is the "maximum observable value". *Amin* and *Amax* might be estimated depending on the nature of data.

$$Ascale = min + \frac{max - min}{Amax - Amin}(A - Amin). \tag{7}$$

In the current application, the intervals used to scale the inputs of MLPs, considering the maximum and minimum ranges for the parameters of the unnatural patterns (±σ3) and given the mean (μ = 5.4) and standard deviation (σ = 0.1), were [5.1, 5.7], which were scaled with confidence intervals of [3.4, 7.4]. In the LVQ network, the data scaling range (training and testing) was [−5, +5]. In this study, all input and output data of MLP networks were scaled in a [−1, +1] interval; however, before the output values were scaled in the [−1, +1] interval, each parameter was scaled to the separate maximum and minimum values. The different scaling values corresponding to the outputs of the MLPs are visible in Table 6. The maximum cumulative error (MCE) for training the LVQ was 0.047 (188 in 4000 training data), and that for testing was 0.0525. Table 7 shows the MCE and training iteration of each MLP network.

Table 7. Module I, evaluation results.

Pattern Type	Direct Identification	Identification (Incomplete/Indirect)	Wrong Identification	Type I Error	Type II Error
Shift	41	6/50 = 0.12	3/50 = 0.06	0.00	-
Trend	48	2/50 = 0.04	0.00	0.00	1/50 = 0.02
Cycle	48	1/50 = 0.02	1/50 = 0.02	0.00	1/50 = 0.02
Systematic	50	0.00	0.00	0.00	0.00
Shift + Trend	50	0.00	0.00	0.00	0.00
Shift + Cycle	50	0.00	0.00	0.00	0.00
Trend + Cycle	44	0.00	6/50 = 0.06	0.00	0.00
Total Error	21/400 = 0.052	9/400 = 0.022	10/400 = 0.025	0.00	2/400 = 0.005

4.1.4. Neural Network Test

After training the network with the training dataset, the network was evaluated by the test dataset. In the training phase, network efficiency was increased by minimizing errors between actual outputs.

In the test phase, solely the input vector was given to the network, where, in this case, the network was validated by predicting the response values for input and output.

• **Module I, Evaluation of LVQ Network**

For each input vector, the LVQ network decides about the production process situation. Therefore, a chance of errors in decision-making issues will arise. In case the network incorrectly recognizes the natural variation of the process abnormally, it commits a type I error. If it does not recognize the abnormal pattern in the process, a type II error takes place. An incorrect identification error occurs when random deviations cause the basic patterns in the early parts of the formation to have similar behavioral characteristics. The same will apply to concurrent patterns. As each random pattern warns of a particular disturbance in the process, incorrect pattern recognition has different costs. Moreover, if a basic abnormal pattern is identified in the form of a concurrent pattern comprising a basic pattern, it is considered indirect identification. On the other hand, if only one of the unnatural patterns is identified during the simultaneous occurrence of two abnormal patterns, the identification is putatively incomplete. The performance of Module I was measured according to the instructions and definitions performed by 400 test vectors, where each of them represents 25 samples of the plaster production process, which represents one of the eight patterns identified by the neural network. We applied each of the samples as input to the network and then compared the network response with the target response and calculated the network error rate. Table 7 presents the merged results for the 400 test vectors. As can be seen in the table below, the maximum LVQ network error in pattern recognition was 0.052 (21 in 400 data), which demonstrates that the proposed model was successful and effective due to the variety of trained patterns in the identification problem.

• **Module II, Evaluation the MLP Networks**

One of the important issues in neural network training is the overfitting problem of the training data. To put it bluntly, the network learns data very well, and it even remembers the noise in the data (disturbances)—excessive compliance—but it has serious problems identifying and generalizing new data [26]. To solve the problem, when the test data error increases while maintaining or decreasing errors related to training data, the training is stopped, and the final parameters are considered with the minimum error of the test data. The performance of the MLP networks in Module II was examined with numerous examples, and the results were satisfactory. As seen in Table 8, the calculated cumulative error of each MLP network was less than 0.02, which indicates that Module II was successful in identifying the parameters.

Table 8. Module II, evaluation results.

Network Name	Maximum Cumulative Error (MCE)	Training Iteration	Minimum Number in Each Training	Hidden Layer Neurons	Output Neurons	Error of MLPs
Shift	10	10	10,537	14	2	0.01
Trend	18	13	16,423	15	2	0.018
Cycle	25	10	35,360	22	2	0.016
Systematic	12	10	48,343	17	2	0.012
Shift + Trend	18	10	48,343	25	3	0.012
Shift + Cycle	25	11	46,703	23	4	0.012
Trend + Cycle	28	10	37,298	21	4	0.014

4.2. Expert Systems

In designing the general framework for the proposed expert system, a rule-based approach was used. ES assists quality control engineers, and it can be used for training operators as well. Therefore, the proposed system runs in three modes: A tutorial mode that offers explanation and training if requested by the user, a status mode that concludes from the evidence and responses provided by the user, and a diagnosis mode that provides inference or reasoning with the rules within the knowledge base.

4.2.1. Knowledge Acquisition

"The knowledge acquisition process includes extracting, transforming, and validating expertise from different information sources for developing a knowledge base repository" [23]. The knowledge used in this research consists of "general knowledge" and "process-specific knowledge". In this study, to assist the fault diagnosis process, "Western Electric" [44] tests were utilized as general knowledge, and, to gather process-specific knowledge, "cause and effect (Ishikawa diagram)" [41] diagrams were prepared to investigate the root-cause. Using cause and effect diagrams, the most problematic reasons in the plaster production process were systematically determined (Figure 9). Next, failure modes and effects analysis (FMEA) (Table 9) was applied as an analytical method that incorporates the technology and experts' knowledge in identifying and prioritizing foreseeable failure modes of the process in order to eliminate or reduce their occurrence [42]. Finally, FMEA analysis results were collected in the knowledge base with the aim of making a complete reference for future issues. As can be seen in Table 9, FMEA uses the risk priority number (RPN) to evaluate the risk level of the process. RPN is calculated by multiplying the scores of three risk factors named occurrence (O), which is the frequency of the failure, severity (S), which is related to the effect of disruption on the system, and detection (D), which refers to the probability of detecting the failure. FMEA uses five scales or scores (1–5) to measure these factors.

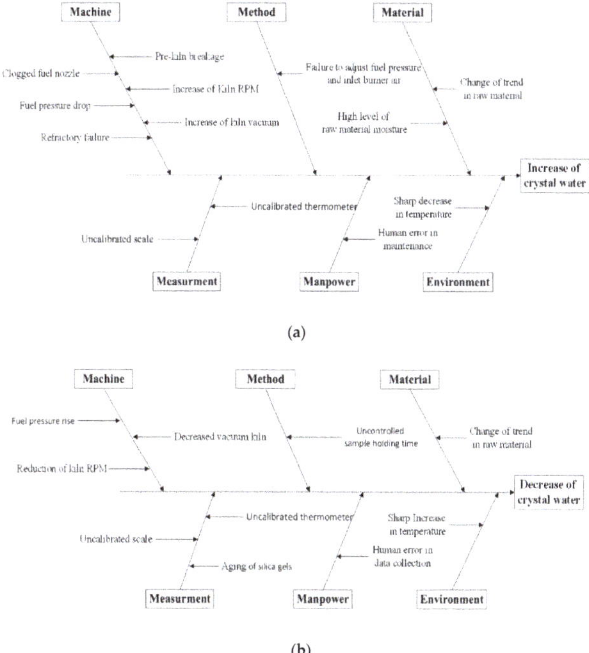

(b)

Figure 9. Cause and effect diagram: (**a**) for increase of crystal water; (**b**) for decrease of crystal water.

Table 9. Failure modes and effects analysis (FMEA) form for the critical parameter (crystal water).

	Failure Modes and Effects Analysis (FMEA)							
	Failure Mode	**Index**	**Cause Effect**	**Corrective Actions**	**O**	**S**	**D**	**RPN** [1]
1	**Increase of Crystal Water**	**Decrease of middle temperature in the kiln body.**	Clogged fuel nozzle.	Cleaning of nozzle, establishment of PM for burner, and installation of filter.	3	5	5	75
2			Failure to adjust fuel pressure and inlet burner air.	Adjustment of fuel and air regulator in a defined period.	3	4	4	48
3			Pre-kiln breakage.	Chang or patching of pre-kiln and thickness monitoring.	2	5	1	10
4			Decrease of kiln temperature according to fuel pressure.	Installation of a shutdown sensor for fuel pressure.	2	3	1	6
5			Increase of Kiln RPM.	Adjustment of RPM via frequency.	1	5	4	20
6			Sharp decrease in temperature.	-	1	2	1	2
7			Erosion in kiln blades	Change of runner blade and periodically monitoring.	1	5	5	25
8			Becoming dirty or clogging of the burner's air nozzle.	Cleaning the air nozzles and instating a multilayer filter.	4	2	3	24
9			Heat transition between kiln and environment because of the lake of refractory.	Establishment of PM and thickness monitoring of refractory.	5	5	5	125
10	**Decrease of Crystal Water**	**Increase of middle temperature in the kiln body.**	Increase of negative pressure of kiln (filter).	Installation of ΔPMeter.	5	5	1	25
11			Fuel pressure rise.	-	1	2	1	2
12			Decreasing of negative pressure of exhaust fan.	-	5	5	3	75
13			Reduction of kiln RPM.	-	1	5	4	20
14			Sharp Increase in temperature.	-	1	2	1	2
15	**Variation in Crystal Water**	-	Change of trend in raw material because of mine.	-	4	5	4	80
16	**Increased Crystal Water**	-	High level of raw material moisture.	-	3	5	1	15
17	**Variation in Crystal Water**	-	Changing of raw material spec.	-	3	5	1	15

[1] RPN = O × S × D.

4.2.2. Knowledge Representation

In this study, a rule-based approach was considered to codify experts' problem-solving knowledge through inference rules: IF <a condition or premise>, THEN <an action or conclusion> rules. In total, 60 rules were used using technical documentation, operations procedures, and interviews with experts, for interpreting control charts (X-bar and R) and providing diagnostic expertise.

4.2.3. Implementation

In this study, the desired expert system was designed using three principal modules. The first module is related to knowledge base development, the second module is relevant to interface design and required questions to reach the answer, and the third module is associated with system run and dialogue to the user. Figure 10 shows a schematic of the proposed neural expert system and its components. Below is a brief description of each system component.

- A knowledge base uses the knowledge of experts and other sources acquired by knowledge engineers to support reliable, complete, and consistent decision-making in a time-critical situation. The knowledge base of the proposed model is an organized collection of facts and heuristics about the plaster production domain, as described briefly below.
    - "Facts" refers to a set of facts relating to the current process state extracted by a knowledge engineer (KE) from the records of the quality management system, preventive maintenance, calibration, brainstorming sessions, and interviews with experts.

- o "Procedures" focuses on manuals, standards, and procedures. Some examples include technical operation instructions, plaster production standards, and intelligent statistical process control (ISPC) tutorials.

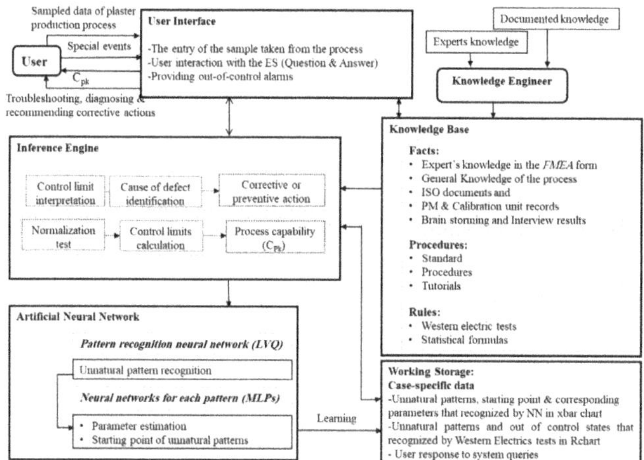

Figure 10. The proposed hybrid fault diagnosis system scheme or ISPC.

"Rules" relate to production rules that represent inferential knowledge learned from experts. In the knowledge base of the proposed model, 60 rules were employed that were extracted from interview with the experts, documents including "*Western Electric*" tests (general knowledge of the process), and the NN's response (case-specific data) extracted by KE and presented in the form if–then. Below is an example of a typical rule, based on specific knowledge of a process.

IF "diagnosis" is "upward trend",

AND "failure mode" is "increase of crystal water",

AND "process index" is "decrease of kiln's temperature",

THEN "specific cause" can be "clogged fuel nozzle",

AND "corrective actions" can be either "cleaning the fuel nozzle, the establishment of preventive maintenance (PM) for the burner, or installing fuel filter".

- The inference engine contains the inference strategies and matches the condition part of rules against facts of a specific case to reach a decision or conclusion. A backward chaining inference engine was used, as it is best suited for diagnosis-type systems, in which the codified program is executed with two groups of rules. The first group defines goals (assumed conclusions) for the properties and checks if their values are supported by the existing data. The second group updates the rules and transmits satisfying goals.
- The working memory acts as a repository for all data including the initial facts of the given case, the user's responses to system queries, and generated facts (e.g., type, change point, and parameters) derived by the inference engine.
- The user interface facilitates communication between the user and the proposed expert system through various input methods including dialog boxes and command prompts.

5. Comparative Analysis and Case Study

5.1. Comparison Study

To decisive the accuracy, consistency, and repeatability of test results, neural network model verification was done. The NN model was verified by comparing the error of the NN algorithm and the error rate of the discriminant analysis (DA) method—a classification counterpart in the statistical approach [1]. The method of DA is to find a rule to separate two or more groups of observations from one another. The most important application of DA is classification. The output of the DA for the test dataset is presented in Table 10.

Table 10. Output of discriminant analysis (DA) for test dataset.

The SAS System (DISCRIM Procedure) Classification Summary for Test Data Set Using Linear Discriminant Function									
Number of Observations and the Percentage of Correctly Classified into the Target Classes									
Target	**1**	**2**	**3**	**4**	**5**	**6**	**7**	**8**	**Total**
1	17 35.42	15 31.25	8 16.67	0 0.00	0 0.00	8 16.67	0 0.00	0 0.00	48 100.00
2	9 15.25	25 42.37	14 23.73	0 0.00	0 0.00	7 11.86	4 6.78	0 0.00	59 100.00
3	10 19.23	12 23.08	23 44.23	1 1.92	0 0.00	6 11.54	0 0.00	0 0.00	52 100.00
4	2 3.77	3 5.66	0 0.00	48 90.57	0 0.00	0 0.00	0 0.00	0 0.00	53 100.00
5	0 0.00	0 0.00	0 0.00	0 0.00	50 100.00	0 0.00	0 0.00	0 0.00	50 100.00
6	7 15.22	11 23.91	9 19.57	0 0.00	0 0.00	19 41.30	0 0.00	0 0.00	46 100.00
7	0 0.00	0 0.00	0 0.00	0 0.00	0 0.00	0 0.00	39 100.00	0 0.00	39 100.00
8	0 0.00	3 5.66	0 0.00	0 0.00	0 0.00	0 0.00	4 7.55	46 86.79	53 100.00
Total	45 11.25	69 17.25	54 13.50	49 12.25	50 12.50	40 10.00	47 11.75	46 11.50	400 100.00
Priors	0.12	0.1475	0.13	0.1325	0.125	0.115	0.0975	0.1325	

For example, in Table 10, the number 0 in the first row and the eighth column indicates that no "natural" pattern was mistakenly placed in the systematic patterns class. In the first row and the first column, the number 17 represents the number of patterns correctly assigned to the "natural" type. Furthermore, the value of 35.42 is the percentage correctly assigned to the "natural" class. In the first row and second column, 15 is the number of "natural" patterns that were mistakenly classified as "shift". The value of 31.25 is also the percentage of the "shift" pattern error in the "natural" class. The value of 31.25 is also the percentage of the "shift" pattern error in the "natural" class. Table 11 shows the errors for each class. This table lists the errors for each category in the "rate" line and the weight for each type in the "priors" row, while "total" (0.3325) indicates the total error of DA method for the test dataset. Tables 12 and 13 provide the output for the training dataset. As shown in the diagrams below (Figures 11 and 12), the NN outperformed the DA method in terms of performance and accuracy.

Table 11. DA error in the test dataset classification.

Error Estimation for Target Classes									
	1	**2**	**3**	**4**	**5**	**6**	**7**	**8**	**Total**
Rate	0.6458	0.5763	0.5577	0.0943	0.0000	0.5870	0.0000	0.1321	0.3325
Priors	0.1200	0.1475	0.1300	0.1325	0.1250	0.1150	0.0975	0.1325	

Table 12. Output of discriminant analysis (DA) for training dataset.

The SAS System (DISCRIM Procedure) Classification Summary for Training Data Set Using Linear Discriminant Function									
Number of Observations and the Percentage of Correctly Classified into the Target Classes									
Target	**1**	**2**	**3**	**4**	**5**	**6**	**7**	**8**	**Total**
1	218	175	124	1	0	0	0	0	518
	42.08	33.78	23.94	0.19	0.00	0.00	0.00	0.00	100.00
2	162	227	116	3	0	21	8	2	539
	30.06	42.12	21.52	0.56	0.00	3.90	1.48	0.37	100.00
3	158	158	163	1	0	21	0	0	501
	31.54	31.54	32.53	0.20	0.00	4.19	0.00	0.00	100.00
4	5	24	6	471	0	0	11	1	518
	0.97	4.63	1.16	90.93	0.00	0.00	2.12	0.19	100.00
5	4	1	0	0	489	0	0	0	499
	0.80	0.20	0.00	0.00	98.00	0.00	0.00	0.00	100.00
6	49	25	42	0	0	314	0	0	430
	11.40	5.81	9.77	0.00	0.00	73.02	0.00	0.00	100.00
7	0	1	0	13	0	1	399	1	415
	0.00	0.24	0.00	3.13	0.00	0.24	69.14	0.24	100.00
8	0	0	0	5	0	56	42	477	580
	0.00	0.00	0.00	0.86	0.00	9.66	7.24	82.24	100.00
Total	596	616	451	494	489	413	460	481	4000
	14.90	15.40	11.28	12.35	12.23	10.33	11.50	12.03	100.00
Priors	0.1295	0.13475	0.12525	0.1295	0.12475	0.1075	0.10375	0.145	

Table 13. DA error in the training dataset classification.

Error Estimation for Target Classes									
	1	**2**	**3**	**4**	**5**	**6**	**7**	**8**	**Total**
Rate	0.5792	0.5788	0.6747	0.0907	0.0200	0.2698	0.0386	0.1776	0.3105
Priors	0.1295	0.1348	0.1253	0.1295	0.1248	0.1075	0.1038	0.1450	

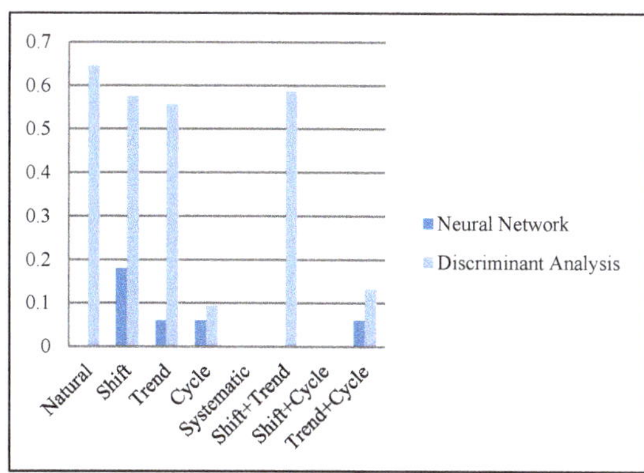

Figure 11. Comparison of DA and neural network (NN) error for each pattern in in test dataset.

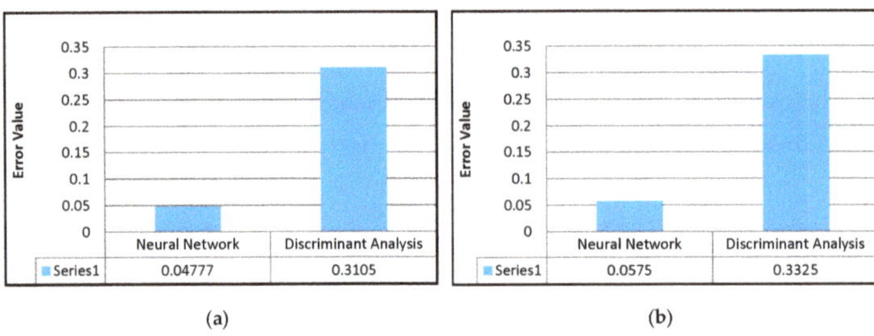

Figure 12. (**a**) Comparison of DA and NN error in test dataset; (**b**) comparison of DA and NN error in training dataset.

5.2. Case Study

In this section, to demonstrate the applicability and capability of the proposed model, a case study in a plaster-producing company is presented. In a traditional statistical process control system, after gathering the data, the following steps are done:

- Plotting the sequence of process measurements (observations).
- Setting UCL and LCL.
- Determining the process capability (Cpk).
- Performing the "normality test" on the data.
- Interpreting both R and X-bar chart for statistical control.

The proposed ISPC model, designed in Matlab, can adequately perform the above operations (Figure 13). Here, pursuant to the plaster-producing experts' opinion, if "Cpk > 1", the chart is considered as the baseline for interpreting the process. As can be seen in the ISPC implementation flow chart, after collecting the process data, the baseline chart should be set by eliminating and replacing points beyond the control limits of new data. As shown in Figure 14, the normality test was done, where the normality assumption was valid, and "Cpk > 1" in control mode. After drawing the baseline chart, the actual data of the process were inserted and checked and analyzed by the desired control charts. As illustrated in Figure 15, although there is no "out of control" mode within the "R-chart", the process was unable to meet specifications due to "Cpk < 1", being equal to 0.81 (Figure 16). On the other hand, by choosing the "X-bar chart" (Figure 17), the user receives the following error message: "X-bar chart is out of control" (Figure 18). Then, the ES using "Western Electric tests" announces that "out of control" modes may have the following reasons: "carelessness in the measurement, machinery stop, or off-spec materials". Later, the user receives a suggestion message from ES to check the unnatural patterns identified by NN (Figure 19). As can be seen in Figure 20, not only was the "downward shift pattern" in the "X-bar chart" identified by the NN, but the "starting point" of the unnatural pattern was estimated (point 6). The "shift magnitude parameter" (−0.161) was also determined. In this scenario, because of the appearance of a "downward shift pattern" and based on user observation, which was "kiln body scarlet", the reason for the deviation was recognized as the "temperature exchange of kiln with the environment due to the loss of refractory and thickness". "Establishment of maintenance and inspection of refractory" was also recommended as corrective or preventive activities. In this scenario, by making corrective actions and following re-sampling the process (Figure 21), "out of control" modes did not appear in the control charts anymore (Figure 22) and, furthermore, "process capability" increased from Cpk = 0.81 to Cpk = 1.15 (Figure 23). The experimental results show that corrective actions could significantly contribute to process recovery. Thus, the proposed fault diagnosis system could be used to support decision-makers of the plaster production.

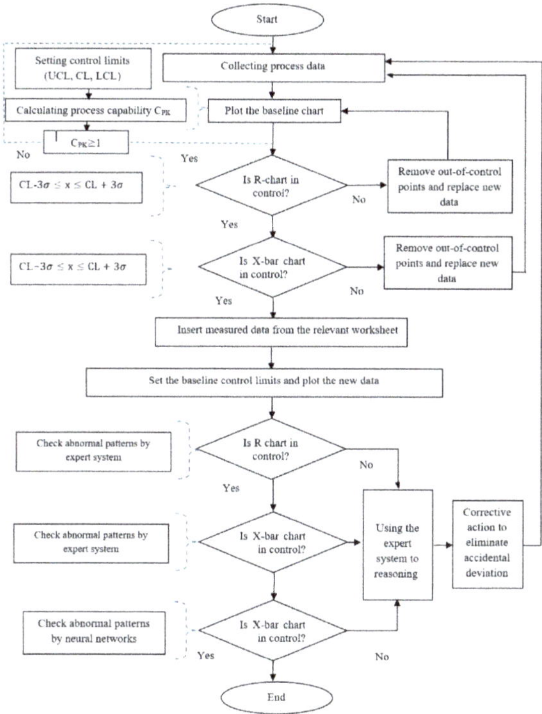

Figure 13. ISPC implementation flow chart.

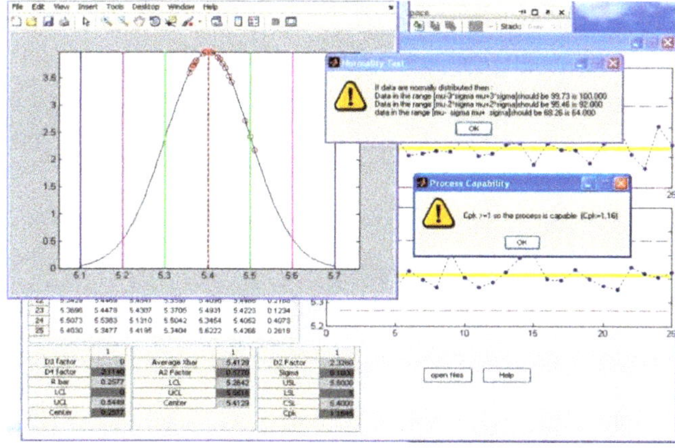

Figure 14. Normality test and process capability calculation.

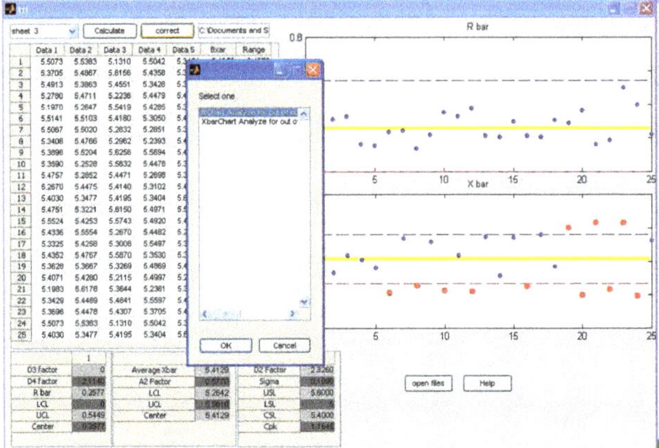

Figure 15. X-bar chart in "out of control" mode.

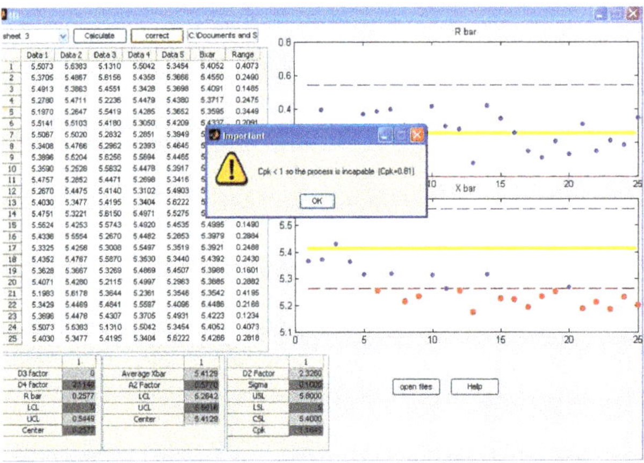

Figure 16. Process capability calculation.

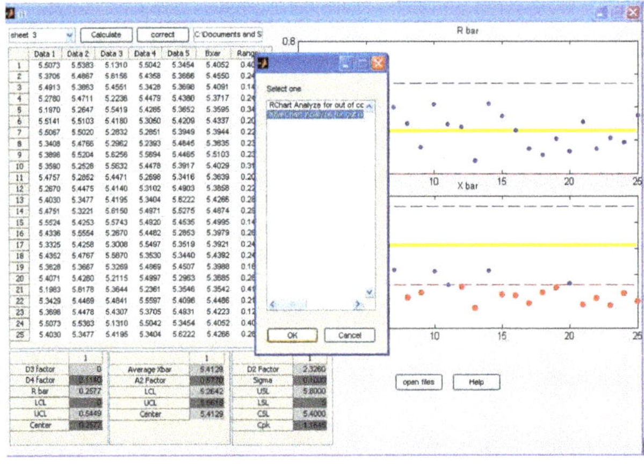

Figure 17. Request for analyzing the X-bar chart.

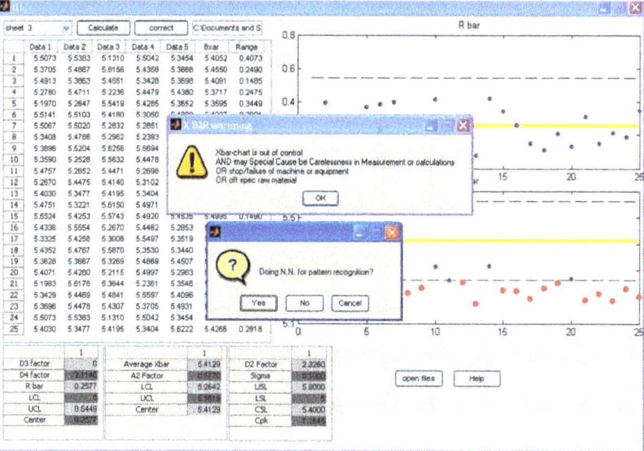

Figure 18. X-bar chart reasoning by Western Electric tests.

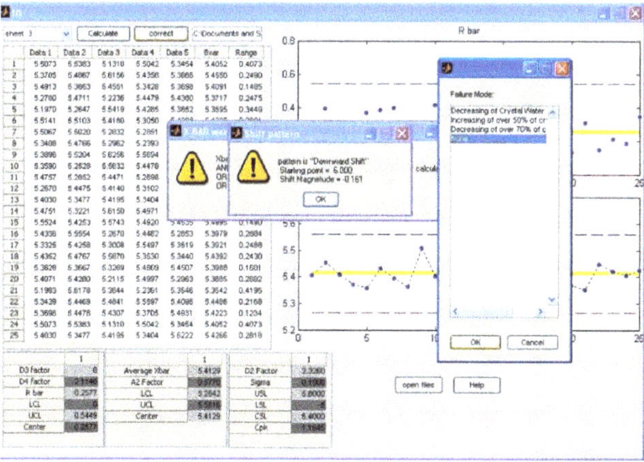

Figure 19. Pattern recognition in X-bar chart by NN.

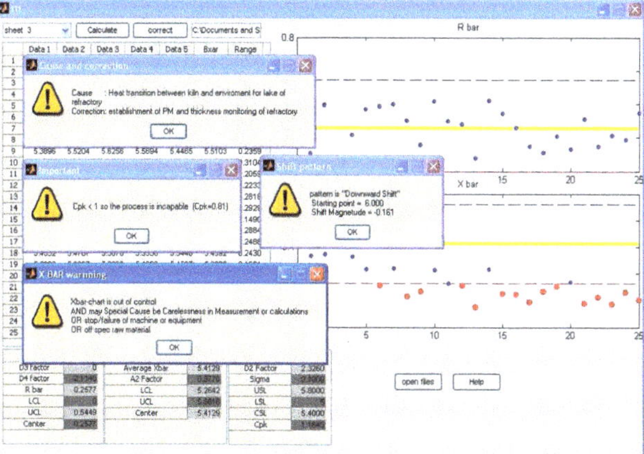

Figure 20. Determining Cpk, unnatural pattern, parameters, and recovery actions.

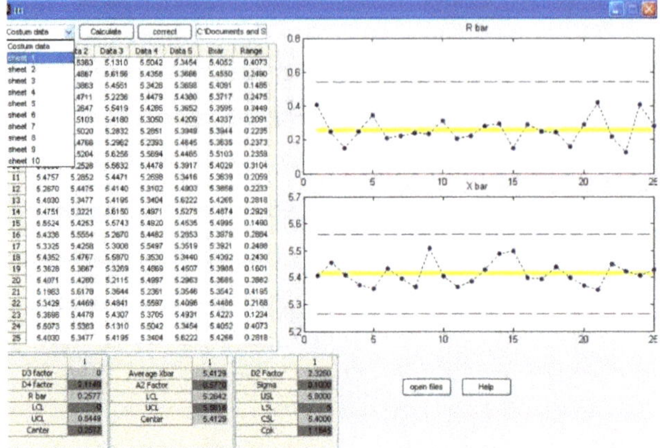

Figure 21. Inserting new dataset.

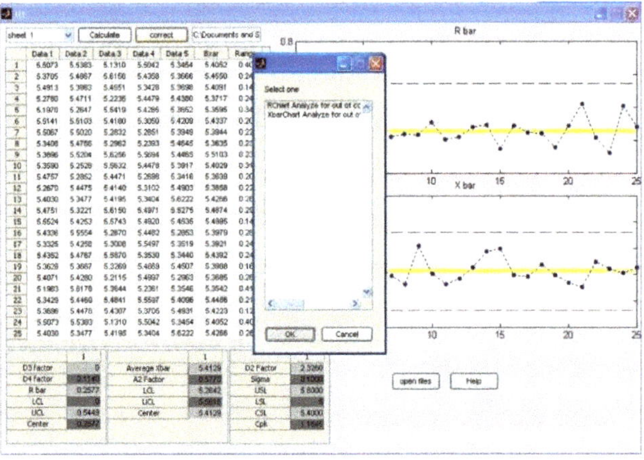

Figure 22. X-bar and R charts in control mode.

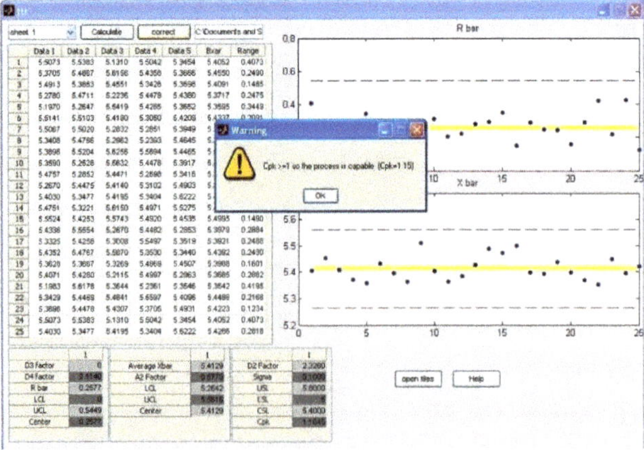

Figure 23. Capability of the process after performing corrective actions.

6. Conclusions

This paper aimed to target one of the most challenging subjects in smart manufacturing, which is the quality control at the shop floor level considering emerging technologies. There are many conceptual models and general recommendations when discussing a new paradigm of quality associated with I4.0, but there are relatively few works in action. The hybrid model proposed in this work supports the troubleshooting of the plaster production process, which is a complex manufacturing system. To have both descriptive and prescriptive approaches, NN and ES were integrated where NN deals with the determination of fault areas, and ES provides the recommendation of corrective actions.

The main achievements and contributions of this work are as follows:

1. Successful implementation of Quality 4.0 to blend traditional quality control models based on CCPs with an intelligent system at the shop floor level.
2. According to Tables 3 and 5, the diagnosis of behavioral patterns coming from Module I is acceptable, and parameters of corresponding patterns estimated by Module II are effective and reliable.
3. It was a multitask project including production, delivery, and encryption of neural network input.
4. Using a wide range of data in training the NNs to assure stable behavior in performance.
5. Using the integrated system, most SPC requirements, such as drawing of basis chart, checking X-bar and R charts for being under control, and calculating Cpks, were achieved.
6. Using LVQ for pattern classification and MLP in parallel. This helps in simultaneously enjoying the competitive power of the LVQ network and the interoperability of multilayer perceptron networks.
7. The result of the case study shows the improvement of process capability, while control charts did not show any out of control mode after following the corrective actions; thus, the capability of the proposed model to serve as a reliable decision support system (DSS) was confirmed.

This paper shows the capability of I4.0 to change the quality paradigm in factories of the future. The key element is the level of intelligence of the system, which leads to smart manufacturing. There is no doubt that emerging technologies will shift quality processes to a different level, while monitoring, fault detection, cause, root analysis, and even corrective actions and strategies would be autonomous.

For further research, there are many potential areas of working, as outlined below.

1. Developing and comparing the result of new models based on adaptive neuro-fuzzy inference systems (ANFIS).
2. Using the particle swarm optimization (PSO) to improve the performance.
3. Using some techniques and algorithms such as deep learning to increase the efficiency of the model.
4. Using collective sensor networks and Internet of things (IoT) platforms to develop a real-time smart quality control system.
5. Connecting smart quality to new services such as predictive maintenance to achieve smart, collaborative devices and support new product and production lines based on information from intelligent quality information.

The models proposed in this paper were independent of software; thus, free software and modern scripting languages such as Python, Ruby, etc. could be utilized for the same purpose.

This paper was a real case of Quality4.0 in action to show the capabilities and applicability of emerging technologies and intelligent algorithms to shift control quality to the new stage, and it represents the initial step of a long journey.

Author Contributions: Conceptualization, J.R.; methodology, J.R. and J.J.; writing—drafting Section 2, J.R. and J.J.; writing—drafting Section 3, J.R. and J.J.; writing—drafting Section 4, J.R.; writing—drafting Section 5, J.R.; writing—drafting Section 6, J.J.; writing—reviewing and editing, J.R. and J.J. All authors read and agreed to the published version of the manuscript.

Funding: This research was funded in part by the Portuguese "Fundação para a Ciência e a Tecnologia" (FCT) in the context of the Center of Technology and Systems CTS/UNINOVA/FCT/NOVA, reference UIDB/00066/2020.

Acknowledgments: This work was supported by the Portuguese Foundation for Science and Technology (FCT) and the Center of Technology and Systems (CTS).

Conflicts of Interest: The authors declare no conflict of interest.

Appendix A

```
for l = 1:Ntraining;
for i = 1:Nhidden
for j = 1:Ninput
D(i,j) = (Data(l,j)-W(j,i))^2;
end
Distance(i,:) = [sqrt(sum(D(i,:),2)),i];
end
MinDis = min(Distance(:,1));
ONN = find(Distance(:,1) == MinDis);
for i = 1:Nhidden
if i == ONN
HiddenOut(i) = 1;
else
HiddenOut(i) = 0;
end
end
NetOut(l,:) = HiddenOut*V;
if NetOut(l,:) == RealOut(l,:)
W(:,ONN) = W(:,ONN) + Lambda*(transpose(Data(l,:))-W(:,ONN));
else
W(:,ONN) = W(:,ONN)-Lambda*(transpose(Data(l,:))-W(:,ONN));
end
Er(l) = sum(abs(RealOut(l,:)-NetOut(l,:)),2);
end
t = t + 1;
Error(t) = sum(Er,2)
if Error(t) < 500
Lambda = 0.99*Lambda;
End
```

Appendix B

```
while Error(l) > Emax
l = l + 1;
Epoch(l) = l;
HO = V0*transpose(InData);
HiddenOut = (1-exp(-A*HO))./(1 + exp(-A*HO));
DiffHiddenOut = (A/2)*(1-HiddenOut.*HiddenOut
HiddenOut(Nshifthidden,:) = 1;
DiffHiddenOut(Nshifthidden,:) = (A/2)*(1-HiddenOut(Nshifthidden,:).*HiddenOut(Nshifthidden,:));
OO = W0*HiddenOut;
Output = (1-exp(-A*OO))./(1 + exp(-A*OO));
DiffOutput = (A/2)*(1-Output.*Output);
DeltaO = (transpose(OutData)-Output).*DiffOutput;
```

```
DeltaH = DiffHiddenOut.*transpose(transpose(DeltaO)*W0);
E = (OutData-transpose(Output)).*(OutData-transpose(Output));
for i = 1:Nshifthidden
for j = 1:Nshiftinput
V(i,j) = V0(i,j) + Lambda*DeltaH(i,:)*InData(:,j);
end
end
for i = 1:Nshiftoutput
for j = 1:Nshifthidden
W(i,j) = W0(i,j) + Lambda*DeltaO(i,:)*HiddenOut(j,:)';
end
end
V0 = V;W0 = W;
Error(l) = 0.5*sum(sum(E,1),2);
```

References

1. Bag, M.; Bengal, W.; Gauri, S.K.; Bengal, W.; Chakraborty, S.; Bengal, W. Recognition of Control Chart Patterns using Discriminant Analysis of Shape Features. In Proceedings of the International Conference on Industrial Engineering and Operations Management, Dhaka, Bangladesh, 9–10 January 2010; pp. 88–93.
2. Lu, C.; Shao, Y.E.; Li, C. Recognition of Concurrent Control Chart Patterns by Integrating ICA and SVM. *Appl. Math. Inf. Sci.* **2014**, *8*, 681–689. [CrossRef]
3. Awadalla, M.; Sadek, M.A. Spiking neural network-based control chart pattern recognition. *Alex. Eng. J.* **2012**, *51*, 27–35. [CrossRef]
4. Ebrahimzadeh, A.; Ranaee, V. Control chart pattern recognition using an optimized neural network and efficient features. *ISA Trans.* **2010**, *49*, 387–393. [CrossRef] [PubMed]
5. Demirli, K.; Vijayakumar, S. Fuzzy logic based assignable cause diagnosis using control chart patterns. *Inf. Sci.* **2010**, *180*, 3258–3272. [CrossRef]
6. Noskievičová, D. Complex Control Chart Interpretation. *Int. J. Eng. Bus. Manag.* **2013**, *5*, 1–7. [CrossRef]
7. Lavangnananda, K.; Sawasdimongkol, P. Neural Network Classifier of Time Series: A Case Study of Symbolic Representation Preprocessing for Control Chart Patterns. In Proceedings of the 2012 8th International Conference on Natural Computation, Chongqing, China, 29–31 May 2012; pp. 344–349.
8. Hachicha, W.; Ghorbel, A. A survey of control-chart pattern-recognition literature (1991–2010) based on a new conceptual classification scheme. *Comput. Ind. Eng.* **2012**, *63*, 204–222. [CrossRef]
9. Haghtalab, S.; Xanthopoulos, P.; Madani, K. Expert Systems with Applications A robust unsupervised consensus control chart pattern recognition framework. *Expert Syst. Appl.* **2015**, *42*, 6767–6776. [CrossRef]
10. Wang, C.; Guo, R.; Chiang, M.; Wong, J.Y. Decision tree based control chart pattern recognition. *Int. J. Prod. Res.* **2008**, *46*, 4889–4901. [CrossRef]
11. Guh, R. Simultaneous process mean and variance monitoring using artificial neural networks. *Comput. Ind. Eng.* **2010**, *58*, 739–753. [CrossRef]
12. Ramezani, J.; Jassbi, J. A hybrid expert decision support system based on artificial neural networks in process control of plaster production—An industry 4.0 perspective. In *Technological Innovation for Smart Systems*; IFIP AICT; Springer: Cham, Switzerland, 2017; pp. 55–71. [CrossRef]
13. Hassan, A. An improved scheme for online recognition of control chart patterns. *Int. J. Comput. Aided Eng. Technol.* **2014**, *3*, 309–321. [CrossRef]
14. Das, P.; Banerjee, I. A hybrid detection system of control chart patterns using cascaded SVM and neural network–based detector. *Neural Comput. Appl.* **2011**, *20*, 287–296. [CrossRef]
15. Demircioglu Diren, D.; Boran, S.; Cil, I. Integration of Machine Learning Techniques and Control Charts for Multivariate Processes. *Sci. Iran.* **2019**. [CrossRef]
16. Lavangnananda, K.; Khamchai, S. Capability of control chart patterns classifiers on various noise levels. *Procedia Comput. Sci.* **2015**, *69*, 26–35. [CrossRef]

17. Fuqua, D.; Razzaghi, T. A cost-sensitive convolution neural network learning for control chart pattern recognition. *Expert Syst. Appl.* **2020**, *150*, 113275. [CrossRef]
18. Biehl, M.; Hammer, B.; Villmann, T. Prototype-based models in machine learning. *WIRE Cogn. Sci.* **2016**, *7*, 92–111. [CrossRef] [PubMed]
19. El-midany, T.T.; El-baz, M.A.; Abd-elwahed, M.S. A proposed framework for control chart pattern recognition in multivariate process using artificial neural networks. *Expert Syst. Appl.* **2010**, *37*, 1035–1042. [CrossRef]
20. Gauri, S.K. Control chart pattern recognition using feature-based learning vector quantization. *Int. J. Adv. Manuf. Technol.* **2010**, *48*, 1061–1073. [CrossRef]
21. Ghiasabadi, A.; Noorossana, R.; Saghaei, A. Identifying change point of a non-random pattern on control chart using artificial neural networks. *Int. J. Adv. Manuf. Technol.* **2013**, *67*, 1623–1630. [CrossRef]
22. Bayat, A.B.; Gharehkhani, A.; Mohajeran, A.; Addeh, J. Control Chart Patterns Recognition Using Optimized Adaptive Neuro-Fuzzy Inference System and Wavelet Analysis. *J. Eng. Technol.* **2013**, *3*, 76–81.
23. Bag, M.; Gauri, S.K.; Chakraborty, S. An expert system for control chart pattern recognition. *Int. J. Adv. Manuf. Technol.* **2012**, *62*, 291–301. [CrossRef]
24. Xanthopoulos, P.; Razzaghi, T. Computers & Industrial Engineering recognition q. *Comput. Ind. Eng.* **2014**, *70*, 134–149.
25. Xie, L.; Gu, N.; Li, D.; Cao, Z.; Tan, M.; Nahavandi, S. Computers & Industrial Engineering Concurrent control chart patterns recognition with singular spectrum analysis and support vector machine. *Comput. Ind. Eng.* **2013**, *64*, 280–289.
26. Lin, S.; Guh, R.; Shiue, Y. Effective recognition of control chart patterns in autocorrelated data using a support vector machine based approach. *Comput. Ind. Eng.* **2011**, *61*, 1123–1134. [CrossRef]
27. Zafar, R.F.; Mahmood, T.; Abbas, N.; Riaz, M.; Hussain, Z. A progressive approach to joint monitoring of process parameters. *Comput. Ind. Eng.* **2018**, *115*, 253–268. [CrossRef]
28. Villmann, T.; Bohnsack, A.; Kaden, M. Can learning vector quantization be an alternative to SVM and deep learning? *J. Artif. Intell. Soft Comput. Res.* **2017**, *7*, 65–81. [CrossRef]
29. Zarandi, M.H.F.; Alaeddini, A. A general fuzzy-statistical clustering approach for estimating the time of change in variable sampling control charts. *Inf. Sci.* **2010**, *180*, 3033–3044. [CrossRef]
30. Radziwill, N. Let's Get Digital: The many ways the fourth industrial revolution is reshaping the way we think about quality. *Qual. Prog.* **2018**, 24–29.
31. LSN Research. Quality 4.0 Impact and Strategy Handbook eBook. blog.lnsresearch.com. 2017. Available online: https://blog.lnsresearch.com/quality40ebook (accessed on 10 May 2019).
32. Nenadál, J. The New EFQM Model: What is Really New and Could Be Considered as a Suitable Tool with Respect to Quality 4.0 Concept? *Qual. Innov. Prosper.* **2020**, *24*, 17–28. [CrossRef]
33. Madsen, D.Ø. The Emergence and Rise of Industry 4.0 Viewed through the Lens of Management Fashion Theory. *Adm. Sci.* **2019**, *9*, 71. [CrossRef]
34. Ramezani, J.; Camarinha-Matos, L.M. A collaborative approach to resilient and antifragile business ecosystems. *Procedia Comput. Sci.* **2019**, *162*, 604–613. [CrossRef]
35. Dallasega, P.; Rauch, E.; Linder, C. Industry 4.0 as an enabler of proximity for construction supply chains: A systematic literature review. *Comput. Ind.* **2018**, *99*, 205–225. [CrossRef]
36. Maskuriy, R.; Selamat, A.; Ali, K.N.; Maresova, P.; Krejcar, O. Industry 4.0 for the Construction Industry—How Ready Is the Industry? *Appl. Sci.* **2019**, *9*, 2819. [CrossRef]
37. Ault, J.H.; Jenkins, J. Control Charts as a Productivity Improvement Tool in Construction. Master's Thesis, Purdue University, West Lafayette, Indiana, 2013.
38. Camarinha-Matos, L.M.; Fornasiero, R.; Ramezani, J.; Ferrada, F. Collaborative Networks: A Pillar of Digital Transformation. *Appl. Sci.* **2019**, *9*, 5431. [CrossRef]
39. Ramezani, J.; Camarinha-Matos, L.M. Novel Approaches to Handle Disruptions in Business Ecosystems. In *Technological Innovation for Industry and Service Systems*; DoCEIS 2019; IFIP AICT; Springer: Cham, Switzerland, 2019; pp. 43–57.
40. Petersen, K.; Vakkalanka, S.; Kuzniarz, L. Guidelines for conducting systematic mapping studies in software engineering: An update. *Inf. Softw. Technol.* **2015**, *64*, 1–18. [CrossRef]
41. Tague, N.R. *The Quality Toolbox, American Society for Quality*; Quality Press: Milwaukee, WI, USA, 2008.
42. Lipol, L.S.; Hag, J. Risk Analysis Method: FMEA/FMECA in the Organizations. *Int. J. Basic Appl. Sci.* **2011**, *11*, 5.

43. Ramchoun, H.; Idrissi, M.A.J.; Ghanou, Y.; Ettaouil, M. New modeling of multilayer perceptron architecture optimization with regularization: An application to pattern classification. *IAENG Int. J. Comput. Sci.* **2017**, *44*, 261–269.
44. Electric, W. *Statistical Quality Control Handbook*; Western Electric Company: New York, NY, USA, 1956.

Article

Modeling of Spiral Wound Membranes for Gas Separations—Part II: Data Reconciliation for Online Monitoring

Diego Queiroz Faria de Menezes [1,*], Marília Caroline Cavalcante de Sá [1], Tahyná Barbalho Fontoura [1], Thiago Koichi Anzai [2], Fabio Cesar Diehl [2], Pedro Henrique Thompson [2] and Jose Carlos Pinto [1]

1 Programa de Engenharia Química/COPPE, Universidade Federal do Rio de Janeiro, Rio de Janeiro CEP 21941-972, RJ, Brazil; marilia@peq.coppe.ufrj.br (M.C.C.d.S.); tahyna@peq.coppe.ufrj.br (T.B.F.); pinto@peq.coppe.ufrj.br (J.C.P.)

2 Centro de Pesquisas Leopoldo Américo Miguez de Mello—CENPES, Petrobras—Petróleo Brasileiro SA, Rio de Janeiro CEP 21941-915, RJ, Brazil; tanzai@petrobras.com.br (T.K.A.); fabio.diehl@petrobras.com.br (F.C.D.); pedrothompson@petrobras.com.br (P.H.T.)

* Correspondence: dmenezes@coppe.ufrj.br; Tel.: +55-21-98807-7489

Received: 24 July 2020; Accepted: 19 August 2020; Published: 25 August 2020

Abstract: The present work presents a methodology based on data reconciliation to monitor membrane separation processes reliably, online and in real time for the first time. The proposed methodology was implemented in accordance with the following steps: data acquisition; data pre-treatment; data characterization; data reconciliation; gross error detection; and critical evaluation of measured data with a soft sensor. The acquisition of data constituted the slowest stage of the monitoring process, as expected in real-time applications. The pre-treatment stage was fundamental to assure the robustness of the code and the initial characterization of collected data was carried out offline. The characterization of the data showed that steady-state modeling of the process would be appropriate, also allowing the implementation of faster numerical procedures for the data reconciliation step. The data reconciliation step performed well, quickly and consistently. Thus, data reconciliation allowed the estimation of unmeasured variables, playing the role of a soft sensor and allowing the future installation of a digital twin. Additionally, monitoring of measurement bias constituted a tool for measurement diagnosis. As shown in the manuscript, the proposed methodology can be successfully implemented online and in real time for monitoring of membrane separation processes, as shown through a real dashboard web application developed for monitoring of an actual industrial site.

Keywords: membrane; data reconciliation; real-time; online; monitoring

1. Introduction

A common problem in oil production is the excess of CO_2 gas present in natural gas streams. The first and most notorious issue is related to the emission of this gas into the environment. However, in addition to the possible environmental problems, the excess of CO_2 in oil streams can cause problems in the process plant, such as freezing due to pressure drop in compression and cooling sections of the plant and corrosion of metal pipelines [1]. According to an ANP (Brazilian National Agency of Petroleum, Natural Gas and Biofuels) resolution, commercial natural gas must contain a maximum of 3% (mol) of CO_2 [2]. Therefore, a possible solution to deal with the produced CO_2 is the reinjection of CO_2 into the oil well, which may also allow the increase of the productivity of the well. This can certainly minimize environmental impacts and problems in natural gas process plants.

Therefore, the CO_2 separation constitutes a fundamental step during the treatment of natural gas in oil production fields.

Different physical/chemical processes can be used to separate CO_2 from natural gas, such as cryogenic distillation, absorption, or membrane reverse osmosis processes [1,3]. Particularly, the removal of CO_2 from natural gas with help of membrane separation processes has been used since 1981 [4]. However, applications were initially limited because of intrinsic economic risks associated with the oil production activity and operation constraints related to membrane separations. Nevertheless, the scenario has been changing due to advantages related to the lower energy consumption, low capital investment, low operating costs, and more compact nature of these pieces of equipment [5–7].

Given the increase of the industrial importance of membrane separation processes, demands for development of mathematical modeling, simulation, optimization, control, statistical data treatment, and online monitoring procedures have also increased, as these techniques are fundamental for design and monitoring of chemical processes. As a consequence, the performance of the analyzed process can be evaluated more precisely and monitored, allowing the detection of failures in line and in real time. Based on these technologies, risks and time required for decision-making can be minimized [8].

Based on the previous paragraphs, the main objective of the present work is to develop and implement a web application that makes possible the online and real-time monitoring of membrane CO_2 separation processes on an industrial scale for the first time, based on rigorous numerical and statistical procedures. The application can also be used to provide information about unmeasured variables (soft sensor) and to diagnose the occurrence of gross error measurements and instrument malfunctioning. The proposed methodology comprises the following stages: (i) pre-treatment and characterization of process data; (ii) data reconciliation of process data to minimize measurement uncertainties, with the aid of mass balance equations; (iii) detection of systematic deviations for identification of process malfunctions; and (iv) observation of unmeasured variables (soft sensor or digital twin). Finally, the proposed data acquisition and visualization system is implemented online for successful monitoring of an actual industrial membrane separation site in real time for the first time.

1.1. Data Rectification

Technological and computer advancement have allowed the wide, easy, and fast access to process data of industrial plants. As a matter of fact, access to actual data are extremely important for real-time monitoring and optimization of production units [9]. The dynamic monitoring of a plant, unit or industrial equipment is increasingly necessary to improve product quality, enhance process safety, and reduce energy costs and waste; however, the acquired information must be reliable and validated with physical process constraints, as the reliability of the data are of paramount importance for any decision-making related to the analyzed process [10]. Nevertheless, process measurements are subject to errors and fluctuations due to intrinsic imprecision, degradation, malfunction, improper installation, poor calibration, and failure of measurement instruments. Additionally, human errors associated with operation and calibration, or gross errors related to the operation of the process, can result in data that do not represent the process reliably. Consequently, measured data are not expected to satisfy physical constraints precisely and are not expected to comply with conservation laws (mass, momentum, and energy balances) [11]. For these reasons, process controllers and data acquisition systems, if not treated properly, can cause the plant to operate at sub-optimal or unsafe operation conditions. In addition, decision-making based on unreliable data can lead to the occurrence of industrial accidents, reduction of product quality, and financial losses [12]. Therefore, the use of data rectification procedures can be essential to improve the quality of the information contained in the data, and consequently provide a margin of reliability for the control and optimization of the process in real time.

Data rectification procedures usually comprise three steps: variable classification; gross error detection (GED); and data reconciliation (DR). Among these three steps, DR and GED are the ones studied most often and applied more frequently in data rectification schemes [13].

1.1.1. Data Reconciliation and Gross Error Detection

The variable classification step determines whether the available information is sufficient to solve the DR problem and identify the sets of observable variables (measured and unmeasured variables, which can be estimated using the other measured variables and the process constraints) and unobservable variables (unmeasured variables that cannot be estimated). This way, the variable classification step makes possible data set size reduction in order to include only the relevant variables that can be observed and used to build the mathematical model of the process, reducing the size of the process database [14].

The GED step is performed to identify and/or eliminate/compensate for the occurrence of deviations that do not follow the admitted statistical distribution of errors. Gross errors can be caused by poor calibration of the measuring instruments, deterioration of the sensors, power surges, among other causes described previously. However, in order to obtain accurate estimates of parameters and variables, the negative influence of gross errors must be minimized or eliminated. The use of robust estimators has been frequently suggested to eliminate the negative effect of gross errors, often implemented simultaneously with DR to avoid the use of iterative and computationally intensive numerical procedures [11].

In the DR stage, measured data are adjusted in a statistically coherent manner by an estimator, which frequently is based on the maximum likelihood principle, with the support of a statistical distribution admitted a priori for measurement fluctuations. According to the DR technique, adjusted data must satisfy the conservation laws and other constraints imposed on the system, maximizing the probability of occurrence of that particular measurement and, simultaneously, respecting the mathematical model of the process. Thus, more reliable estimates can be obtained for the variables and parameters of the process [13,15]. Traditionally, the normal distribution is assumed to be valid, which results in the Weighted Least Squares (WLS) estimator [16]. For illustrative purposes, data rectification applications in the industry can be implemented as shown in Figure 1.

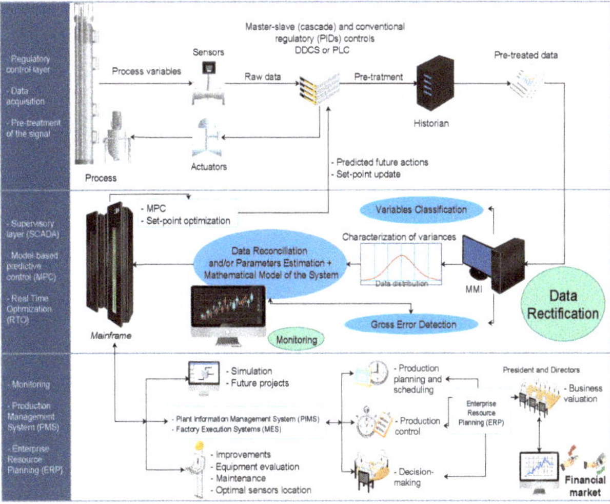

Figure 1. Illustrative representation of industrial data rectification applications [12,17].

Kuehn and Davidson [18] were the pioneers in using DR procedures in chemical engineering processes. Since then, many works have proposed the use of DR procedures for monitoring of industrial processes, although the vast majority of the published material investigates simulated processes that operate at steady-state conditions. Investigations of actual dynamic processes in real time and using actual data are scarce and have never been performed for industrial membrane separations [14,19]. Therefore, the present work contributes to the development of DR procedures through the successful implementation of an original application in an actual industrial environment and using real data in real time to perform the proposed analyses. In addition, the present work shows that similar DR applications can be implemented in many industrial membrane separation environments using simple computational resources in real time.

1.2. Membrane Separation Process

Membranes constitute excellent alternatives for gas separations due to their low installation and maintenance costs. In the industrial environment, membranes are usually organized in modules with spiral-wound or hollow fiber geometries. Hollow fiber separation units are normally applied to relatively smaller fluxes when compared to spiral-wound modules. On the other hand, spiral-wound modules are cheaper, capable of handling higher operating pressures, and are more resistant to scaling, as particles present in the feed gas stream can block the fine membrane fibers [20,21]. Because of that, the spiral-wound units are largely used in industrial gas separation processes. In a previous work of our group, a mathematical model based on a phenomenological approach for a leaf of a spiral-wound membrane was developed. The model was validated in four case studies of common gas separations, with very good performance and robustness. Furthermore, it allowed the prediction of flow rates and concentrations along the membrane leaf, which are important features for the understanding of membrane operation processes. In addition, a discretization method was proposed to solve the model, which proved to be faster and more efficient than the shooting method [22]. It is also important to emphasize that industrial spiral-wound membrane separation units for CO_2 applications are formed by several leaves, which are wounded onto a central perforated collecting tube, forming one modular separation element. These elements can be arranged in series to build a membrane separation tube. Then, tubes can be organized in parallel to form a bank. Finally, the banks can be aligned in parallel to compose a membrane separation train, while the trains can be arranged in parallel to form a stage [21]. Figure 2 shows an example of this kind of unit.

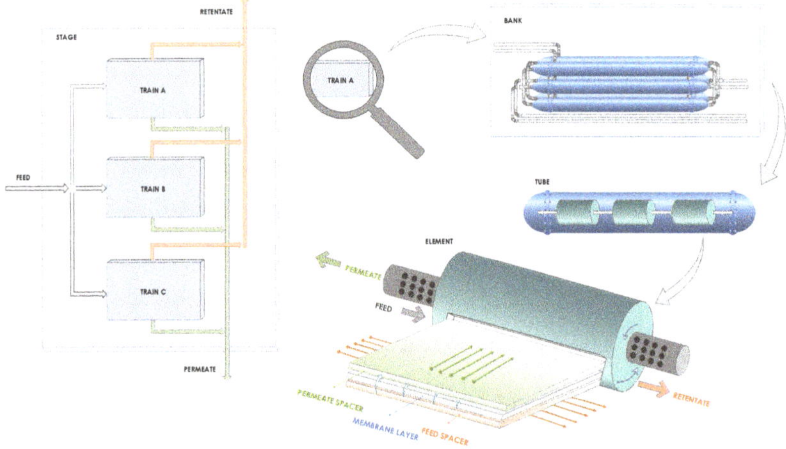

Figure 2. Schematic representation of a spiral-wound membrane unit.

1.2.1. Data Reconciliation in the Membrane Separation Process

Currently, very few papers are somehow involved with Data Reconciliation in Membrane Separation Processes. However, the work developed by Lashkari and Kruczek [23] showed that it is possible to reconcile the properties of the membrane separation process, using data related to the process dead time, which is affected by the resistance that the membrane offers to the gas flow. In fact, this work developed procedures to estimate effective permeabilities and diffusion coefficients, allowing the modeling of position-dependent resistance effects, using a lab-scale unit as an example.

Crivellari [19] proposed a model to simulate the separation of CO_2 from natural gas through a counter-current spiral wound polymer membrane. The model was used to analyze the influence of distinct variables on process operation conditions and was based on phenomenological balance equations. The model was validated with data collected from the literature and some industrial sites. Moreover, a DR procedure was used to treat the available data and estimate the model parameters; however, the study was implemented offline and did not allow any sort of real-time monitoring of the process operation.

Based on the previous discussions, the present work pioneers the use of DR procedures to monitor membrane separation processes reliably, online and in real time. For this purpose, a membrane plant located in one of the Petrobras Offshore Units was used during this paper. Finally, it is worth noting that, for reasons of industrial confidentiality, numerical results are presented in normalized form.

2. Methods

The methodology implemented in the present work comprises six stages: data acquisition; data pre-treatment; data characterization; data reconciliation; gross error detection, and process monitoring (soft sensor or digital twin). The first stages of the implemented procedure involved pre-treatment and characterization of the data. As a matter of fact, proper understanding of some characteristics of the data are fundamental for adequate implementation of the data reconciliation stage [24]. The initial characterization of the data was performed offline and using historical data available in the data acquisition system of the industrial site. The available data were used to determine appropriate sampling periods (based on process response times) and calculate measurement variances (used to formulate the estimation problem) and variable correlations (to characterize independence of measuring devices). Variable classification was also performed to determine the sets of observable and unobservable variables (with the help of the proposed model, as described below) [25].

Using the available data and the model equations, the DR procedure (as described in the following paragraphs) was solved offline to validate the proposed procedure and determine some performance indexes. Particularly, a statistical metrics was used to describe the magnitudes of the deviations between measured and reconciled variables. Then, the model was used offline for calculation of unmeasured variables, providing the soft sensor (or digital twin) response. Finally, the proposed procedures were implemented online and in real time.

The numerical procedures and codes were developed and implemented in Python 3.7.6 (Python Software Foundation, Beaverton, OR, USA) and the details of the proposed methodology are explained in the following sections.

2.1. Data Acquisition

Data acquisition was performed through direct access to an industrial database, using standard Plant Information (PI) resources. After performing the numerical operations, a file was saved with the measured, reconciled, estimated, and calculated variables. Storage was performed during monitoring, to avoid accumulation of data in the computer memory and save the relevant information in real time.

The "pandaspi" library was utilized to provide communication between Python and PI, transferring the information directly to a data frame [26,27]. By using these resources, the data acquisition process became very simple and practical, as access to the data depended only on the login, password, tags of the desired variables, the size of the sample window, and the sampling frequency. The time interval selected for offline analyses was equivalent to two weeks with a sampling frequency of 5 min, which provided a sufficiently high number of points for the execution of the pre-treatment step. An additional number of data points did not provide any significant improvement of the preliminary analyses in the considered case so that this should not be regarded as a drawback of the proposed analysis.

2.2. Data Pre-Treatment

Data pre-treatment comprised the following steps: reading the Excel Workbookfiles in UTF-8 encoding; pre-treatment of raw data spreadsheets; and data storage in Hierarchy Data Format version 5 (HDF5).

During real-time monitoring, the data reading step was performed as described in the previous section. Data storage in HDF5 format was also performed as described previously. When the file is saved in HDF5 format, reading presents better performance, since reading data directly from the Excel spreadsheet can be too slow for real-time applications [28,29].

The pre-treatment stage organized the raw data from the PI into data frames (Pandas library) [26,27]. The treatment followed the following steps: standardize the indices (day/hour/minute of the samples); variables that contain some string must be replaced (such as on/off by 0/1 and error notices by NaN, not a number); chronologically sort the data and replace missing data (NaN) with neighbors (back and forward fill). This last step is crucial to assure that the acquired data window does not contain missing data, which can make the calculation of variances difficult in the acquisition window. Unit conversions, normalization of concentrations and calculation of standard deviations and variances completed this step.

Data storage must be carried out after data reconciliation and energy balance calculations. The file was saved with the measured, reconciled, estimated, and calculated variables. As already explained, storage was performed during monitoring, to avoid accumulation of data in the computer memory and save the relevant information in real time.

2.3. Data Characterization

During this step, the procedures that must precede the data reconciliation task, such as the visualization and statistical characterization of the data, must be carried out. The characterization of the data was performed in accordance with the following steps:

- Visualization of variables;
- Selection of variables of interest;
- Quantification and visualization of missing data (NaN);
- Construction of the boxplots of the variables of interest [30];
- Analysis of the variance spectra [31];
- Calculation of variances and correlations.

The main pursued objectives during this stage were the proper characterization of the data quality, the analysis of the operation dynamics, and the characterization of the stationarity of the phenomenological process model. The analyses of missing data and boxplot properties can indicate the quality of collected data during the selected time period and the number of gross errors in the data base.

The boxplot is a graph used to assess the empirical distribution of data. The boxplot is a non-parametric analytical technique, which shows the measurement variations within a statistical population without making any assumption about the underlying statistical distribution. The box is usually built with the first (Q_1) and third (Q_3) quartiles (50% of the data) and the median (Q_2). The lower and upper lines extend, respectively, from the lower quartile to the lowest value not lower than the lower limit (LL), and from the upper quartile to the highest value not higher than the upper limit (UL). The limits can be calculated according to Equations (1) and (2):

$$LL = \max\left[\min(data),\quad Q_1 - 1.5(Q_3 - Q_1)\right] \tag{1}$$

$$UL = \min\left[\max(data),\quad Q_1 - 1.5(Q_3 - Q_1)\right] \tag{2}$$

The value 1.5 is tuned to capture 99.7% of the data between the lower and upper limits, assuming the normal distribution [30]. In summary, the boxplot identifies the regions in the variable domain where 50% and 99.7% of the data are located. The points outside these limits are tagged as outliers. Boxplots were plotted using the Seaborn library available for Python [32].

The analysis of the variance spectra shows how the process variance depends on the size of the sampling window. This type of spectrum provides information about the various sources that contribute to the signal of a variable, including noise/measurement errors (short window sizes) and intrinsic process variations (large window sizes). The variance spectrum can be defined as a set of variances calculated while some variable related to them evolves [31]. The spectrum of variances for short sampling windows are controlled by the variances of the measuring instrument. This way, the best estimate for the variance of the measuring device can be calculated using the variance spectrum with short sampling windows. However, the use of very short sampling windows may not reveal the actual variability of the data, due to poor measurement quality. The spectrum for sufficiently large sampling windows captures the variability of the entire process, including operational changes. More details about the usefulness of this technique for characterization of process data are provided by Feital and Pinto [31].

Analyzing the correlations, autocorrelations, and cross-correlations between pairs of variables of interest allows the characterization of stationarity, seasonality, and regions of operation of the process [33,34]. Observing correlations between input and output flows can indicate process stationarity. The absence of correlation can indicate the occurrence of significant nonlinearity or dynamics in the process response [35]. For dynamic responses, cross-correlation can capture the lags between the actions on the input variables and the steady-state of the output variables. Obviously, the identification of dynamic correlations among the many variables of the system may indicate the necessity to build and implement dynamic models for more accurate representation of the available data. For this reason, proper characterization of stationarity can be important for more successful implementations of monitoring procedures [36].

2.4. Data Reconciliation

The DR procedure consists of solving an optimization problem characterized by an Objective Function (OF) that must be minimized while respecting certain restrictions (model). The OF of the DR is often proposed as the maximum likelihood estimator resulting from a statistical distribution of measurement errors, which is commonly adopted as the normal distribution. After application of the principle of maximum likelihood, the normal distribution results in the WLS estimator. This way, the problem originally formulated by Kuehn and Davidson [18] can be written in accordance with Equations (3)–(7) [17]:

$$\underline{\hat{z}} = \min_{\underline{\hat{z}}} \frac{1}{2}\left[\underline{z} - \underline{\hat{z}}\right]^T \underline{\underline{V}}^{-1}\left[\underline{z} - \underline{\hat{z}}\right] \tag{3}$$

subject to:

$$\underline{h}(\hat{\underline{z}}, \underline{u}) = \underline{0} \tag{4}$$

$$\underline{g}(\hat{\underline{z}}, \underline{u}) \geq \underline{0} \tag{5}$$

$$\hat{\underline{z}}^L \leq \hat{\underline{z}} \leq \hat{\underline{z}}^U \tag{6}$$

$$\underline{u}^L \leq \underline{u} \leq \underline{u}^U \tag{7}$$

where $\hat{\underline{z}}$ is the vector of the reconciled variables; $\underline{z}$ is the vector of the measured variables; $\underline{\underline{V}}$ is the matrix of variances for measurement errors; $\underline{u}$ is the vector of the unmeasured variables (observable); $\underline{h}()$ is the vector of linear or nonlinear algebraic constraint equations; $\underline{g}()$ is the vector of the inequalities of linear or nonlinear algebraic restrictions; $\hat{\underline{z}}^L$ and $\hat{\underline{z}}^U$ are the upper and lower parameter vectors of the $\hat{\underline{z}}$ vector and $\underline{u}^L$ and $\underline{u}^U$ are the upper and lower parameter vectors of the $\underline{u}$ vector.

For DR problems where the model constraints are linear and all variables are measured, the analytical resolution of the problem can be obtained with help of Lagrange Multipliers [37]. However, the set of individual mass balance equations generate a nonlinear system of equations that involve unmeasured variables, as described below. For this reason, a successive linearization procedure was used to solve this problem [38].

The observability analysis of the system can also be performed during the successive linearization procedure, using QR factorization. The procedure consists of describing the nonlinear model in terms of two matrices: a matrix related to the measured variables and a second matrix related to the unmeasured variables. Therefore, the system is observable if the rank of the matrix of unmeasured variables is equal to the number of unmeasured variables [39].

The following points were considered during the implementation of the proposed model in the industrial site, illustrated in Figure 3:

- Period for preliminary characterization of database: two weeks with sample frequency of 5 min;
- Measured variables (z):
  - Flowrates: Total Feed (F), Total Retentate (R), Train Retentate A, B and C (R_A, R_B and R_C) $[kNm^3/h]$;
  - Components: C_1(methane), C_2(ethane), C_3(propane), C_6(hexane), C_7(heptane), C_8(octane), CO_2(carbon dioxide), iC_4(i-butane), iC_5(i-pentane), N_2(nitrogen), nC_4(n-butane) and nC_5(n-pentane) in the 3 streams.
- Unmeasured variables (u):
  - Flowrate: Total permeate (P) $[kNm^3/h]$.

Model (Component Mass Balance—Steady-State):

$$\underline{h}(\hat{\underline{z}}, \underline{u}) = \begin{cases} \underline{0} = F\underline{y}_F - R\underline{y}_R - P\underline{y}_P \\ 0 = R - R_A - R_B - R_C \\ 0 = 1 - \sum_{i=1}^{nc} \underline{y}_{F,i} \\ 0 = 1 - \sum_{i=1}^{nc} \underline{y}_{R,i} \\ 0 = 1 - \sum_{i=1}^{nc} \underline{y}_{P,i} \end{cases} \tag{8}$$

- Number of points in the study phase (nt) = 3457
- Number of components (nc) = 12
- Number of measured variables at each sampling point (Nm) = $3nc + 5$ = 41
- Number of unmeasured variables at each sampling point (Nu) = 1
- Number of total equations at each sampling point (Nv): $Nm + Nu$ = 42

- Number of constraint equations (Nce) = $nc + 4$ = 16
- Number of optimization variables at each sampling point (Nopt): $Nv - Nce$ = 26
- Degrees of freedom at each sampling point (DF): $Nm - Nopt = Nm - (Nm + Nu - Nce)$ = 15

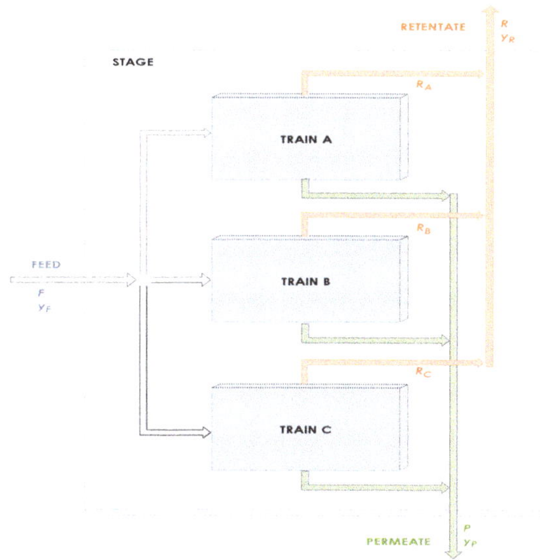

Figure 3. Individual mass balances in the envelope around the process.

2.5. Gross Error Detection

Gross errors are those originating from non-random events, having little or no connection with the measured value. They may be related to measurements (such as malfunctioning of instruments) or process (such as leaks). Consequently, gross errors invalidate the classic statistical basis of traditional DR methods and undermine the systematic analysis of the data, demanding the implementation of Gross Error Detection procedures for compensation or elimination of gross errors, which may precede the DR step. Gross errors can be further classified as **bias**, **outliers**, and **drifts** [40].

The metrics applied in this step to identify the occurrence of gross errors is related to the statistical test of standard hypotheses for GED [12]. This test is widely used and simple, and also forms the basis for all other classic GED procedures. The test is applied to the data set to determine whether the measurements follow a symmetrical distribution of the measurement errors.

Reilly and Carpani [41] were the first to study the detection of gross errors in process engineering. The authors proposed the use of a statistical test of the type χ^2 (chi-square) based on the residues of the process model, which was called the Global Test (GT).

In the present work, GED was based on the GT, using the sampling window to observe samples and identify the occurrence of gross measurement errors. According to this procedure, the data in the moving window are used to compare the errors of the Weighted Least Squares estimator with the deviations from median values within the window and using the χ^2 function to characterize significant deviations between these computed variances [42].

The outlier effect during DR is removed through compensation. After identification of a possible outlier, the variable value was manipulated to adjust the expected variance. Thus, a moving variance window was implemented to monitor changes in operation, measurement errors, and failures. The statistical test was performed in sampling windows containing at least 20 samples. With this, the median and the standard deviation were calculated and compared with the value of the variable at the current point. Therefore, if the value was more than seven standard deviations apart from the median, the measurement would be regarded as an outlier and the variance adjustment would be performed.

To monitor the possible occurrence of bias, a dynamic bar graph illustrating the magnitude of errors for each variable was implemented, using the following metrics:

$$Bias_{i,t} = \frac{med(|\underline{z}_{i,t} - \hat{\underline{z}}_{i,t}|)}{dp_{i,t}} \tag{9}$$

$$dp_{i,t} = NMAD(|\underline{z}_{i,t} - \hat{\underline{z}}_{i,t}|) \tag{10}$$

where $NMAD$ is Normalized Median Absolute Deviation.

2.6. Monitoring

Instrumenting the whole process can be very costly and may lead to acquisition of unnecessary and obsolete information. In addition, in some cases, it can be impossible to measure the desired variable. For this reason, some information that is essential for process monitoring can be assessed through models. Therefore, a soft sensor can estimate variables with a mathematical model, in real time, using available plant data, as measured by existing instrumentation. Particularly, the use of plant variables can provide opportunities to improve the performance of a plant [43].

In the analyzed process, the permeate flow was not measured. It must be emphasized that this is not unusual at real industrial sites. However, through DR, this information can be obtained with the help of the model, after the application of the DR procedure. The complete set of temperature and pressure was not available either, which is not unusual at the plant site. In the analyzed case, the pressure of the feed stream and the temperature of the permeate stream were not measured. Therefore, the full implementation of the energy balance in the proposed DR procedure was not viable due to the lack of observability. For this reason, the pressure of the feed stream was evaluated through calculation of the pressure loss in the separation stage, based on design data and the pressure data of the retained stream. In this case, after characterization of the pressure loss, the energy balance equation can be used to estimate the permeate stream temperature, with the aid of Equation (11), calculating the enthalpies of the process streams with the Peng–Robinson equation of state [44]:

$$0 = F.H(T_F, P_F, \underline{y}_F) - R.H(T_R, P_R, \underline{y}_R) - P.H(T_P, P_P, \underline{y}_P) \tag{11}$$

It must be noted that the execution of the numerical procedure in this analyzed case was extremely fast (order of milliseconds), so that the computer hardware exerted little influence on the application. The slowest step of the numerical procedure was data acquisition (and the bottleneck was data transfer and connection speed), so that instrumentation hardware and data handling software constituted the most sensitive parts for this particular real-time application. Despite that, it took only few seconds for data downloading to be complete; consequently, the online and real-time implementation could be performed in a standard notebook equipped with an Intel Core i7 8th gen processor (Intel Corporation, Santa Clara, CA, USA).

3. Results and Discussion

3.1. Data Characterization

The Missingno library was used to analyze the missing data [45]. In Figure 4, one can observe the data density, visualizing completely the pattern of missing data in the whole set, with the columns representing the variables and lines representing data points of the time series. In addition, a frequency bar that indicates the number of variables measured at each particular timeline can be seen on the right side of the graph.

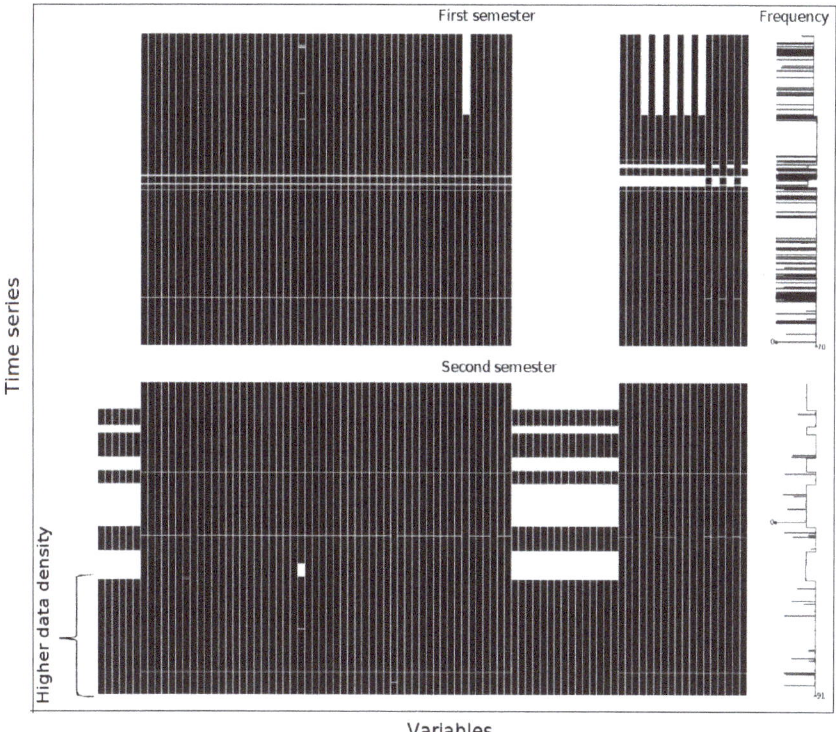

Figure 4. Missing data analysis.

The analysis of Figure 4 illustrates how one can select the best data period for the execution of the initial data processing stage. The period with the highest data density allows the analysis of the plant operation with better reliability, avoiding blind periods of instrumentation or shutdown of operation. Therefore, the selection of the period is fundamental for the data characterization stage.

Figure 5 illustrates the boxplot of chromatographic analyses associated with the four main components of the feed stream. This analysis allows the preliminary evaluation of the precision of the instrumentation and/or the variability of the operation. A process with distinct operating points generates multimodal distributions, which requires more involving analysis to qualify the precision of the measurement, such as violin plot (a combination of boxplot and kernel density estimate) [46]. However, during stationary operation periods, the boxplot analysis showed that chromatographic measurements presented good precision, which is also illustrated by the small number of outliers with respect to the total number of 3673 samples.

Figure 6 shows that the "gross errors" followed a downward trend, with highest concentration below the modal value. This is because a drop in the process flowrate was observed during this period. Therefore, in this case, the analysis interprets that the operating changes are "gross error", when they are not. On the other hand, the good behavior of the data indicates that the flowrate was measured with good precision. Instrumentation accuracy was also analyzed after DR, with help of the bias analysis.

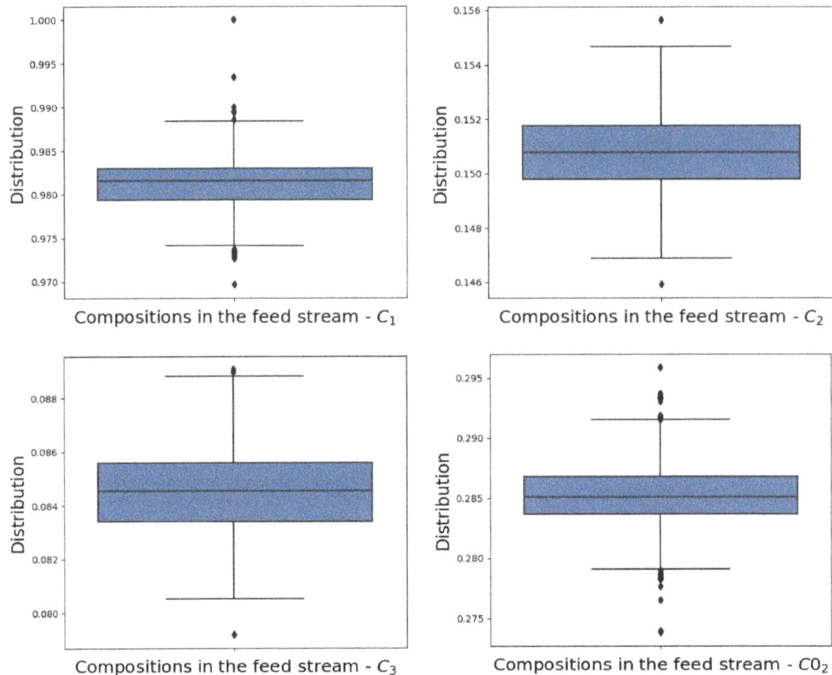

Figure 5. Analysis of boxplots for compositions C_1, C_2, C_3, and CO_2 in the feed stream.

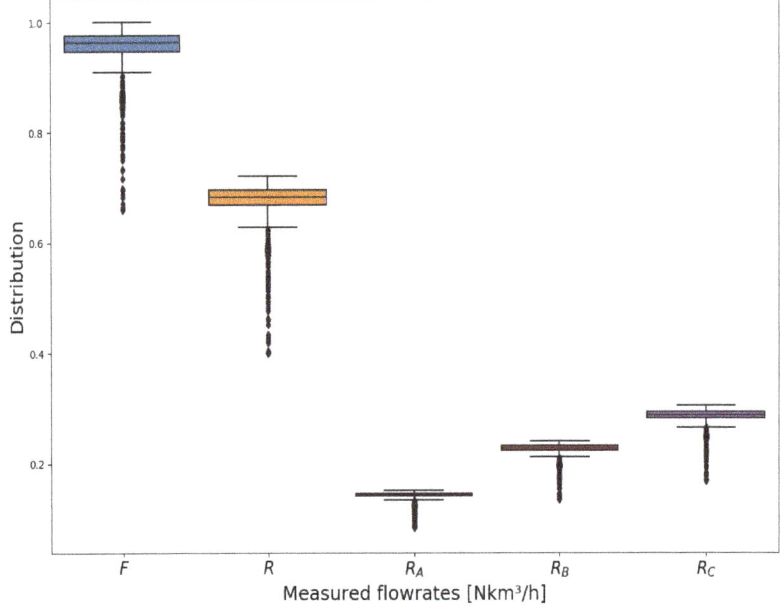

Figure 6. Boxplot analysis of flowrates.

Figure 7 illustrates the time series for the compositions of C_1, C_2, C_3, and CO_2 in the feed stream, while Figure 8 shows the feed and retentate flowrates, for the same time ranges analyzed in the boxplot.

Therefore, it becomes evident in the case of flowrates that the supposed gross errors actually indicated a change in operation and not a failure of the sensor. In the case of composition, outliers can possibly be assigned to gross errors, although only the DR can allow the proposition of reliable statements about the alleged gross errors.

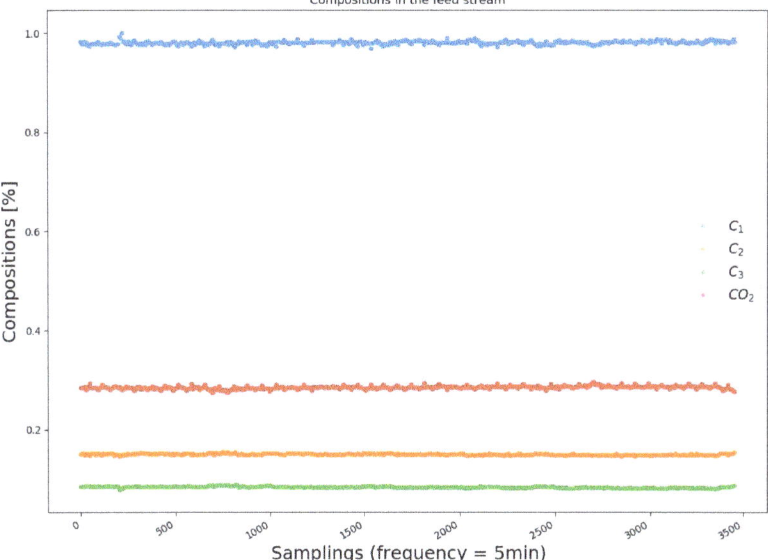

Figure 7. C_1, C_2, C_3, and CO_2 compositions in the feed stream.

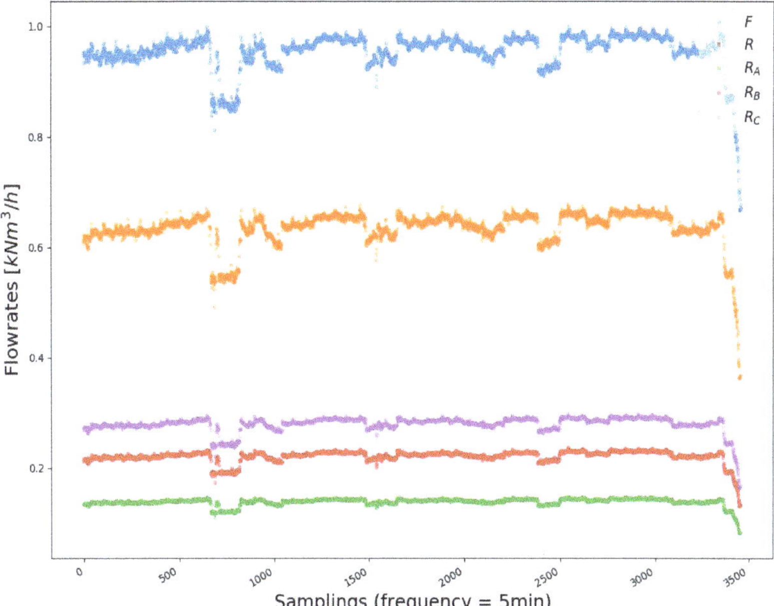

Figure 8. Measured flowrates.

An important analysis is related to the observation of the correlations between pairs of variables. Correlations can indicate absence or presence of process stationarity. Figure 9 shows a strong linear correlation between feed, retentate, and permeate flowrates. Strong linear correlation between inlet and outlet flowrates can be an indication of process stationarity [47].

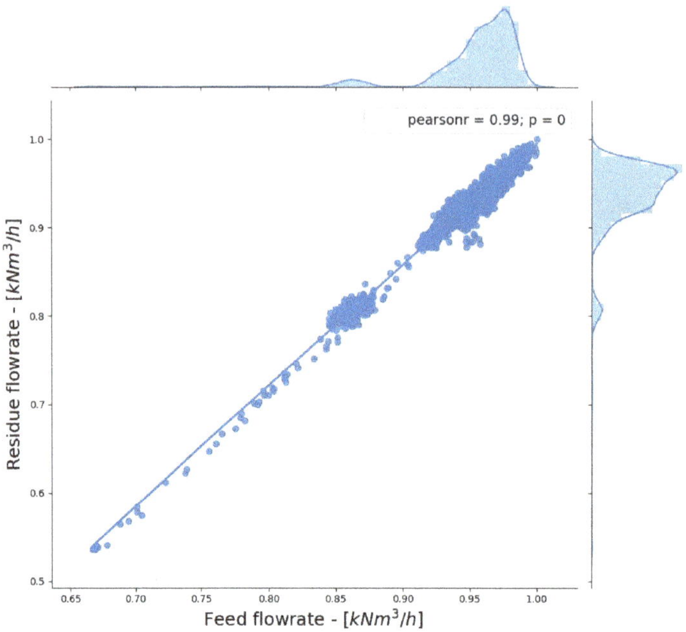

Figure 9. Correlation analysis for feed and retentate flowrates.

Figure 10 illustrates the Autocorrelation Function (ACF) and Partial Autocorrelation Functions (PACF) for feed and residue streams. This analysis provides the diagnosis of temporal dependence between the lags of individual variables, which in this case were evaluated for lags ranging from 0 to 50 lags. As shown in Figure 10, the ACF decayed continuously and just one lag caused the appearance of strong correlation (close to 1) in the PACF. Therefore, the process presents very short dynamic memory, indicating the quasi steady-state behavior and constituting an auto-regressive process of order 1 [33].

Figure 11 illustrates the Cross-Correlations Function (CCF) between feed and retentate flowrates up to 50 lag. Cross-correlations decayed slowly for different pairs of variables, indicating that the process operated at quasi steady-state conditions and that responses were much faster than the characteristic sampling times. Therefore, the analyzed membrane separation process could be considered to operate at steady-state. This validated the use of the steady-state mass and energy balance equations in the DR problem. Given the small volumes of most membrane separation modules and the large flowrates of typical industrial plants, this conclusion can probably be extended to other industrial sites, allowing the more general use of the analysed procedures in other industrial facilities.

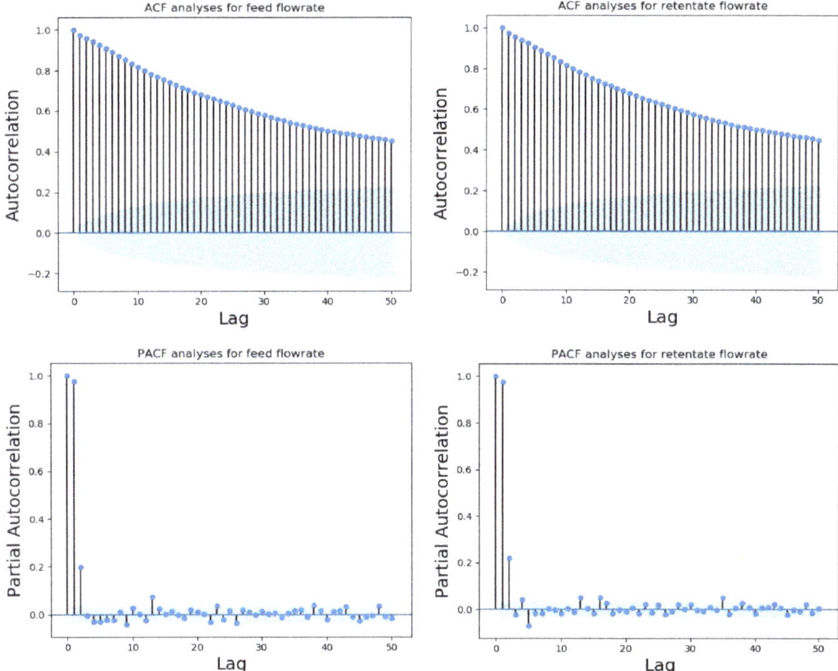

Figure 10. ACF (Autocorrelation Function) and PACF analyses (Partial Autocorrelation Functions) for feed and retentate flowrates.

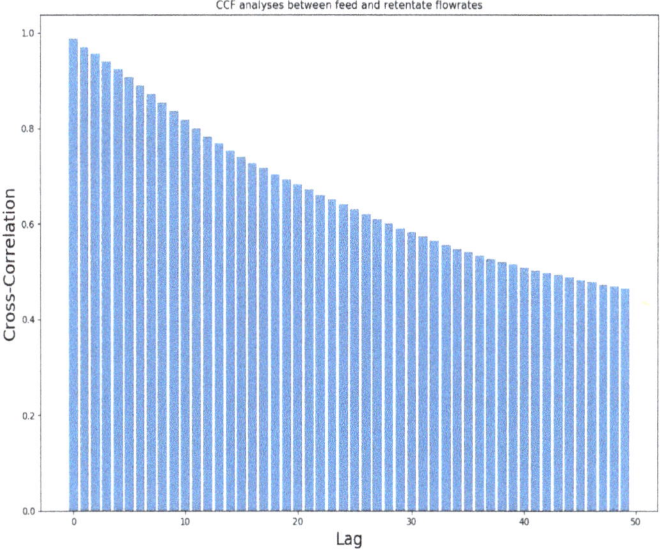

Figure 11. CCF analyses between feed and retentate flowrates.

3.2. Data Reconciliation

Before starting the DR procedure, the system was analyzed with the help of Variable Classification techniques and the system was classified as observable, indicating that the measured variables can be reconciled and that unmeasured variables (permeate flowrate) can be estimated.

The first DR results were obtained through offline simulations, using a sampling period of two weeks with a sampling interval of 5 min. The data reconciliation performed very well and the problem was solved at average computational speed of 1.7 ms/sample. This result clearly showed that the application could be implemented online and in real time due to the sampling interval of 5 min.

Figure 12 illustrates, as an example, the measured and reconciled data for chromatographic measurements of the four main components in the feed stream. One of the advantages of DR is to restore the resolution of the amplitude signal of the measured value, which can be seen in Figure 12, especially for samplings of the C_2 component.

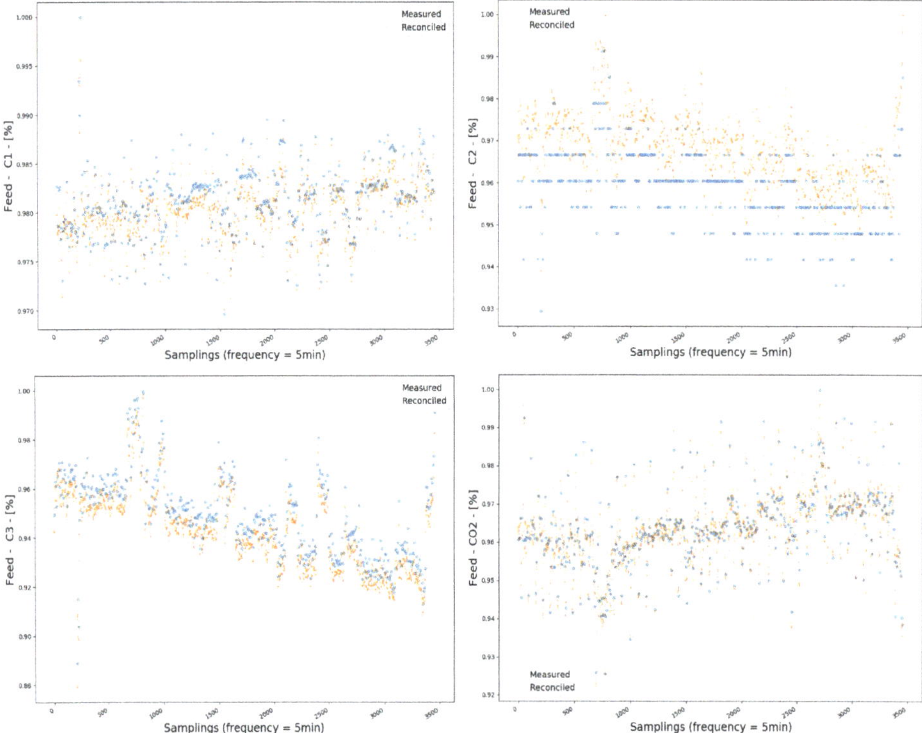

Figure 12. Offline data reconciliation for compositions C_1, C_2, C_3, and CO_2 in the feed stream.

Figure 13 illustrates the measured, reconciled, and estimated data for the flowrates. Another major advantage of DR is to identify the occurrence of systematic deviations in measurements, often caused by miscalibrated instruments. Figure 13 shows the occurrence of bias for the feed and residue flowrates.

Figure 14 shows the sum of residuals with respect to the mathematical model. The total squared residual is a measure of corrections that were needed in order to reconciled variables to satisfy the mass and energy balance equations.

Based on the previous results, it could be concluded that the DR procedure presented good performance and advantageous aspects for monitoring of the process. Monitoring processes with statistically treated information, detection of measurement bias, and identification of poor instrumentation performances constitute good tools for diagnosing the states of the analyzed process and respective instrumentation.

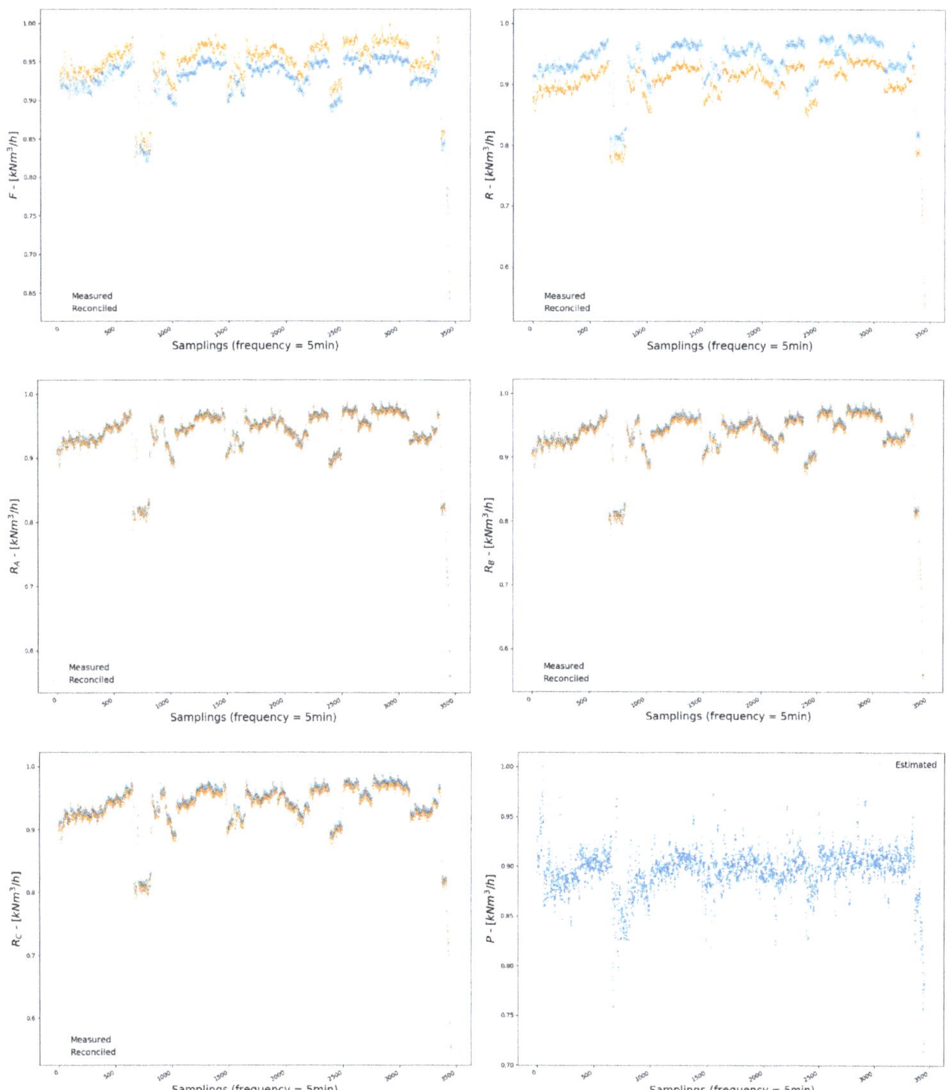

Figure 13. Offline data reconciliation of flowrates.

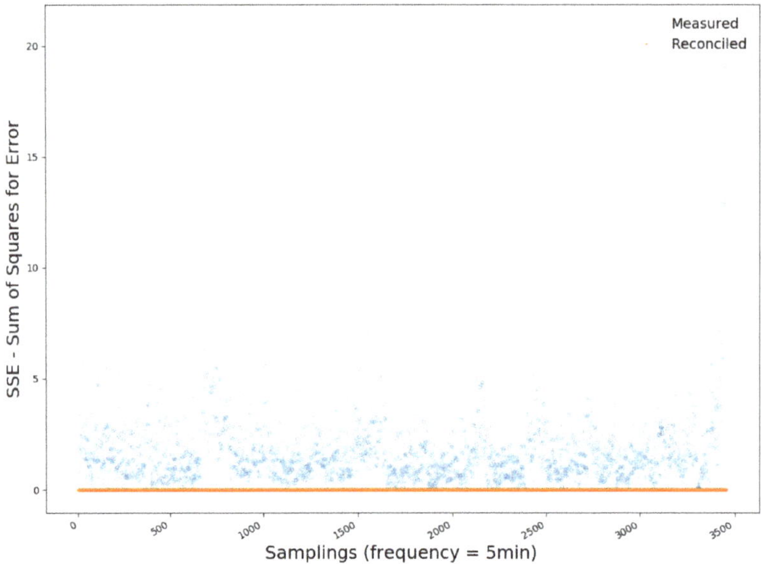

Figure 14. Sum of squares of model deviations (residuals) during offline data reconciliation.

3.3. Gross Error Detection

Procedures for removal of gross errors were implemented for online and real-time DR. These procedures were based on statistical tests using a moving variance window. Figures 15–20 illustrate a case of gross errors in which the procedure proved to be robust. However, it is possible to observe that the gross errors that affected the compositions of the retentate flowrate influenced the reconciliation of the feed flowrate. This occurred because of the well-known "smearing effect" when the DR procedure was performed with the WLS estimator (non-robust), even when variance adjustment was performed [48]. As gross error measurements were observed for a short period of time, it was not possible to detect the main source of the problem in the analyzed data set. Nevertheless, the occurrence of the problem was reported for maintenance teams for evaluation of measurement consistency.

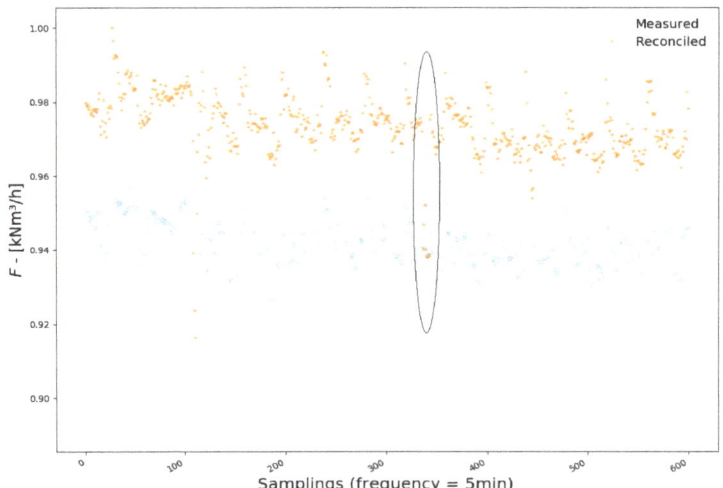

Figure 15. "Smearing" effect during the DR of feed flowrates.

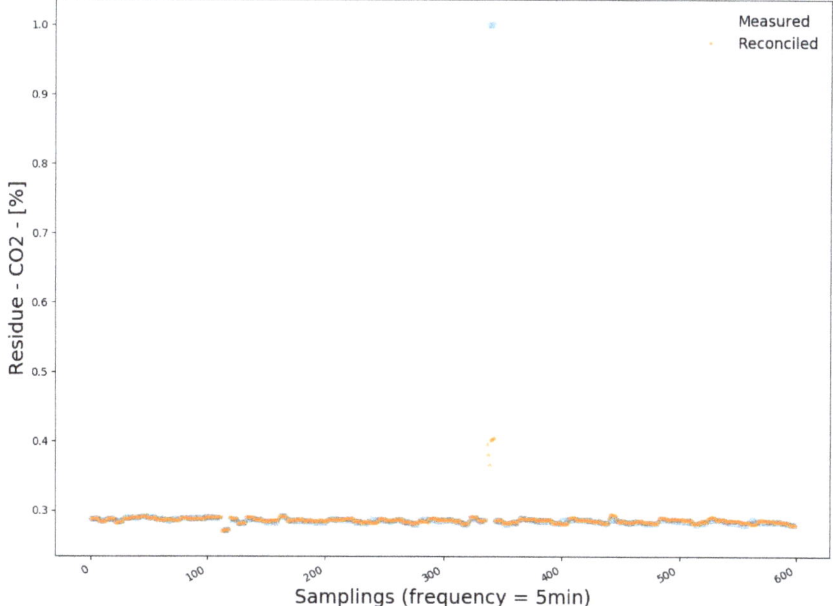

Figure 16. Gross Error Detection—CO_2 in the feed stream.

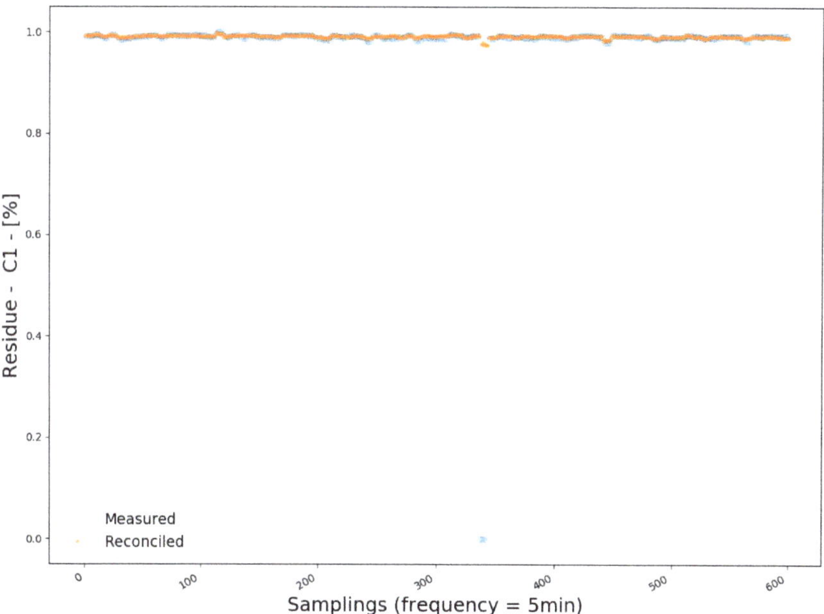

Figure 17. Gross Error Detection—C_1 in the feed stream.

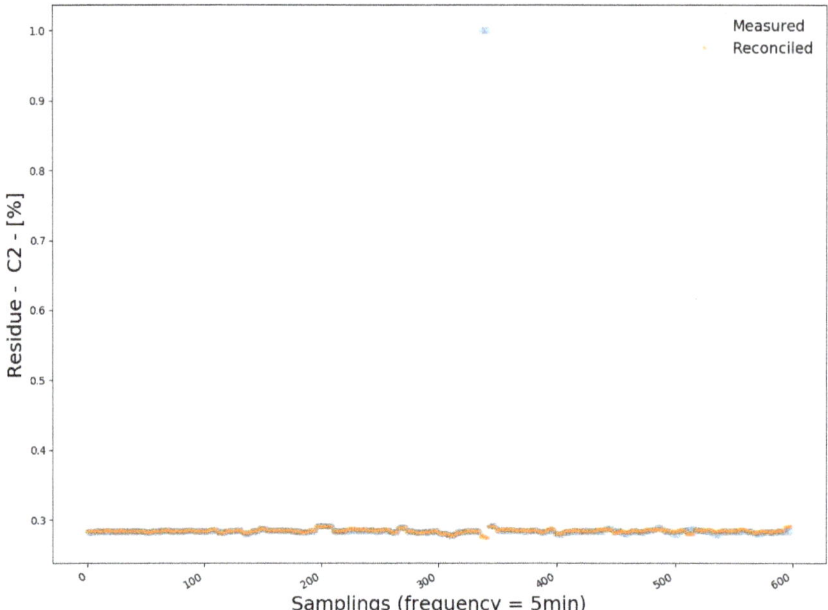

Figure 18. Gross Error Detection—C_2 in the feed stream.

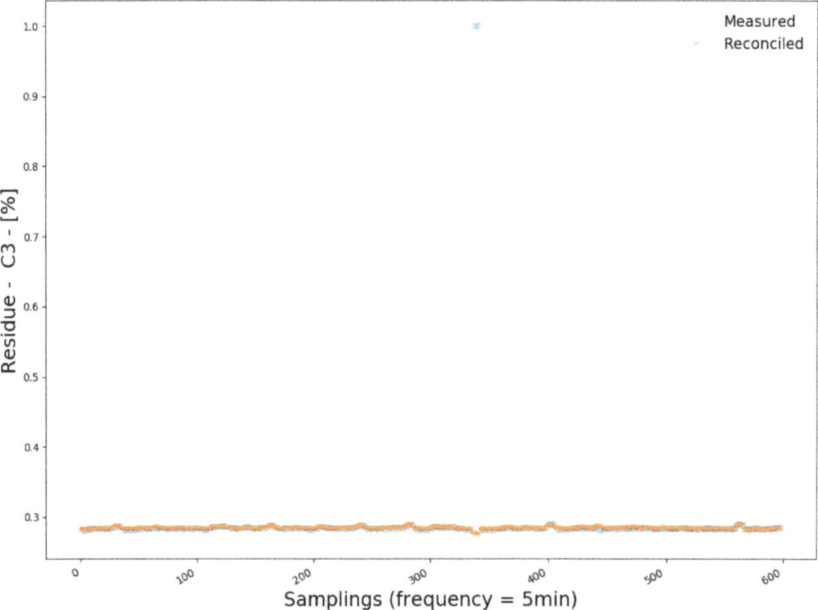

Figure 19. Gross Error Detection—C_3 in the feed stream.

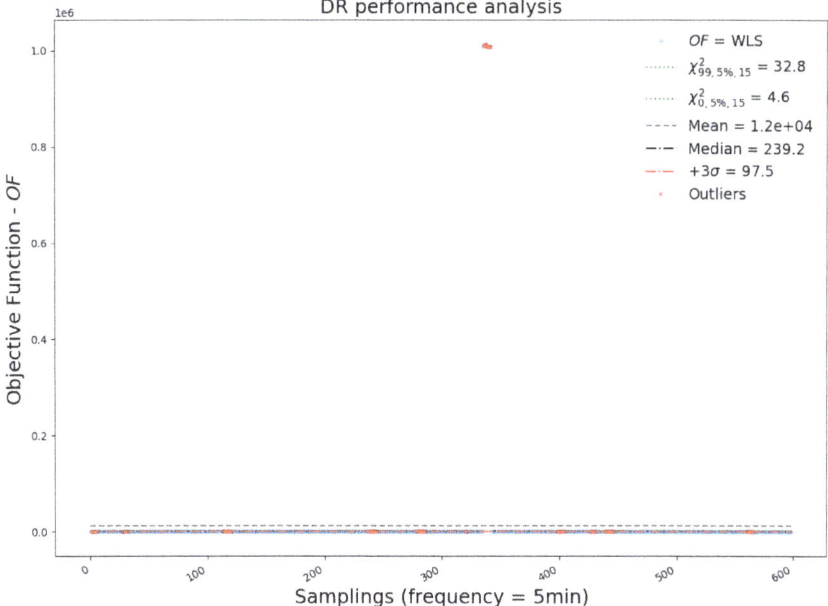

Figure 20. Gross Error Detection—DR performance analysis.

It is important to note that the statistical tests were implemented only for the compositions. The fact is that operational changes hindered the test because in many cases the test interpreted operational changes as outliers. Figure 21 illustrates that data reconciliation was effective and performed well after several operational changes.

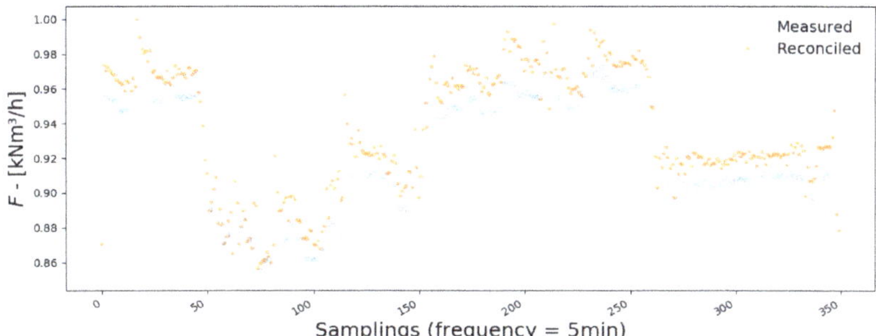

Figure 21. Monitoring through data reconciliation.

An important advantage of the moving variance window was to avoid the interruption of the online DR procedure due to measurement problems. These failures occur more frequently with compositions measured online through gas chromatography. These measurement failures cause missing data and, consequently, occurrence of series of constant values. As a result, the signal loses variability, preventing the realization of DR. Figures 22 and 23 illustrate cases of variable freezing caused by missing data.

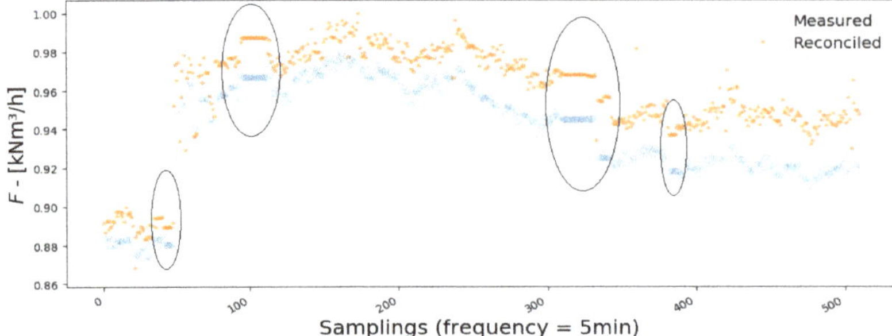

Figure 22. Measurement failures: missing data and frozen values of feed flowrate.

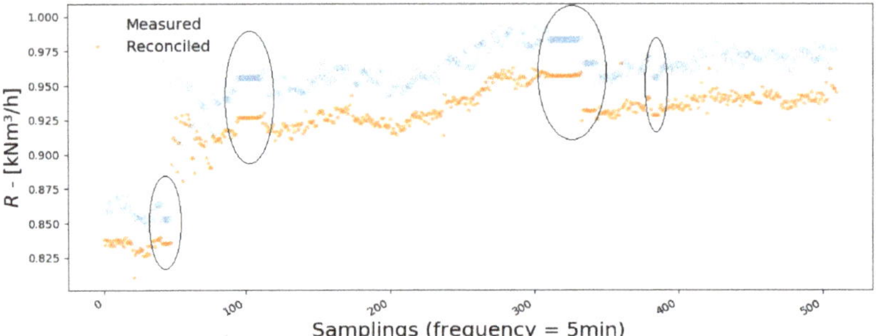

Figure 23. Measurement failures: missing data and frozen values of residue flowrate.

The analysis of bias can be performed through dynamic bar graph monitoring, illustrating the magnitude of the errors of each variable. Figure 24 informs the magnitudes of the systematic deviations from the median, that is, how many times the reconciled variable deviated from the measured median value. Systematic deviations that are larger than three times the value of the standard deviation can be regarded as a bias. Therefore, analyzing Figure 24, five variables with measurement biases could be observed: N2 (feed); N2 (residue); C8 (permeate); feed flowrate; and residue flowrate. Generally, biases can indicate the occurrence of unbalanced measurements, calibration problems, and instrument malfunctioning. For this reason, the obtained results were relayed to maintenance teams for evaluation of the instrumentation performances.

Figure 25 analyzes the performance of the DR, presenting the value of the OF, which represents the degree of correction of the reconciliation. The two green dotted lines represent the region where a normal distribution is expected for the errors of all measured variables. The region above the red dotted line indicates the samples that were subject to large corrections during the reconciliation step. Therefore, it is reasonable to consider the possible occurrence of outliers when the obtained value of OF deviated more than three standard deviations from the median value.

A test to observe the influences of biases on the analysis and performance of DR was performed. Figures 26 and 27 illustrate the same analyses performed for the same time window, as presented in Figures 24 and 25, but after identification and compensation of outliers. It can be observed that outliers significantly affected the average OF value. Based on Figures 24–27, it can be said that the analysis of bias and outliers performed very well. Figure 27 also illustrates the benefits of bias and outlier adjustments, as objective function values were reduced significantly and shed light on the existence of persistent outlier measurements. This reinforces the importance of bias and gross error diagnosing and the necessity to involve maintenance teams for evaluation of the instrumentation performances.

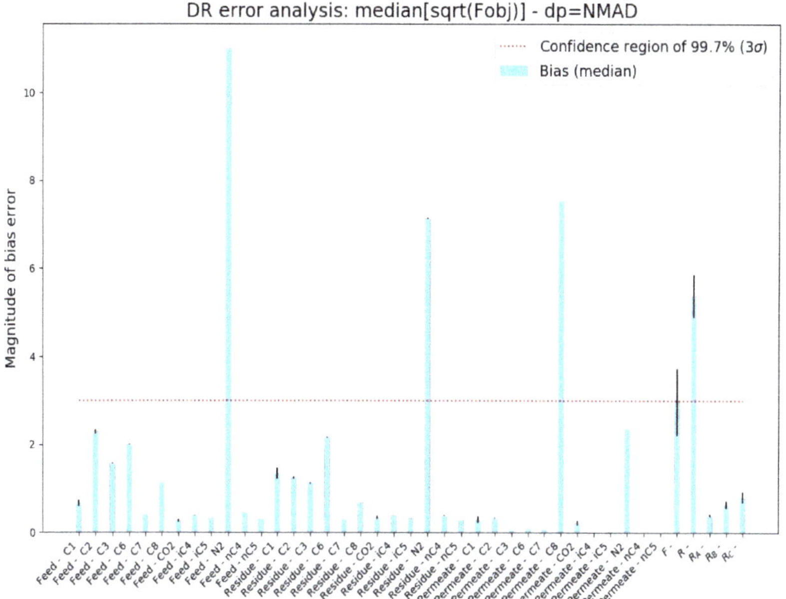

Figure 24. DR analysis without bias compensation.

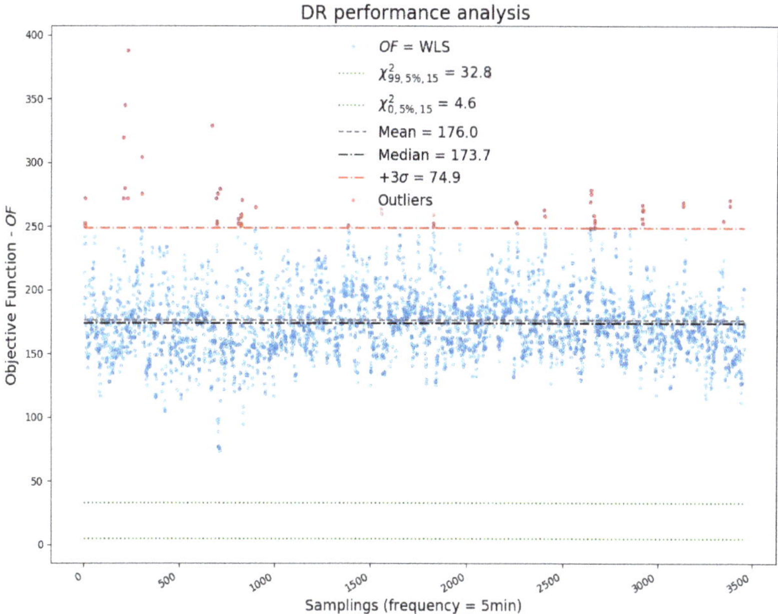

Figure 25. DR performance analysis without bias compensation.

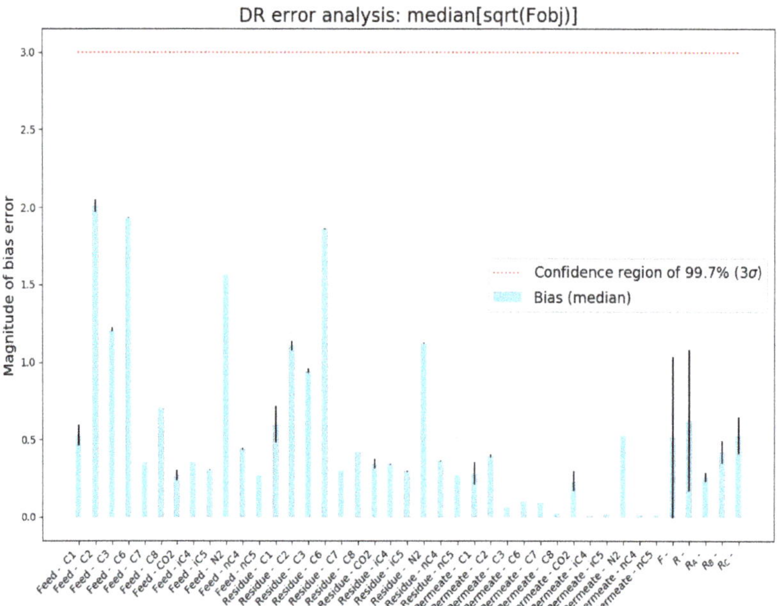

Figure 26. DR analysis with bias compensation.

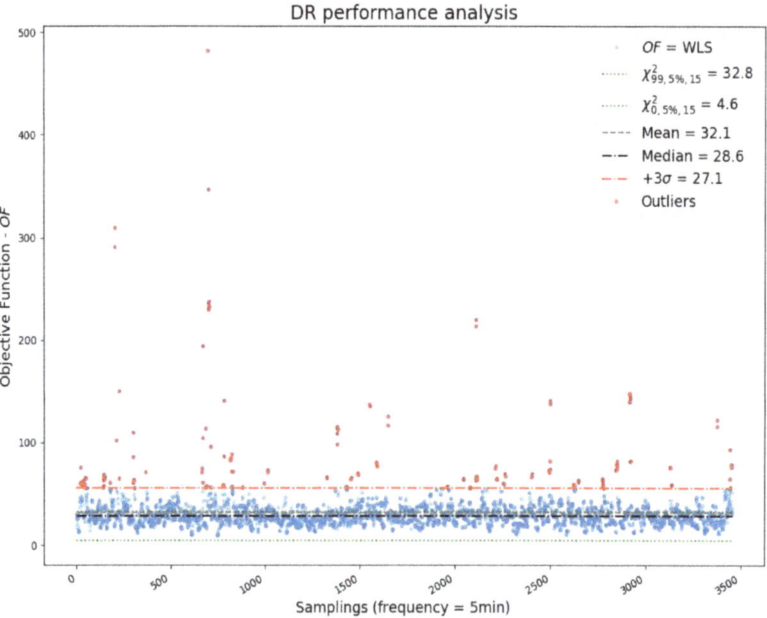

Figure 27. DR performance analysis with bias compensation.

3.4. Monitoring

In each new cycle of data acquisition, the code runs in sequence the pre-treatment, statistical tests and outlier compensation, data reconciliation with the permeate flowrate estimation, and finally the energy balance to calculate the temperature of the permeate flowrate. The first inferred variable was the permeate flowrate, estimated within the data reconciliation procedure. Figure 28 illustrates the real-time monitoring of the inferred variable.

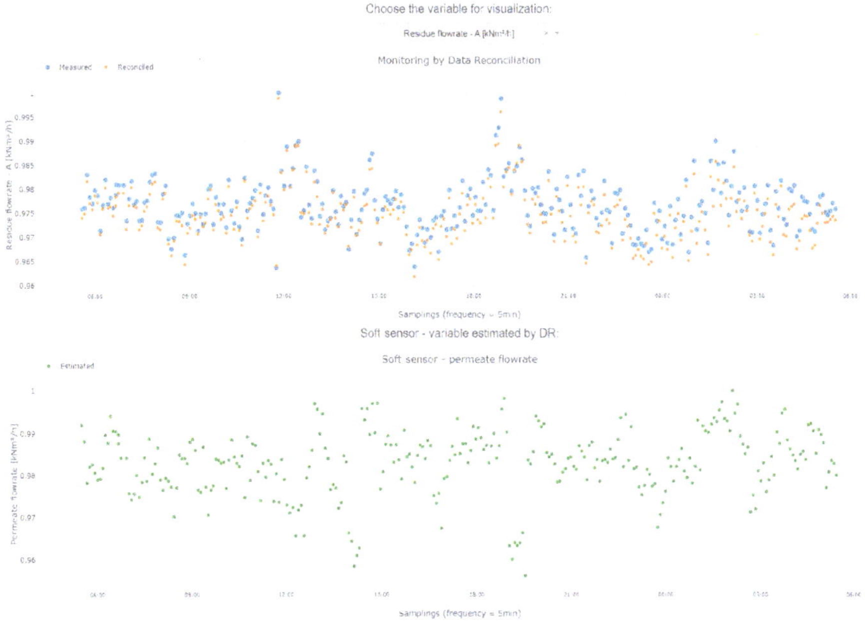

Figure 28. Real-time soft sensor—permeate flowrate.

The second inferred variable was the temperature of the permeate. At this stage, it was not possible to reconcile data due to the lack of redundancy of measured variables. Thus, this variable was inferred without the proper statistical treatment by the DR stage. Figure 29 illustrates part of the web application (web-app) where the user interacts with the interface. The variables can be selected through a dropdown menu. In addition, the application provides graphs of gross error analysis (Figure 30), visualization of inferred variables, and a button to start and stop monitoring. Figure 31 illustrates the three temperatures of each stream, during the testing period of the web-app. The permeate temperature was calculated with help of the energy balance and the regions without data are the days when the web-app was paused. All monitoring data are saved and can be read and analyzed offline, as in the case of Figure 31.

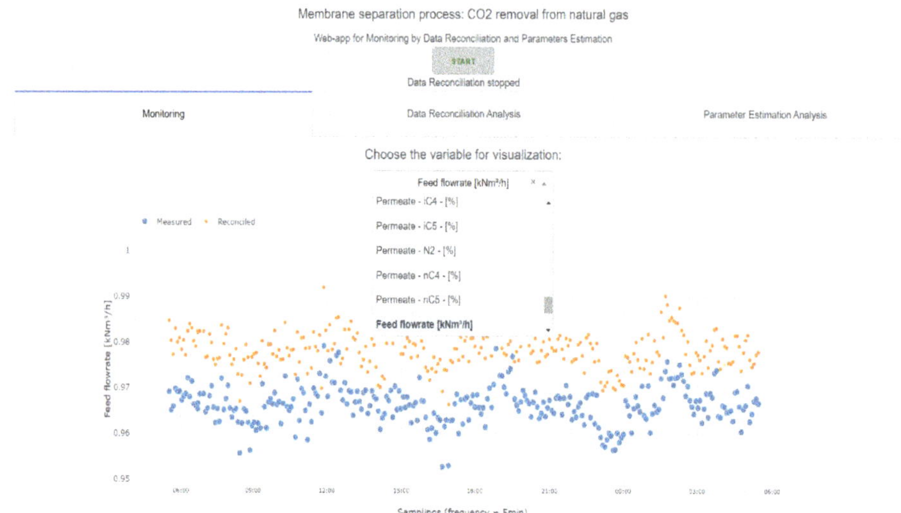

Figure 29. Part of the web-app: variables measured and reconciled.

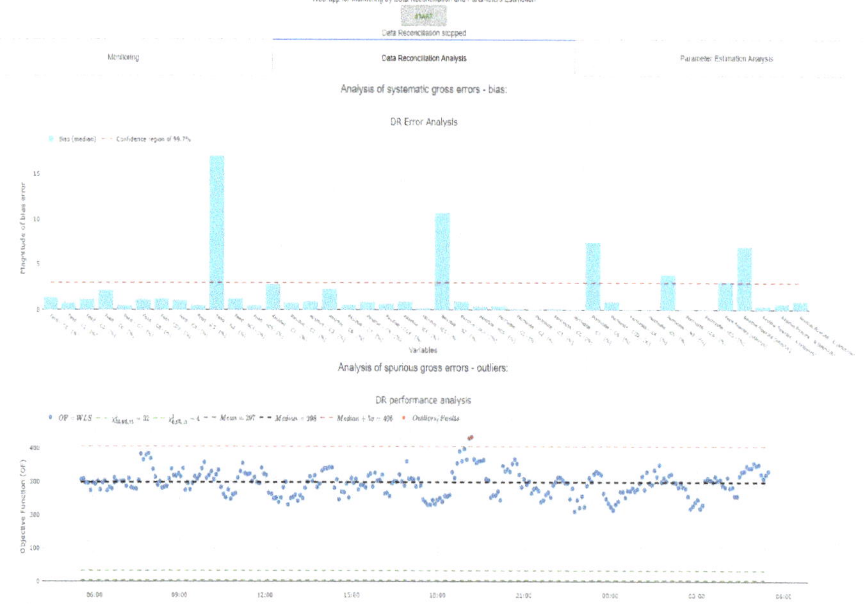

Figure 30. Part of the web-app: data reconciliation analysis.

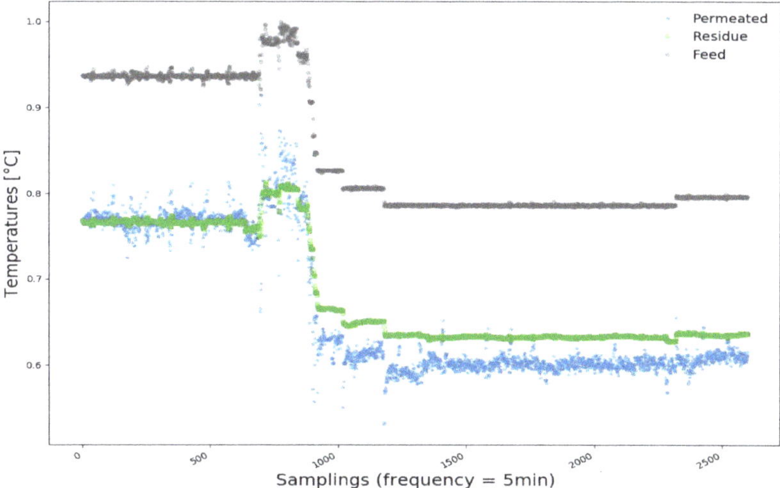

Figure 31. Offline data analysis: Temperatures in the testing period.

Therefore, the temperature inferred by the calculation of the energy balance showed good accuracy in relation to reported offline measurements, which demonstrates the importance of DR for the treatment of the variables used in the soft sensor.

4. Conclusions

A methodology was developed and implemented for the first time in the form of a web application to allow the monitoring of membrane separation processes online and in real time, making use of statistical techniques for treatment of process data. The proposed methodology comprises the following stages: (i) pre-treatment and characterization of process data; (ii) data reconciliation of process data to minimize measurement uncertainties, with the aid of mass balance equations; (iii) detection of systematic deviations for identification of process malfunctions; and (iv) observation of unmeasured variables (working as a soft sensor or digital twin). The pre-treatment and data characterization steps were fundamental for the understanding and correct formulation of the problem. The characterization step can find wide application, as this procedure can be applied in any chemical process. This step is essential for the appropriate selection of data reconciliation techniques and gross error detection procedures. After that, the proposed data reconciliation and gross error detection steps showed robustness, good performance, and speed. The proposed scheme was based on detailed steady-state balance equations, validated after proper characterization of actual operation data. The numerical procedures were validated offline and then implemented online and in real time for the first time, allowing the successful identification of measurement biases and outliers and providing estimates for unmeasured data. The developed procedures can be used for online and real-time detection of process faults and process diagnosing. In addition, the procedure provides reliable data for future stages of simulations and parameter estimation, allowing the implementation of digital twins, as the model proposed in part I of this research project. Production Management System and Enterprise Resource Planning steps can also benefit from availability of more reliable data, and variables inferred by a soft sensor. Therefore, the main advantages of the procedure are reliable data handling, diagnosis of gross errors/failures, and real-time monitoring of the process.

Author Contributions: D.Q.F.d.M.: Conceptualization, Methodology, Software, Validation, Data curation, Writing—original draft, Writing—review and editing; M.C.C.d.S.: Software, Validation, Formal analysis; T.B.F.: Software, Validation, Formal analysis; T.K.A.: Resources and Funding acquisition; F.C.D.: Resources and Funding acquisition; P.H.T.: Resources and Funding acquisition; J.C.P.: Supervision, Project administration, Writing—review and editing. All authors have read and agreed to the published version of the manuscript.

Funding: This work was supported by the Petróleo Brasileiro SA (Petrobras); Conselho Nacional de Desenvolvimento Científico e Tecnológico (CNPq) and Coordenação de Aperfeiçoamento de Pessoal de Nível Superior (CAPES).

Acknowledgments: The authors thank Petrobras, CNPq, and CAPES for the financial support to this work, as well as for covering the costs to publish in open access.

Conflicts of Interest: The authors declare that they have no known competing financial interests or personal relationships that could have appeared to influence the work reported in this paper.

Abbreviations

The following abbreviations are used in this manuscript:

GED	Gross error detection
DR	Data reconciliation
WLS	Weighted least squares
PI	Plant information
HDF5	Hierarchy Data Format version 5
NaN	Not a number
LL	Lower limit
UL	Upper limit
OF	Objective function
GT	Global test
NMAD	Normalized median absolute deviation
ACF	Autocorrelation functions
PACF	Partial autocorrelation functions
CCF	Cross-Correlations Function

References

1. Mokhatab, S.; Poe, W.A.; Mak, J.Y. *Handbook of Natural Gas Transmission and Processing: Principles and Practices*; Gulf Professional Publishing: Houston, TX, USA, 2018.
2. Agência Nacional do Petróleo, Gás Natural e Biocombustíveis. *ANP, ANP RESOLUTION N° 16, D*; Diário Oficial da União: Brasilia, Brazil, 2008. Available online: http://www.anp.gov.br/ (accessed on 11 July 2020).
3. Speight, J.G. *Natural Gas: A Basic Handbook*; Gulf Professional Publishing: Houston, TX, USA, 2018.
4. Henis, J.M.; Tripodi, M.K. Composite hollow fiber membranes for gas separation: The resistance model approach. *J. Membr. Sci.* **1981**, *8*, 233–246. [CrossRef]
5. Al-Obaidi, M.A.; Kara-Zaïtri, C.; Mujtaba, I.M. Simulation and sensitivity analysis of spiral wound reverse osmosis process for the removal of dimethylphenol from wastewater using 2D dynamic model. *J. Clean. Prod.* **2018**, *193*, 140–157. [CrossRef]
6. Singh, V.; Jain, P.; Das, C. Performance of spiral wound ultrafiltration membrane module for with and without permeate recycle: Experimental and theoretical consideration. *Desalination* **2013**, *322*, 94–103. [CrossRef]
7. Kovvali, A.S.; Vemury, S.; Admassu, W. Modeling of multicomponent countercurrent gas permeators. *Ind. Eng. Chem. Res.* **1994**, *33*, 896–903. [CrossRef]
8. Soares, R.d.P. Desenvolvimento de um Simulador Genérico de Processos Dinâmicos. Master's Thesis, Universidade Federal do Rio Grande do Sul, Porto Alegre, RS, Brazil, 2003.
9. Nicholson, B.; López-Negrete, R.; Biegler, L.T. On-line state estimation of nonlinear dynamic systems with gross errors. *Comput. Chem. Eng.* **2014**, *70*, 149–159. [CrossRef]
10. Câmara, M.M.; Soares, R.M.; Feital, T.; Anzai, T.K.; Diehl, F.C.; Thompson, P.H.; Pinto, J.C. Numerical Aspects of Data Reconciliation in Industrial Applications. *Processes* **2017**, *5*, 56. [CrossRef]
11. Prata, D.M.; Schwaab, M.; Lima, E.L.; Pinto, J.C. Simultaneous robust data reconciliation and gross error detection through particle swarm optimization for an industrial polypropylene reactor. *Chem. Eng. Sci.* **2010**, *65*, 4943–4954. [CrossRef]
12. Farias, A.C. Avaliação de Estratégias para Reconciliação de Dados e Detecção de Erros Grosseiros. Master's Thesis, Universidade Federal do Rio Grande do Sul, Porto Alegre, RS, Brazil, 2009.

13. de Menezes, D.Q.F. Reconciliação Robusta de Dados em Colunas de Destilação. Master's Thesis, Universidade Federal Fluminense, Niterói, RJ, Brazil, 2015.
14. Prata, D.M. Reconciliação Robusta de Dados para Monitoramento em Tempo Real. Ph.D. Thesis, COPPE—Universidade Federal do Rio de Janeiro, Rio de Janeiro, RJ, Brazil, 2009.
15. Prata, D.M.; Schwaab, M.; Lima, E.L.; Pinto, J.C. Nonlinear dynamic data reconciliation and parameter estimation through particle swarm optimization: Application for an industrial polypropylene reactor. *Chem. Eng. Sci.* **2009**, *64*, 3953–3967. [CrossRef]
16. Benqlilou, C. Data Reconciliation as a Framework for Chemical Processes Optimization and Control. Ph.D. Thesis, Universitat Politècnica de Catalunya, Barcelona, Spain, 2004.
17. Narasimhan, S.; Jordache, C. *Data Reconciliation and Gross Error Detection: An Intelligent Use of Process Data*; Gulf Professional Publishing: Houston, TX, USA, 1999.
18. Kuehn, D.R.; Davidson, H. Computer control II. Mathematics of control. *Chem. Eng. Prog.* **1961**, *57*, 44–47.
19. Crivellari, G.P. Modelagem Matemática e Simulação de um Permeador de Gases para Separação de CO_2 de Gás Natural. Master's Thesis, Universidade de São Paulo, São Paulo, SP, Brazil, 2016.
20. Dortmundt, D.; Doshi, K. *Recent Developments in CO_2 Removal Membrane Technology*; UOP LLC: Des PLaines, IL, USA, 1999; pp. 1–30.
21. Rackley, S.A. *Carbon Capture and Storage*; Butterworth-Heinemann: Oxford, UK, 2017.
22. Dias, A.C.S.; De Sá, M.C.C.; Fontoura, T.B.; Menezes, D.Q.; Anzai, T.K.; Diehl, F.C.; Thompson, P.H.; Pinto, J.C. Modeling of spiral wound membranes for gas separations. Part I: An iterative 2D permeation model. *J. Membr. Sci.* **2020**, *612*, 118278. [CrossRef]
23. Lashkari, S.; Kruczek, B. Reconciliation of membrane properties from the data influenced by resistance to accumulation of gasses in constant volume systems. *Desalination* **2012**, *287*, 178–189. [CrossRef]
24. Feital, T.; Prata, D.M.; Pinto, J.C. Comparison of methods for estimation of the covariance matrix of measurement errors. *Can. J. Chem. Eng.* **2014**, *92*, 2228–2245. [CrossRef]
25. Stanley, G.; Mah, R. Observability and redundancy classification in process networks: Theorems and algorithms. *Chem. Eng. Sci.* **1981**, *36*, 1941–1954. [CrossRef]
26. McKinney, W. Pandas, Python Data Analysis Library. 2020. Available online: http://pandas.pydata.org (accessed on 11 July 2020).
27. McKinney, W. Pandas: A foundational Python library for data analysis and statistics. *Python High Perform. Sci. Comput.* **2011**, *14*. Available online: https://www.dlr.de/sc/en/desktopdefault.aspx/tabid-7649/13008_read-32724/ (accessed on 11 July 2020).
28. Kuriakose, J. Using HDF5 with Python. 2017. Available online: https://medium.com/ (accessed on 11 July 2020).
29. Zaitsev, I. The Best Format to Save Pandas Data. 2019. Available online: https://towardsdatascience.com/ (accessed on 11 July 2020).
30. Galarnyk, M. Understanding Boxplots. 2018. Available online: https://towardsdatascience.com/ (accessed on 11 July 2020).
31. Feital, T.; Pinto, J.C. Use of variance spectra for in-line validation of process measurements in continuous processes. *Can. J. Chem. Eng.* **2015**, *93*, 1426–1437. [CrossRef]
32. Waskom, M. Seaborn: Statistical Data Visualization. Python 2.7 and 3.5. 2020. Available online: https://seaborn.pydata.org (accessed on 11 July 2020).
33. Box, G.E.; Jenkins, G.M.; Reinsel, G.C.; Ljung, G.M. *Time Series Analysis: Forecasting and Control*; John Wiley & Sons: Hoboken, NJ, USA, 2015.
34. Peixeiro, M. The Complete Guide to Time Series Analysis and Forecasting. 2019. Available online: https://towardsdatascience.com (accessed on 11 July 2020).
35. Evsukoff, A.G. *Inteligência Computacional—Fundamentos e Aplicações*; E-papers: Rio de Janeiro, RJ, Brazil, 2020.
36. Vanhatalo, E.; Kulahci, M.; Bergquist, B. On the structure of dynamic principal component analysis used in statistical process monitoring. *Chemom. Intell. Lab. Syst.* **2017**, *167*, 1–11. [CrossRef]
37. Himmelblau, D.M. *Fault Detection and Diagnosis in Chemical and Petrochemical Processes*; Elsevier Science Ltd.: Amsterdam, The Netherlands, 1978; Volume 8.
38. Romagnoli, J.A.; Sanchez, M.C. *Data Processing and Reconciliation for Chemical Process Operations*; Academic Press: Cambridge, MA, USA, 1999; Volume 2.

39. Veverka, V.V.; Madron, F. *Material and Energy Balancing in the Process Industries: From Microscopic Balances to Large Plants*; Elsevier: Amsterdam, The Netherlands, 1997; Volume 7.
40. Chen, J.; Romagnoli, J.A. A strategy for simultaneous dynamic data reconciliation and outlier detection. *Comput. Chem. Eng.* **1998**, *22*, 559–562. [CrossRef]
41. Reilly, P.; Carpani, R. Application of statistical theory of adjustment to material balances. In Proceedings of the 13th Canadian Chemical Engineering Congress, Montreal, QC, Canada, 1963; pp. 21–23.
42. Schwaab, M. *Análise de Dados Experimentais: I. Fundamentos de Estatística e Estimação de Parâmetros*; E-papers: Rio de Janeiro, RJ, Brazil, 2007.
43. Lotufo, F.A.; Garcia, C. Sensores Virtuais ou Soft Sensors: Uma Introdução. In Proceedings of the 7th Brazilian Conference on Dynamics, Control and Applications, Presidente Prudente, Brazil, 5–9 May 2008; pp. 1–9.
44. Seader, J.D.; Henley, E.J.; Roper, D.K. *Separation Process Principles*, 3rd ed.; Wiley: New York, NY, USA, 2010.
45. Bilogur, A. Missingno: A missing data visualization suite. *J. Open Source Softw.* **2018**, *3*, 547. [CrossRef]
46. Lewinson, E. Violin Plots Explained—Learn How to Use Violin Plots and What Are Their Advantages over Box Plots! 2019. Available online: https://towardsdatascience.com/ (accessed on 11 July 2020).
47. Câmara, M.M.; Quelhas, A.D.; Pinto, J.C. Performance evaluation of real industrial RTO systems. *Processes* **2016**, *4*, 44. [CrossRef]
48. Özyurt, D.B.; Pike, R.W. Theory and practice of simultaneous data reconciliation and gross error detection for chemical processes. *Comput. Chem. Eng.* **2004**, *28*, 381–402. [CrossRef]

Article

Copper Oxide Spectral Emission Detection in Chalcopyrite and Copper Concentrate Combustion

Gonzalo Reyes [1,*], Walter Diaz [1], Carlos Toro [2], Eduardo Balladares [1], Sergio Torres [3], Roberto Parra [1] and Alejandro Vásquez [1]

1 Metallurgical Engineering Department, University of Concepción, Concepción CCP4070386, Chile; walterdiaz@udec.cl (W.D.); eballada@udec.cl (E.B.); rparra@udec.cl (R.P.); alejanvasquez@udec.cl (A.V.)
2 Dirección de Investigación, Technological University of Chile INACAP, Avenida El Condor 720, Ciudad Empresarial, Huechuraba, Santiago RM858000, Chile; ctoron@inacap.cl
3 Electrical Engineering Department, University of Concepción, Concepción CCP4070386, Chile; sertorre@udec.cl
* Correspondence: gonzaloreyes@udec.cl

Abstract: In this research, the spectral detection of copper oxide is reported from different combustion tests of chalcopyrite particles and copper concentrates. Combustion experiments were performed in a bench reactor. In all the tests, the radiation emitted from the sulfide particle reactions was captured in the VIS–NIR range. The obtained spectral data were processed by using the airPLS (adaptive iteratively reweighted penalized least squares) algorithm to remove their baseline, and principal component analysis (PCA) and the multivariate curve resolution method alternate least squares (MCR-ALS) methods were applied to identify the emission lines or spectral bands of copper oxides. The extracted spectral pattern is directly correlated with the emission profile reported in the literature, evidencing the potential of using spectral analysis techniques on copper sulfide combustion spectra.

Keywords: combustion; optical sensors; spectroscopy measurements; signal detection; digital processing; principal component analysis; multivariate data analysis; curve resolution

Citation: Reyes, G.; Diaz, W.; Toro, C.; Balladares, E.; Torres, S.; Parra, R.; Vásquez, A. Copper Oxide Spectral Emission Detection in Chalcopyrite and Copper Concentrate Combustion. *Processes* **2021**, *9*, 188. https://doi.org/10.3390/pr9020188

Received: 21 December 2020
Accepted: 16 January 2021
Published: 20 January 2021

1. Introduction

The need for better process monitoring, control, and optimization of industrial reactors, and to secure environmental sustainability, has driven the development of new technologies in metallurgical processes. For example, many conventional reactors in the metallurgical industry have undergone improvements and optimization in their design and operation. Advanced sensing techniques, such as those presented in this work, increase understanding of the physical chemistry phenomena that take place in the process, facilitating these process improvements.

Flash smelting technology produces over 50% of the primary copper in the world [1] and it is widely used by the top producers of primary copper, such as China, Japan, Chile, and Russia, who represent more than 60% of the copper production through the pyrometallurgy processes. The flash furnace used in this type of process originated in Finland at the end of World War II and presented an excellent alternative to the energy shortage that existed in post-war Europe since, at that time, the intensive demand for thermal energy was mainly supplied by the combustion of hydrocarbons [2]. Over the years, flash smelting furnaces have become the most widely used technology in new smelting companies due to their ability to take advantage of the heat released through combustion reactions.

At present, new processes and technologies have emerged as competitors to flash combustion [3,4]. The flash furnace continues to be widely used for both its energy efficiency and its environmental performance. Since these furnaces appeared on the market, numerous studies have been reported to improve the understanding of physicochemical phenomena and thus optimize their performance. One example is the work of Jorgensen in 1981 [5], who measured the temperature of pyrite particles reacting inside a laminar flow reactor, all of which was facilitated by optical pyrometry techniques and particularly by the two-color pyrometry method.

Over the years, it led to the improvement of measurement techniques as can be seen in the work of Tuffrey et al. [6], who also measured the temperature of pyrite, registering much higher values (maximum 3127 °C) than those reported by Jorgensen. The latter was attributed to the higher speed of the pyrometer data acquisition system used by Tuffrey. Subsequently, many studies on combustion kinetics, reaction mechanisms, and particle size measurement, among others, have applied two-color pyrometry as a support in their experiences [7–10].

The latest works reported on spectral measurements of sulfide combustion in laboratory tests have been carried out by a group of researchers from the Metallurgical and Electrical Engineering Departments of the Universidad de Concepción, Chile. This group has made use of optical pyrometry techniques to retrieve the flame temperature of concentrates and pure mineral species under combustion conditions. In these investigations, different spectral ranges of all the captured radiation were used and a cooled optical fiber was specially designed for high-temperature applications [11,12].

The application of spectroscopy at an industrial level is scarce, and one of the few examples is the mentioned by Sun et al. [13], who measured a coal-fired flame in a boiler, while the only reported measurements in an industrial flash furnace were performed at the Chagres smelter, Anglo American [12]. These tests were carried out by introducing a cooled optical probe inside the furnace through a sight glass positioned on the roof of the reaction tower, which allowed this optical system to capture flame combustion similarly to those found in laboratory experiments [14,15].

2. Spectral Emissions of Iron and Copper Oxides

In the last 50 years, different researchers in the field of emission spectroscopy have found the ranges of wavelengths in which certain chemical elements and molecules emit radiation, thus determining their spectral profiles. Among them are the emission spectra profiles of iron oxide and copper oxide. Although in these investigations, these emission profiles have not been obtained from sulfide combustion tests, they can be used as a reference when determining the presence of these oxides in a copper and iron sulfide emission spectrum, as Toro et al. did in their research on high-grade pyrite combustion spectra [16].

2.1. Spectral Emission of Iron Oxide

In the case of iron oxide (FeO), West and Broida [17] conducted a study in the visible spectral range (VIS). In their work, they reported that the FeO molecule emits a continuous spectrum between 500 and 700 nm. The emission profile contains wide spectral bands centered at 570, 590, and 620 nm. In the case of the wavelength of 590 nm, it is usually not perceived due to the interference of the sodium emission [18]. Toro et al. applied multivariate techniques on high-grade pyrite emission spectra. Through principal component analysis (PCA) and multivariate curve resolution method alternate least squares (MCR-ALS) analysis, the combustion spectrum was deconvolved obtaining pure emission profiles of Na, FeO, and Fe_3O_4 [16].

2.2. Spectral Emission of CuO_x

The study presented by Gole in the early 1990s was the first to report the emission profile of copper oxide, with spectral bands located in the range of 580 to 720 nm [19]. For their part, Knapp et al. proposed a spectral profile of copper oxide and aluminum oxide in their study of the emission spectroscopy of the termite combustion flame [20]. In this study, they presented four peaks located at 606, 616, 629 and 640 nm, which represent CuO_x emissions.

This research shows the application of an optoelectronic technique to measure the combustion flame produced by copper concentrates. The obtained spectral information was related to the physicochemical processes. The industrial process control could be performed with the application of this optoelectronic technique, allowing the operators to visualize important indicators of the processes online and in real-time.

The objective of this research is to study the emission spectra of chalcopyrite and a real copper concentrate when it is exposed to combustion conditions using an optical measurement system.

3. Methodology

The combustion tests were performed to detect spectral characteristics of interest associated with copper oxides. The emission spectrum of a flame represented by $I(\lambda, T) = I_c(\lambda, T) + I_d(\lambda, T) + I_{mol}(\lambda, T) + n$, where λ is a wavelength sample, T is the flame temperature, I_c is the baseline, I_{mol} and I_d are components associated with molecular and elementals emissions (discontinuous), respectively [21], and n is a noise component related to the electronics and detectors themselves. The spectral range and the number of sampled wavelengths were defined by the spectrometer. Research methods and considerations are represented below.

Experimental Setup and Sample Characteristic

Briefly, the combustion experiments were performed in a bench-scale setup consisting of a drop-tube reactor under laminar flow conditions and heated by a controlled electrical furnace at 1273 K (Figure 1). The spectral data acquisition was composed of cooled optical fiber (Avantes®) and a VIS–NIR spectrometer (Ocean Optics USB4000®) which is sensitive in the spectral range between 344 and 1034 nm and can deliver 3648 wavelengths samples in such a range.

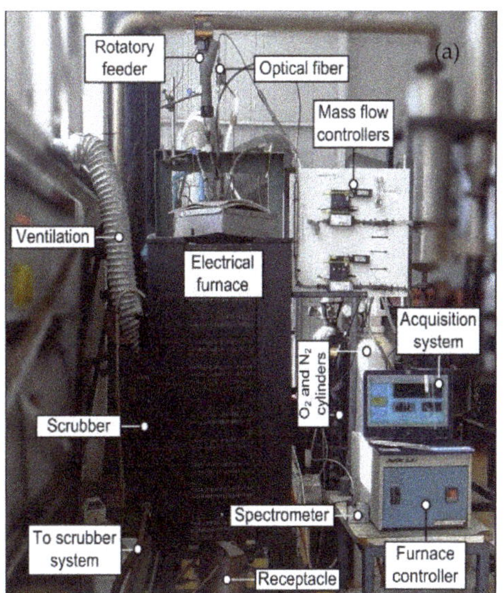

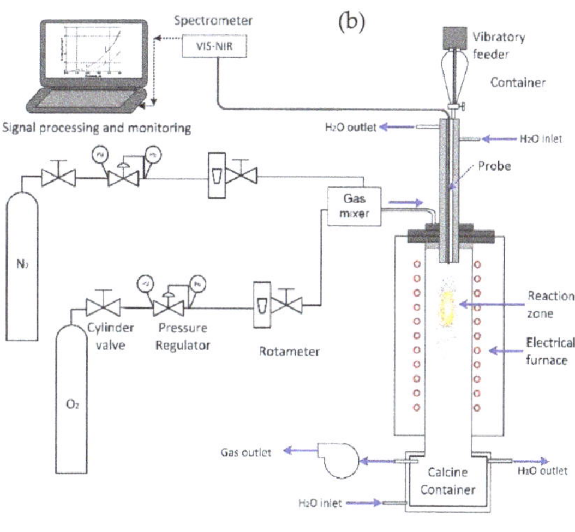

Figure 1. (**a**) Experimental setup [15] and (**b**) schematic diagram.

The measurement and interpretation of the emitted radiation from the cloud of particles in the reactor is a difficult task, especially for copper concentrates, since it involves many chemical and physical processes and interactions between the particles and the particles with their surroundings. In Figure 2, a sample of the incandescent cloud generated during combustion is depicted. This figure also summarizes the spectral data acquisition and the implemented methodologies to process and analyze the spectral information.

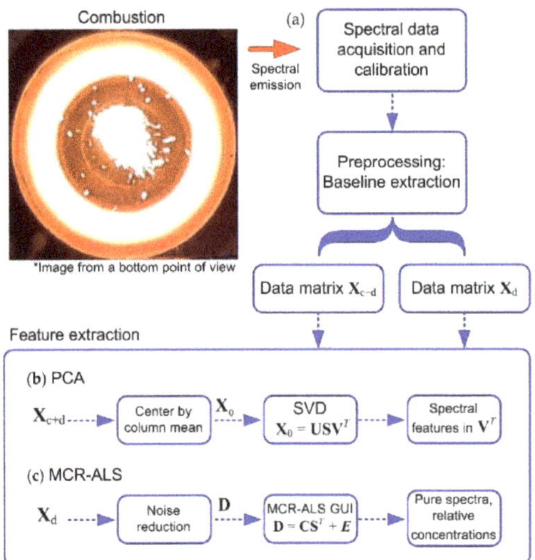

Figure 2. Data acquisition and analysis pipeline. (**a**) Spectral data acquisition and preprocessing. (**b**) Principal component analysis (PCA) features extraction. (**c**) Multivariate curve resolution method alternate least squares (MCR-ALS) application for separation of pure spectral signals. Note that the depicted combustion image is only for visualization purposes, and the optical fiber is located at the central centered position.

In this research, spectral data processing with multivariate data analysis methods were implemented to extract important characteristics related to the formation of copper oxides. PCA was used for an exploratory analysis on the data matrix containing all the spectral information, whilst MCR-ALS was used to deconvolve the original emission spectrum based on its pure spectral components. The airPLS (adaptive iteratively reweighted penalized least squares) algorithm baseline correction [16,22] was used to unmix the continuous and discontinuous spectral components to ensure that the analysis was on the chemical behavior of the combustion and not on its energy. The data analysis was carried out in MATLAB™ (MathWorks, Inc., Natick, MA, USA) with the PLS Toolbox 8.9 (Eigenvector Research, Inc., Manson, WA, USA) and MCR-ALS GUI 2.0.

The chalcopyrite sample was purchased from Ward's Science® (Rochester, NY, USA), while the concentrate was donated by a Chilean mining company. The predominant mineralogical composition of the concentrates is detailed in Table 1.

Table 1. Copper concentrate mineralogical composition.

Mineral	Conc. A (wt %)	Conc. B (wt %)
$CuFeS_2$	32.71	66.7
FeS_2	45.52	16.52
Cu_5FeS_4	3.23	8.35
CuS	0.85	0.45
Others	17.69	7.98

In this research, eight sets of tests were carried out. Six of them consisted of the combustion of chalcopyrite at different particle sizes, corresponding to 105 to 149, 74 to 105, 53 to 74, 44 to 53, 37 to 44, and <37 µm. From now on, these samples will be referred to as CpyA, CpyB, CpyC, CpyD, CpyE, and CpyF respective to the previous order. The concentrates used had a granulometric distribution with a p80 of ~36 μm for sample Conc. A and ~47 μm for sample Conc. B. A Sympatec Helos-Succel™ particle size analyzer based on a diffraction laser was used for the particle size analysis. All the laboratory experiments were carried out under similar conditions. In addition, we worked with oxygen supply three times over the stoichiometric quantity to ensure the total oxidation of the chalcopyrite and the sulfurized species in the concentrates to form copper oxides. The calcines obtained in each test were analyzed using scanning electronic technology. For more details about the methodology and special considerations, see Toro et al. [16].

4. Results

4.1. Spectral Measurement from Chalcopyrite

The first step was to calculate the average spectra obtained in each test, which are depicted in Figure 3. The average calibrated spectrum of chalcopyrite combustion for each size is shown in Figure 3. It is observed that the behavior of the spectral irradiance was different for each size of chalcopyrite analyzed. It is possible to observe that for the finer sizes, the intensity of the spectrum is greater; this is because finer particles are completely oxidized, releasing more energy. A cloud of fine particles represents a greater surface area compared to a cloud of "coarse" particles, which also has a higher oxygen consumption which in turn translates into greater heat generation. For their part, the larger particles take time to burn in the reaction zone, resulting in a lower intensity emission spectrum. Even so, this behavior is not possible to observe in the smallest sample size (CpyF), which may be related to an agglomeration problem. The smallest size tends to form clumps in the cloud combustion, increasing its particle size and thus behaving as coarse particles.

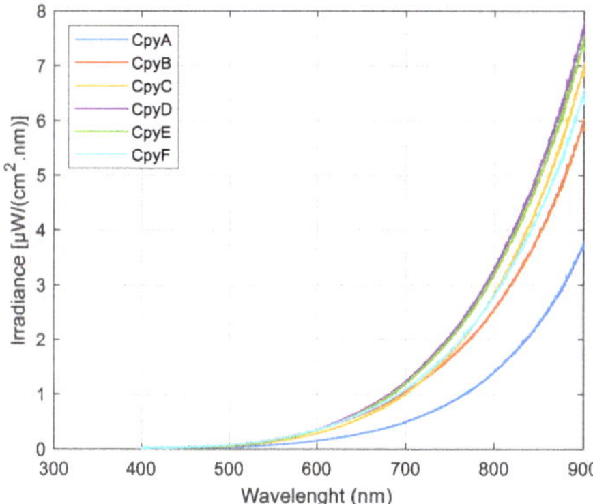

Figure 3. Average emission spectrum of chalcopyrite combustion at different sizes.

It is not possible to observe emission peaks in the chalcopyrite spectra compared to those reported by Toro et al. [16] in their combustion experiments with pyrite. This is likely because chalcopyrite in an oxidizing environment releases less energy than pyrite under the same conditions, which results in a lower temperature combustion flame, making it more difficult to identify discontinuities. Some of the reactions that occurred during the oxidation of chalcopyrite and pyrite are shown in Table 2, a value of $\Delta H < 0$ indicates that the reaction is exothermic (releases thermal energy), while a value of $\Delta H > 0$ indicates a reaction is endothermic (absorbs thermal energy).

Table 2. Typical reactions during chalcopyrite and pyrite combustion [1].

Reactions	$\Delta H^o_{1000\,^\circ C}$ [Kcal/mol]
$CuFeS_{2(s)} \rightarrow \frac{1}{2}Cu_2S_{(s)} + FeS_{(s)} + \frac{1}{4}S_{2(g)}$	12.19
$FeS_{2(s)} \rightarrow FeS_{(s)} + \frac{1}{2}S_{2(g)}$	34.79
$FeS_{(s)} + \frac{1}{2}O_{2(g)} \rightarrow FeO_{(s)} + \frac{1}{2}S_{2(g)}$	−26.36
$FeS_{(s)} + \frac{5}{3}O_{2(g)} \rightarrow \frac{1}{3}Fe_3O_{4(s)} + SO_{2(g)}$	−136.46
$Cu_2S_{(s)} + O_{2(g)} \rightarrow 2CuO_{(s)} + \frac{1}{2}S_{2(g)}$	−40.82
$S_{2(g)} + 2O_{2(g)} \rightarrow 2SO_{2(g)}$	−172.52

[1] Values are calculated with HSC Chemistry®.

In this work, multivariate techniques were implemented to detect weak spectral emission profiles. The application of PCA and the analysis of loadings of the combustion spectra of chalcopyrite, allowed us to find peaks at 606 and 616 nm, which are associated with copper oxides [19,20]. These peaks were observed slightly from the loading of the PC3 of the CpyD sample, being more visible in the CpyF sample (Figure 4d–f). Moreover, the peaks associated with Na and K emissions were observed, around 589 and 767 nm, respectively.

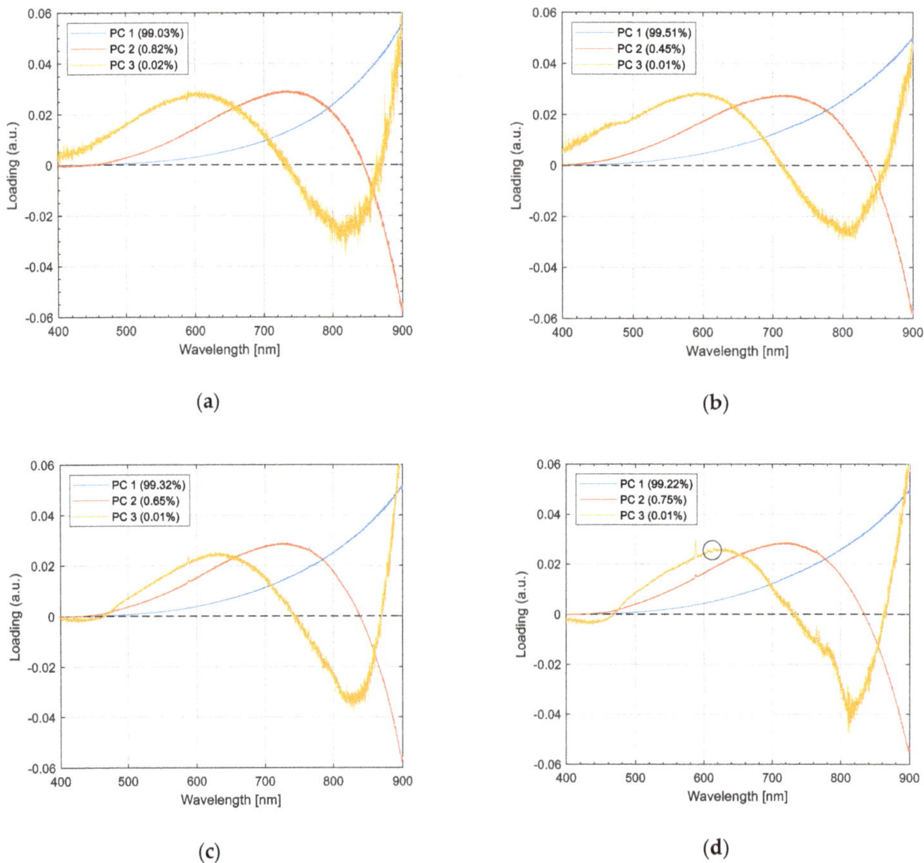

Figure 4. *Cont.*

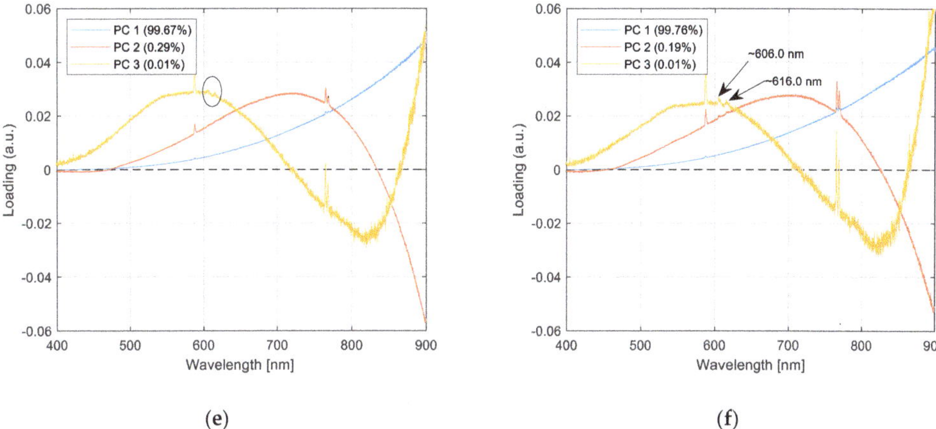

(e) (f)

Figure 4. PCA analysis on the combustion spectrum of chalcopyrite of different sizes. (**a**) CpyA, (**b**) CpyB, (**c**) CpyC, (**d**) CpyD, (**e**) CpyE, (**f**) CpyF.

4.2. Spectral Measurement from Copper Concentrates

In Figure 5, the average spectral signals of the combustion of copper concentrates are represented. It can be noted that the signal associated with the sample Conc. A presents greater intensity of irradiance compared to the spectral signals of the samples of chalcopyrite and Conc. B. In addition, the spectral lines of emission of Na and K are presented with greater intensity.

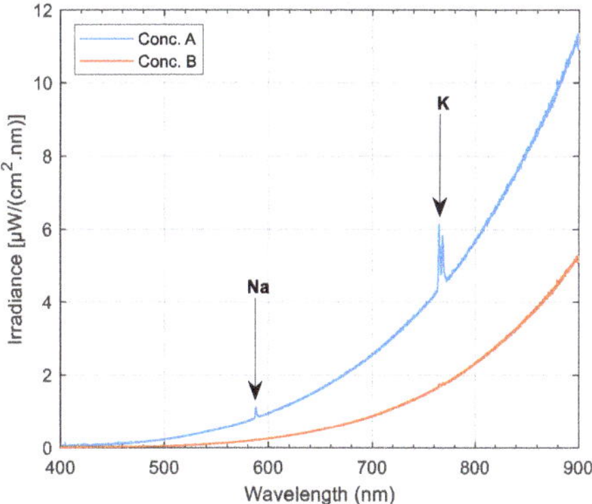

Figure 5. Average emission spectrum from the combustion of copper concentrates.

Through exploratory analysis with PCA to each set of spectral data of the concentrates and the subsequent graphic representation of the first three loadings (Figure 6a,b), it is revealed that the emission spectral lines of Conc. A are more intense than those of Conc. B and, in addition, there are certain characteristics at ~606 and ~616 nm that are associated with copper oxides, while close to the potassium emission, two spectral emissions appear at 779.1 and 793.9 nm, the same ones that appear in the pyrite combustion spectral signals reported by Toro et al. [16].

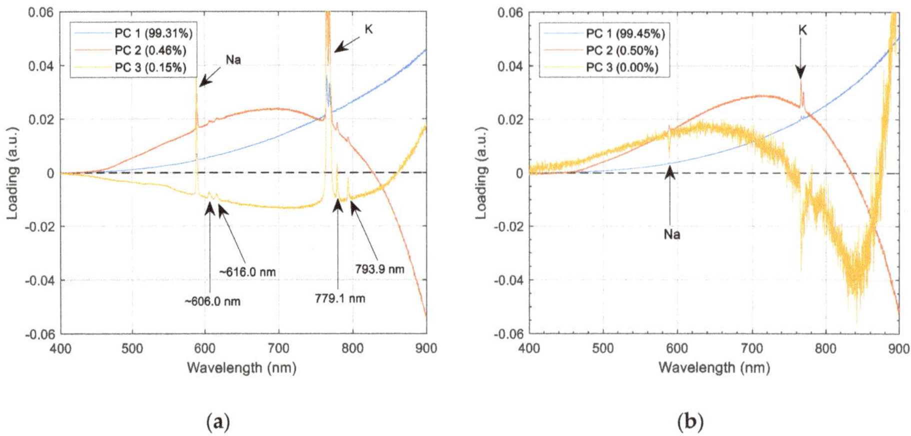

(a) (b)

Figure 6. PCA loadings of the (**a**) Conc. A and (**b**) Conc. B average emission spectrum.

4.3. MCR-ALS Application

From the PCA analysis of the spectral data of chalcopyrite and copper concentrates, the spectral data of CpyF and Conc. A were selected, which have discontinuous sections in which the spectral characteristics are related to copper oxides. The intensity of these spectral characteristics is the result of the temperature reached by the combustion cloud of each sample. Table 3 indicates the average combustion temperatures estimated by two-color pyrometry, and according to [23], a good selection of the sampling wavelengths to apply this method are 650 and 750 nm.

Table 3. Average estimated temperatures of the combustion flames.

Sample	Temperature (°C)
CpyA	1002.5
CpyB	1047.2
CpyC	1061.5
CpyD	1109.5
CpyE	1160.2
CpyF	1310.9
Conc. A	1485.2
Conc. B	1316.8

Continuing with the analysis, the airPLS baseline extraction algorithm was applied to the two selected datasets, thereby separating the dataset to extract the continuous and discontinuous emissions. Since the spectral evidence of copper presence is associated with wavelengths of ~606 and ~616 nm, we proceeded to limit the discontinuous section in the spectral range between 540 and 650 nm, while the noise was reduced through the algorithm of Savitzky–Golay (SG) [24]. Figure 7 summarizes the discontinuities profile of CpyF and Conc. A.

Thus, to obtain a spectral profile associated with copper oxides, the spectra of the sample Conc. A were selected because they present greater irradiance intensity.

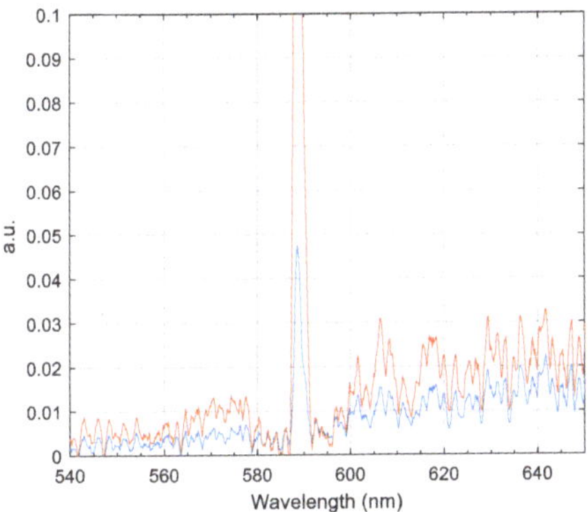

Figure 7. Discontinuities profile of Conc. A (orange) and CpyF (blue).

MCR-ALS can perform multiset analysis of data organized in a single matrix [25] and, therefore, for this study, we proceeded to create a matrix in the range of 540 to 650 nm that contains the discontinuous samples of the Conc. A sample and the spectral signals of sodium (Na), FeO, and Fe_3O_4 were those obtained by Toro et al. [16] (Figure 8). The aim of this was to make the MCR-ALS algorithm separate the signals of the known species (FeO, Fe_3O_4, and sodium) from those that are not, including the copper oxide profile.

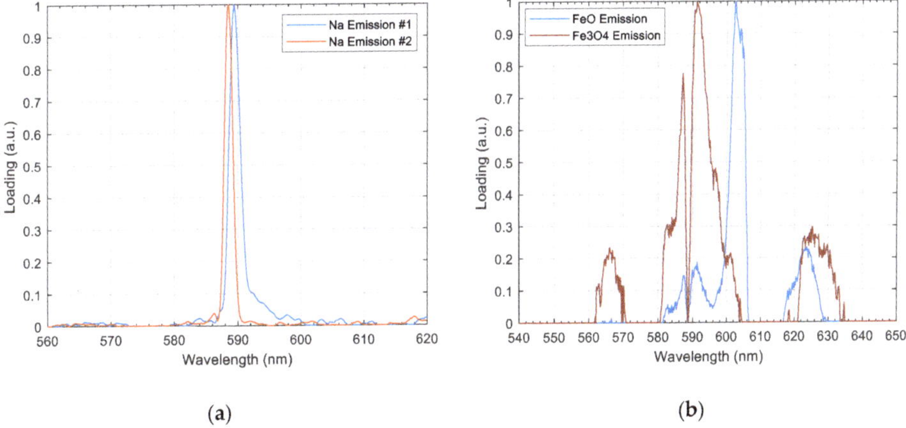

Figure 8. Spectral profiles (**a**) associated with sodium and (**b**) associated with FeO and Fe_3O_4 [16].

The MCR-ALS method was applied to the new matrix, as detailed in Figure 9, with 10 components and a "non-negativity" restriction on spectral profiles and concentrations.

With these settings on the MCR-ALS GUI, 92.5244% of the data variance and an error of fit (PCA) of 3.0201% were achieved with 70 iterations [25]. As expected, within the 10 obtained profiles were those associated with emissions of sodium (Na), FeO, and Fe_3O_4, in addition to a profile whose characteristics resemble the profile reported by Knapp, as seen in Figure 10.

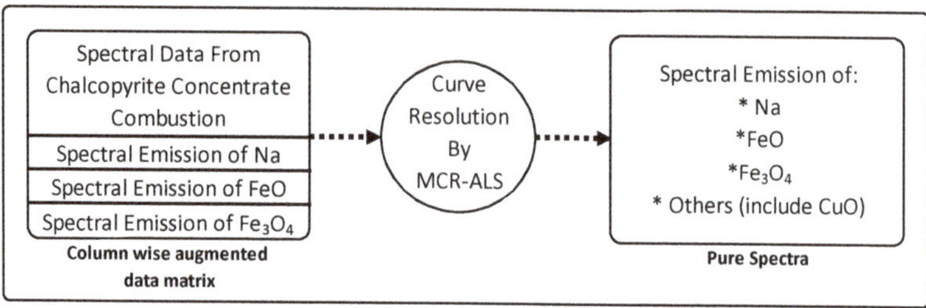

Figure 9. Data processing schemes using MCR-ALS.

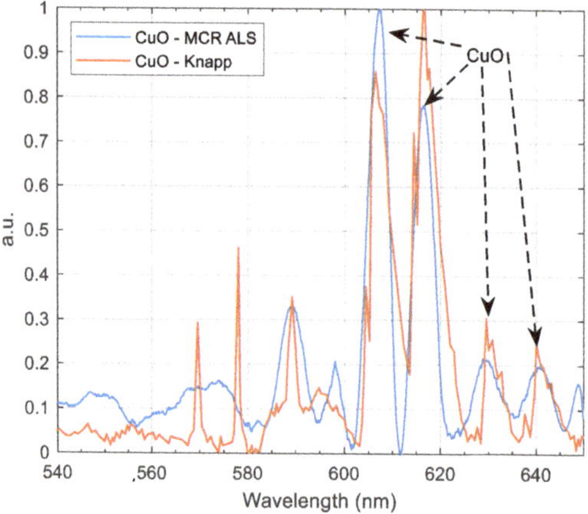

Figure 10. Spectral emission measurements obtained via MCR-ALS and reference spectral characteristics.

On the other hand, the phase analysis of the calcine of the sample Conc. A is presented in Table 4, a majority composition of Fe and Cu oxides was observed according to what was expected, which was a product of the working conditions. The mineralogical composition of the Conc. A calcine was determined using an integrated automated mineralogy solution, QEMSCAN system (quantitative evaluation of minerals by scanning electron microscopy), produced by the Fei Company (Hillsboro, OR, USA). The Cu_2S generated by the decomposition of species such as bornite or chalcopyrite can be oxidized to form CuO or Cu_2O. The formation of these copper oxides in the flame follows the following transition $Cu_2S \rightarrow Cu_2O \rightarrow CuO$, this can be corroborated with the analysis of the stability diagram Cu-S-O in Figure 11.

Table 4. Calcine of the sample Conc. A mineralogical composition.

Minerals	wt %
Fe_2O_3/Fe_3O_4	29.87
FeS_2	1.09
CuS	0.84
CuO/Cu_2O	39.30
$CuFeS_2$	0.11
FeS	0.79
Cu_2S	2.30
Cu_5FeS_4	0.95
SiO_2	4.32
FeO	6.96

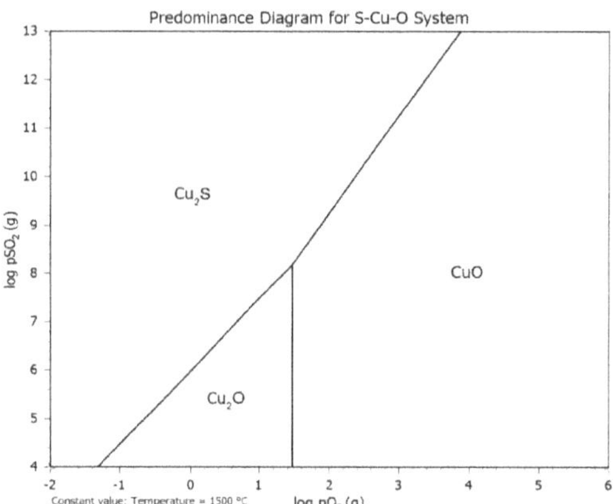

Figure 11. Cu-O-S stability diagram at 1500 °C constructed with HSC® software.

This small thermodynamic analysis ensures the formation of CuO within the working conditions, so the depicted CuO_X emission profile in the combustion spectra of a copper concentrate is effectively the emission profile of copper oxides. The detection of these oxides in the process is important because it will allow having control over copper oxidation in the flame, allowing the operator to adjust process variables preventively. This would avoid greater copper losses in the slag.

5. Conclusions

In this work, a series of multivariate analysis techniques were applied to spectral data obtained during the combustion of chalcopyrite and copper concentrates. The results show that the intensity of the combustion and the spectral characteristics present in the signal depend on the temperature reached by the combustion cloud. The copper concentrates, according to their pyrite content, can reach an intensity such that it allows visualizing some spectral lines, such as those associated with sodium and potassium in the visible range and, even so, the spectral data of all the combustion tests required an exploratory analysis to reveal the presence of spectral lines that are not visible in the average spectrum. The sample that provided the most information on the possible oxidized phases that were generated in the combustion cloud was a high pyrite concentrate sample. The spectral information provided was not only limited to copper oxide emission lines at ~606 and ~616 nm but, instead, there are two emission lines at 779.1 and 793.9 nm that may be associated with iron species, according to some authors. Finally, the application of the MCR-ALS method to a spectral dataset measured from the combustion of a high pyrite concentrate and previous knowledge of spectral bands of Na, FeO, and Fe_3O_4 patterns allowed us to find a spectral profile associated with copper oxides in the range of 540 and 650 nm.

This research shows that the use of spectral measurement techniques is a powerful tool to relate the spectral emission with physicochemical processes in a combustion flame of copper concentrates. The online and real-time identification of copper oxide formation in a flash furnace flame could be a great advance for the non-ferrous mineral in the pyrometallurgical industry, improving process control by measuring a few process variables such as copper content in mate or slag.

Author Contributions: Conceptualization, W.D., G.R. and C.T.; methodology, W.D., G.R., and C.T.; software, C.T.; validation, W.D., C.T.; formal analysis, W.D., G.R., C.T., and S.T.; investigation, W.D. and G.R.; resources, R.P., E.B. and S.T.; data curation, W.D.; writing—original draft preparation, W.D. and G.R.; writing—review and editing, C.T. and A.V.; visualization, W.D.; supervision, C.T., S.T., E.B. and R.P.; project administration, E.B.; funding acquisition, E.B. All authors have read and agreed to the published version of the manuscript.

Article

AOC-OPTICS: Automatic Online Classification for Condition Monitoring of Rolling Bearing

Hassane Hotait [1], Xavier Chiementin [1,*] and Lanto Rasolofondraibe [2,*]

[1] ITheMM, Institut de Thermique, Mécanique et Matériaux, University of Reims Champagne Ardenne, Moulin de la Housse, 51687 Reims Cedex 2, France; hassane.hotait@etudiant.univ-reims.fr

[2] CReSTIC, University of Reims Champagne Ardenne, Moulin de la Housse, 51687 Reims Cedex 2, France

* Correspondence: xavier.chiementin@univ-reims.fr (X.C.); lanto.rasolofondraibe@univ-reims.fr (L.R.)

Received: 3 April 2020; Accepted: 15 May 2020; Published: 20 May 2020

Abstract: Bearings are essential components in rotating machines. They ensure the rotation and power transmission. So, these components are essential elements for industrial machines. Thus, real-time monitoring is required to detect a possible anomaly, diagnose the failure of rolling bearing and follow its evolution. This paper presents a methodology for automatic online implementation of fault diagnosis of rolling bearings, by AOC-OPTICS (automatic online classification monitoring based on ordering points to identify clustering structure, OPTICS). The algorithm consists of three phases namely: initialization, detection and follow-up. These phases use the combination of features extraction methods, smart ranking, features weighting and classification by the OPTICS method. Two methods have been integrated in the dimension reduction step to improve the efficiency of detection and the followed of the defect (relief method and t-distributed stochastic neighbor embedding method). Thus, the determination of the internal parameters of the OPTICS method is improved. A regression model and exponential model are used to track the fault. The analytical simulations discuss the influence of parameters automation. Experimental validation shows detection with 100% accuracy and regression models of monitoring reaching $R^2 = 0.992$.

Keywords: rolling bearing; condition monitoring; classification; OPTICS

1. Introduction

The automation of techniques takes place around the world in the manufacturing and processing of industrial sectors [1,2]. In the industrial and rotary machines, the main idea of automation is the monitoring without input parameters. The error is human, and the limitation of inexact input parameters affects the accuracy of monitoring, so it was interesting to make an autonomous method. There is a growing demand for real-time monitoring in the rotary machines to facilitate advanced maintenance programs [3]. Rotary machines are most often made of a significant and critical component: the rolling bearings [4].

The monitoring of rolling bearings gets the scientist's attention; so many methods applied to detect defects such as support vector machine [5], Bayesian network [6] and clustering [7]. Numerous literature reviews are available on monitoring methods [8,9]. From all these used methods, clustering analysis is one of the most remarkable approaches [10–12]. The density-based method is one of them, the clusters of dense regions of data separated from the less dense [11]. The method OPTICS (ordering points to identify clustering structure), subdivided from density-based, has the basic idea to separate clusters by density [13]. In addition, it has the advantage to attain clusters with varied data density. Clustering by OPTICS methods is an unsupervised learning method directly implemented to vibration data. Being thus can be applied directly in the industrial environments without trained by data measured on a machine under a fault condition [7]. Further advantages of the method are its ease of programming and the accomplishment of a good trade-off and achieved the best performances.

In addition, it is fast for small data, used with different density to detect and attain arbitrary and sphere-shaped clusters [14].

Within the framework of bearing monitoring, the OPTICS method integrated dynamic classification processes for real-time monitoring [15]. The algorithm proposes to make a detection of faults from two time features (rms and kurtosis). The monitoring is then carried out using three geometric values, the contour, the distance and the density. However, the process was incomplete and not completely automated, which required an expert.

The extracted features play an essential role in the classification, for that many methods used to eliminate unwanted and unimportant features. The relief method is used to select features for the classification of biomedical data. It eliminates the irrelevant features and to prepare data of rolling bearings for the classification [16]. The Chi-square is another method that has the same aim of the relief to reduce and rank features, this method used ranking features to detect the defect in the rolling bearing [17]. After selecting features and reducing them by eliminating the uncorrelated ones, the importance of dimension reduction comes before starting the classification. In the literature, many methods have applied for dimension reduction, principal component analysis (PCA) [18] and kernel principal component analysis (KPCA) [19], to detect the defect in rolling bearings. A recently developed nonlinear dimensionality reduction technique shows its efficiency in the detection of a fault in rotary based on t-distributed stochastic neighbor embedding (t-SNE) [20].

The parameters specific to OPTICS are also subject to automation. The lack of automation concerns the choice of features according to a library and the internal parameters of OPTICS: ε (cluster radius), *MinPts* (the minimum number of data points needed to cluster) and the distance metric used to calculate instances between arrays [15]. The determination of the parameter values of the OPTICS algorithm can be a challenging task because the parameter values affect the accuracy and precision of the clustering. Many researchers have discussed this topic, and they were looking for ways to solve it. An automated algorithm AE-DBSCAN, proposed by [21], defines ε like the K-nearest neighbor for this *MinPts*. Regarding the choice of distance, [22] show the hardness approximation of data with Euclidean distance in k-means clustering. Manhattan outperforms the Euclidean distance with the k-means method. The aim is to automate calculation of all the parameters and to offer a complete real-time monitoring solution dedicated to the bearings.

This paper proposes an Online One Class Monitoring based on OPTICS Classification for Rolling Bearing, automatic online classification monitoring based on ordering points to identify clustering structure (AOC-OPTICS). The input parameters are limited to the initialization time of the method and the number of signals collected at a monitoring time t. It integrates the detection and monitoring of the evolution of a fault. After an initialization phase, the detection is carried out by a multidimensional analysis with extraction, ranking (relief method) selection (t-SNE) and classification (OPTICS) of one class clustering type. The follow-up is carried out when creating a new class. In this phase, geometric parameters from this class are proposed and discussed due to regressions models.

This paper is organized as follows. Section 1 introduces the context of the monitoring of the rotating elements and presents the bibliographical review on the contributions and the limits of the classification methods. Section 2 describes the OPTICS method and highlights the parameters to be automated. Section 3 presents the general methodology for automatic monitoring and follow-up of the healthy state of a bearing. Section 4 assesses the relevance of the methodology on data simulating the initiation and growth of a defect on the outer ring of a bearing. The automated parameters and their influences are discussed. Tracking parameters are defined, and mathematical laws are established. Section 5 corresponds to an experimental validation on a test bench. Section 6 concludes this study.

2. Classification Method OPTICS

OPTICS (ordering points to identify clustering structure) is a hierarchical clustering algorithm that relies on a density notion [13]. The application of this method is not limited to one field. It used in many fields and areas of biology, astronomy, topology, and recently for the detection of the defect in rolling bearings in rotary machines [15]. This method is capable of regrouping the base of data into an order of points with different parameter settings, and then detecting a meaningful difference of data with varied density by producing a request of data that is spatially closed to each other and can become a neighbor. It can separate considerable objects from noise and identify all possible levels of clusters. The main idea for the OPTICS algorithm is that for each point of a cluster the neighborhood of a given radius (ε) has to contain at least a minimum number of points ($MinPts$), where ε and $MinPts$ are input parameters. The concept of OPTICS algorithm starts by adding points to the clustered data in arbitrary shape and then to continue by adding points iteratively for developing the final cluster. The addition of points close to each other respecting the ε-neighbor order continues until getting the entire group.

The two-components of OPTICS are the core distance, C_d, and the reachability distance, R_d, Equations (1) and (2). If the number of points in the vicinity of an object, $N_\varepsilon(p)$, is less than $MinPts$, C_d is the distance from p to its $Minpts^{th}$ neighbour, $MinPts_{distance(p)}$. In this case, p is a core-object. The reachability distance of an object o, R_d, is the maximum of the Core Distance of p and the Euclidean distance between o and p. Figure 1a is a representation of the reachability distance and the core distance objects.

$$C_d(\varepsilon, MinPts(p)) = \begin{cases} Undefined & if\ N_\varepsilon(p) < MinPts \\ MinPts_{distance}(p) & else \end{cases} \tag{1}$$

$$R_d(\varepsilon, MinPts(p,o)) = \begin{cases} Undefined & if\ N_\varepsilon(p) < MinPts \\ \max(C_d; distance(o,p)) & else \end{cases} \tag{2}$$

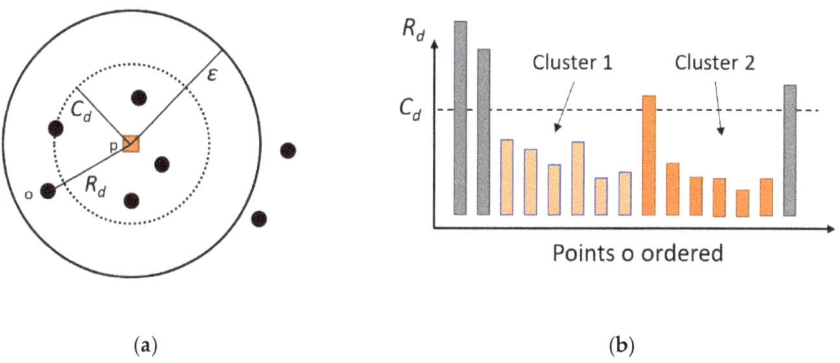

Figure 1. (**a**) Representation of the core distance and reachability distance for $MinPts = 4$. (**b**) Reachability plot.

The number of classes is determined from the reachability plot, Figure 1b. It corresponds to the number of valleys of the graphic representation R_d as a function of the points o ordered.

3. Method

3.1. Global Architecture

The AOC-OPTICS method is developed for monitoring the state of health of a bearing throughout its entire life. It is based on the physical manifestations involved in the deterioration of a bearing. Thus 3 automated phases were proposed, Figure 2. Phase 1 considers that when a bearing is fitted, it is healthy during an interval T_h. This phase allows the initialization of the method. Phase 2 corresponds to the failure detection phase. It is effective if the failure is not detected. Data agglomeration is used for early and reliable fault detection. The third phase corresponds to the follow-up of the evolution of the fault. In view of the evolutionary nature of the fault, the third phase is a follow-up loop of this state by second class geometrical values. It runs until the bearing fails. Each phase is described in the following sections and Table 1 shows the associated pseudo code.

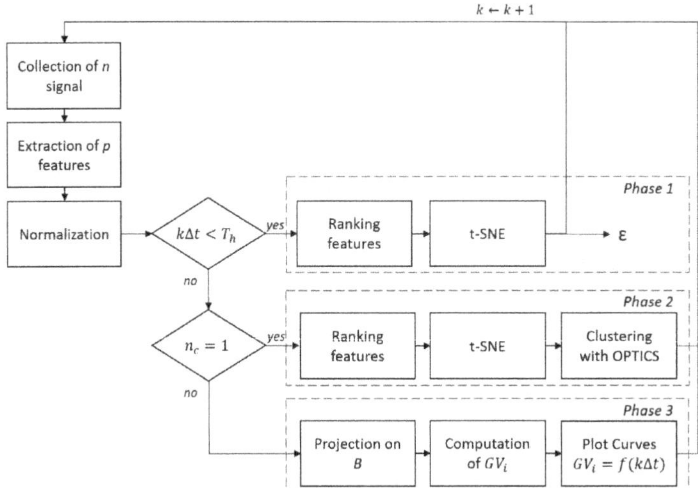

Figure 2. Flowchart of AOC-OPTICS.

Table 1. Pseudocode for automatic online classification monitoring based on ordering points to identify clustering structure (AOC-OPTICS).

Inputs	$T_h, n, \Delta t$ Δt is the interval time between two data collection n is the number of signals collected at time $k\Delta t$ T_h is the time of initialization monitoring
Outputs	n_c, Plot $GV = f(k\Delta t)$ GV are the geometric values n_c is the number of class (=1 for a healthy condition, =2 for a healthy and faulty conditions)
Phase 1 Initialization	$k = 0, n_c = 1$ While $k\Delta t < T_h$ Collection of n signals Computation of p features $[HI]_{p,(k+1)n}$ $k \leftarrow k+1$ End While Normalization $[HI]^{norm}_{p,(k+1)n} with [HI]^{norm}_{i,(k+1)n} = \left(HI_i - \overline{HI_i}\right)/std(HI_i)$ Ranking features Application of t-SNE: $tSNE\left([HI]^{norm}_{p,(k+1)n}\right) = [FI]_{3,(k+1)n}$ Compute ε
Phase 2 Detection	While $n_c = 1$ Collection of n signals Computation of p features $[HI]_{p,(k+1)n}$ Normalization $[HI]^{norm}_{p,(k+1)n} with [HI]^{norm}_{i,(k+1)n} = \left(HI_i - \overline{HI_i}\right)/std(HI_i)$ Ranking features Application of t-SNE: $tSNE\left([HI]^{norm}_{p,(k+1)n}\right) = [FI]_{3,(k+1)n}$ on a basis B Application of OPTICS -> $n_c = 1ou2$ $k \leftarrow k+1$ EndWhile

Table 1. *Cont.*

Phase 3 **Follow**	While $k\Delta t < T_{end}$ where T_{end} is the failure time. Collection of n signals Computation of p features $[HI]_{p,(k+1)n}$ Normalization $[HI]^{norm}_{p,(k+1)n} with [HI]^{norm}_{i,(k+1)n} = \left(HI_i - \overline{HI_i}\right)/std(HI_i)$ Projection on basis B -> $[FI]_{3,(k+1)n}$ Computation of geometrics values GV Curve $GV = f(k\Delta t)$ $k \leftarrow k+1$ EndWhile

3.2. *Phase 1, Initialization*

The first phase is executed for a duration T_h, which is assumed to be a healthy phase of the bearing. For every iteration k, n signals were collected. $p = 17$ features were extracted in the time, spectral and/or time-frequency domains. The use of a multidomain feature in the detection of defect bearing can offer an efficacy diagnosis for different defects of rolling bearings, with variated speed and load. The time domain provides nine characteristic features as descriptive statistics. The statistical indicators are widely used to their relations with significant bearing damages [23]. The frequency-domain allows one to localize and detect the nature of the bearing defect [24]. Six indicators are computed. The time scale domain uses the wavelet method to extract two features [25], Table 2. These indicators are stored in a matrix $[HI]$ where each column corresponds to a signal and each row to an indicator.

Table 2. Computed features. x is the sequence of samples obtained after digitizing the time domain signals, x_i is a signals series for i = 1, 2..., N. $W_S(f_k)$ corresponds to the spectral density of the max coefficients of the continuous wavelet transform. $s(k)$ is a spectrum for $k = 1, 2 \ldots K$, K is the number of spectrum lines, f_K is the frequency value of the kth spectrum value.

TIME DOMAIN	FREQUENCY DOMAIN	TIME SCALE DOMAIN
Root mean square $RMS = \left(\frac{1}{N}\sum_{i=1}^{N}\left(x_i^2\right)\right)^{\frac{1}{2}};$ Kurtosis $Ku = \frac{1}{N}\sum_{i=1}^{N}\left(\frac{x_i-\overline{x}}{\sigma}\right)^4;$ Peak to Peak $x_{PEAK} = max(x_i) - min(x_i)$ Crest factor $x_{CF} = \frac{X_{PEAK}}{X_{RMS}};$ Skewness $Ske = \frac{1}{N}\sum_{i=1}^{N}\frac{(x_i-\overline{x})}{(N-1)\sigma^3}^{3};$ Impulse Factor $(IF) = \frac{Max(\lvert x_i\rvert)}{\frac{1}{N}\sum_{i=1}^{N}\lvert x_i\rvert}$ Standard deviation $(Std) = \frac{1}{N}\sum_{i=1}^{N}(x_i-\overline{x})^{\frac{1}{2}};$ Talaf = log $(Ku + RMS/RMS(1))$; Tikhat = log $((Ku)\,\hat{}\,x_{CF}+RMS$ $/RMS(1))\,\hat{}\,x_{PEAK})$;	Frequency Root mean square $f_{rms} = \left(\sum_{k=1}^{K}\left(\frac{s(k)}{K}\right)\right)^{\frac{1}{2}};$ Weighted Frequency Root mean square $f_{rmsb} = \left(\frac{\sum_{k=1}^{K} f_k^2 s(k)}{\sum_{k=1}^{K} s(k)}\right)^{\frac{1}{2}}$ Frequency Center $f_c = \frac{\sum_{k=1}^{K} f_k s(k)}{\sum_{k=1}^{K} s(k)};$ Weighted Standard deviation frequency $f_{stdb} == \left(\frac{\sum_{k=1}^{K}(f_k-f_c)^2 s(k)}{\sum_{k=1}^{K} s(k)}\right)^{\frac{1}{2}}$ Mean $f_{rmsf} = \frac{1}{K}\sum_{k=1}^{K} s(k)$ Power envelope $PW = \frac{1}{K}\sum_{k=1}^{K} s^2(k)$	Effective value of the frequencies $x_{WRMS} = \left(\frac{\sum_{j=1}^{K} f_k W_S(f_k)}{\sum_{k=1}^{K} W_S(f_k)}\right)^{\frac{1}{2}};$ Average value of the envelope Amplitudes $x_{PCWT} = \frac{\sum_{j=1}^{K} W_S(f_k)}{K-1};$

At the time T_h, the indicator matrix $[HI]_i$ is normalized $[HI]_i^{norm}$, Equation (3). Normalization aims to transform the computed to be on a similar scale.

$$[HI]^{norm}_{i,(k+1)n} = \frac{HI_i - \overline{HI_i}}{std(HI_i)} \tag{3}$$

A ranking step is applied. The ranking features are a significant method for eliminating the unimportant features before reduction the dimension. The massive amount of data calculates features take a long time. To reduce this long process, the method of ranking features is implemented to minimize the number of features, which can make the calculation faster, without touching the accuracy of detecting the defect. For the AOC-OPTICS, two methods are compared in Section 4.3 to eliminate the unnecessary features, with the different amounts of features: the relief method and the chi-square method [26].

Although the nuisance of dimensionality poses serious problems, processing data with high dimensions has an advantage that the data can give more information. The reduction method t-distributed stochastic neighbor embedding (t-SNE) is a powerful dimensional reduction tool, which can reduce functionality dimensions and increase the recognition rate to an overwhelming majority. The dimension reduced to be in three components, which will give more accuracy than two dimensions. The difference accuracy between the dimensions noticed in the representation of amplitude. Due to the use of the three-component in this paper, figures are shown in three dimensions.

Finally, the calculation of ε is done after a reduction in dimension. ε corresponds to the maximum distance between the center of the class, c_h and the $MinPts^{th}$ neighbor, Equation (4).

$$\varepsilon = distance\left(c_h,\ MinPts^{th}\ neighbor\right) \tag{4}$$

The resulting class is a so-called healthy class, noted C_h, with center c_h. This class corresponds to a reference state.

3.3. Phase 2, Detection

The second phase is a step to detect the mechanical failure. The objective of this phase is to detect a new state called the defective class, noted C_f. At each new iteration k, the indicators are extracted, normalized, sorted and reduced as in the previous phase. These features $[FI]_{3,(k+1)n}$ are tested by the OPTICS method to detect or not a second class. If only one class is obtained, which corresponds to the reference state, the algorithm remains in the detection phase, this new data feeds the reference state. If two classes are detected, this new class C_f, is obtained in a plan B, which will be kept for the follow-up phase.

3.4. Phase 3, Follow-up

The third step is carried out in plan B, which is determined in the previous phase. It is important to keep the same plan in order to visualize the evolution of the characteristics. This plan is the best plan to follow the evolution of the bearing failure. With each new series of data, the indicators were extracted, standardized and projected in plan B. From these features, $[FI]_{3,(k+1)n}$, five geometrical parameters GV_i were calculated to monitor over time.

The Calinski-Harabasz index, GV_1, is based on the density and the separated clusters, Equation (5). p is the features number. c_f, c_h are the center of the class C_f, C_h respectively. n_c is the number of clusters. d is the Euclidean distance between x, c_i .

$$GV_1 = \frac{\sum_i pd^2\left(c_f, c_h\right)/(n_c - 1)}{\frac{\sum_{x \epsilon C_h} d^2(x, c_h)}{p - n_c} + \frac{\sum_{x \epsilon C_f} d^2(x, c_f)}{p - n_c}} \tag{5}$$

The Davies–Bouldin index, GV_2, measures the average of similarity between each cluster. The lower index means a better cluster configuration. R_{ij} is the similarity measure of two clusters i *and* j. n_c is the number of clusters.

$$GV_2 = \frac{1}{n_c}\sum_{i=1}^{n_c} R_i \text{ with } R_i = \max_{j=1\ldots n_c,\ i \neq j}\left(R_{ij}\right),\ i = 1\ldots n_c \tag{6}$$

This third parameter, GV_3, calculates the distance between the center cluster of the initial phase C_h with the centre of the fault cluster C_f, where d_M is the Manhattan distance.

$$GV_3 = d_M\left(c_f, c_h\right) \tag{7}$$

Finally, the contour, GV_4, of the cluster is calculated from a convex hull, which is the smallest convex set that contains the points. The density, GV_5, is the number of points of the cluster, C_f for a volume V_f.

4. Numerical Investigation

4.1. Simulated Model

A mathematical model verifies the methodology. It is corresponding to the bearing vibratory signature, with an outer race defect (x_{BPFO}). Equations (8)–(10) [27] describes the used model to present the effect of a rolling element at each passage in the faulty outer race according to time t. The passage of balls in the defect of the outer race creates impacts at the frequency f_{BPFO}. This impact generates an impulse response of the structure with a natural frequency f_0 and a damping μ. Frequency f_{BPFO} depends on the rotation speed of the motor, f_r, and bearing's geometry, Equation (10).

Thus, the model is defined by four-parameters: amplitude A, damping factor μ, rotational speed f_r, the amplitude of the noise signal $b(t)$. The exponential formula implanted in place of amplitude A. Roller bearing simulated is a type SKF 6206 whose characteristics listed in Table 3. Every signal contains 16384 samples (N) with a sampling rate of 51.2 kHz.

$$x_{BPFO}(t) = \sum_{k=1}^{N} A.exp\left(-2\pi\mu f_0\left(t - \frac{k}{f_{BPFO}}\right)\right)\cdot sin\left(2\pi f_0\left(t - \frac{k}{f_{BPFO}}\right)\right) + b(t) \tag{8}$$

$$A = \frac{e^{4\omega} - 1}{e^4} \tag{9}$$

$$f_{BPFO} = \frac{n_b}{2} f_r\left(1 + \frac{d_{ball}}{D_m}\cdot \cos(\alpha)\right) \tag{10}$$

Table 3. Bearing dimensions SKF6206.

D	Outer diameter	62 mm
D_m	Pitch diameter	46 mm
n_b	Number of balls	9
d_{ball}	Ball diameter	9.525 mm
α	Angle	0°

To simulate the appearance and evolution of the defect, the database was made of created fifty-one different values of the amplitude A, noted A_i with $i = 1\ldots51$. For each value, twenty signals were generated with a Gaussian variability of ±5% for the three parameters f_r, μ and f_o, Table 4. Thus, the database was made on 51 signals × 20 signals ordered by increasing values. The amplitude for $A_{i=1-10}$ was constantly equal to zero, which had no variation for amplitude A. The deviation started from eleven to fifty-one, introducing Equation (9) in Equation (8), to create signals, Table 4.

Table 4. Simulation characteristics.

	$A_{i=1-10}$	$A_{i=11-51}$
A	0	$\left(e^{4\omega}-1\right)/e^4$
ω	0	0:0.025:1
f_r (rpm)	1000 ± 5%	1000 ± 5%
μ	0.05 ± 5%	0.05 ± 5%
f_o (kHz)	10 ± 5%	10 ± 5%
$b(t)$	0.1:0.2:0.5	0.1:0.2:0.5

4.2. Effect of Internal Parameters of The OPTICS Method

OPTICS uses two parameters ε and *MinPts*. ε was calculated in the initialization phase, after collecting all the data. ε depends on the *MinPts* value. For simulation, ε had a value in a range (0.909–0.920) for a range *MinPts* =(2–20). Thus, this value varied only slightly during the initialization phase. Its value for the $MinPts^{th}$ neighbour was kept for the rest of the algorithm. Figure 3 confirms the value of ε. After the initialization phase, ε increased abruptly.

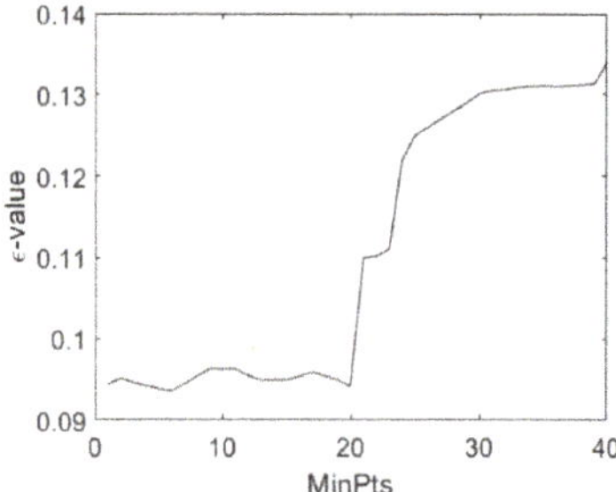

Figure 3. ε as a function of *MinPts* for $0.1b(t)$ level noise.

Minpts was related to the number of signals for an instant. Table 5 shows the effects of *MinPts* for three levels of noise. This table aimed to represent the effectiveness of the automatized ε and *MinPts*, with the initial state that is the Euclidean distance that exists in the OPTICS algorithm, and all the features (seventeen). From this table, the optimal value of *MinPts* was $n/2$. This value of *MinPts* made it possible to detect the fault before the others.

Table 5. Effect of *Minpts* on detection time, with Euclidian distance, 17 features, $\varepsilon = 0.094$.

Minpts	0.1b (t)	0.3b (t)	0.5b (t)
$n/8$	A_{12}	A_{13}	A_{14}
$n/4$	A_{11}	A_{13}	A_{14}
$n/2$	A_{11}	A_{12}	A_{12}

The selection of the distance measure affects the results of clustering algorithms. In this section, the advantages and disadvantages of every distance method used are shown in Table 6. The Euclidean distance used in the OPTICS algorithm in clustering, to calculate the distance between two vectors, was significantly difficult to iterate even an approximate of the precise values of data. Table 7 below shows the effect of distance implanted in the AOC-OPTICS method for three noise levels. The Manhattan distance could lead to the detection of the defect with global accuracy for the different signal to noise ratio equal to 96.7%, and then the second one was the Mahalanobis distance that detected at 88.2%, for the other distances the accuracy equaled 85.5%.

Table 6. Advantages and disadvantages of different uses distance. x and y are features vectors.

Distance	Formula		Comments
Euclidean	$\left(\Sigma(x_i - y_i)^2\right)^{\frac{1}{2}}$	Advantage	(1) Accessible to counting and suitable for datasets with separated clusters [28]. (2) Fast for small data [28].
		Disadvantage	(1) Susceptible to outliers [28]. (2) Results greatly influenced by variables [29]. (3) Failure in classification massive data.
Mahalanobis	$\left((x - y)'C^{-1}(x - y)\right)^{\frac{1}{2}}$ C is the covariance matrix.	Advantage	(1) Suitable for correlated data. (2) Provide curved and linear boundaries. (3) The distance is a distortion caused by a linear combination of attributes. (4) Takes account of the shape of the clusters by employing within-group correlation [30].
		Disadvantage	(1) If the noise has a high effect, it can lead to covers of the data provided and misclassification [31]. (2) It is not able to calculate the inverse of the correlation matrix when the variables highly correlated [32]. (3) When the dimension is proportional, eigenvalues of covariance equal zero, then distance cannot be calculated [33].
Cityblock Or Manhattan	$\Sigma\lvert x_i - y_i\rvert$	Advantage	(1) Shows better performance with the datasets in terms of less computation time. (2) Easily generalized to higher dimensions. (3) Having triangular inequality and offering better data contrast than Euclidean distance [30]. (4) Relatively good data contrast in high dimensions.
		Disadvantage	(1) Sensitive to outliers [34].
Minkowski Order q	$\left(\sum_{i=1}^{p}\lvert x_i - y_i\rvert^q\right)^{\frac{1}{q}}$	Advantage	(1) Useful in high dimensions of data [34]. (2) Useful for datasets with compact or isolated clusters.
		Disadvantage	(1) The terms of computation is expensive [35].
Chebychev	$max_i(\lvert x_i - y_i\rvert)$	Advantage	(1) It takes less time to count distances between data sets [36].
		Disadvantage	(1) More sensitive to the scales of the feature magnitude, the inherent weakness can be resolved by normalization of all features before the classification task [37].

Table 7. Effect of distances, with $MinPts = n/2$, 17 features, $\varepsilon = 0.094$.

	0.1 $b(t)$	0.3 $b(t)$	0.5 $b(t)$	Global Accuracy
Euclidean	A_{11}	A_{12}	A_{12}	85.5%
Mahalanobis	A_{11}	A_{12}	A_{13}	88.2%
Manhattan	A_{10}	A_{10}	A_{11}	96.7%
Minkowski	A_{11}	A_{12}	A_{12}	85.5%
Chebychev	A_{11}	A_{12}	A_{12}	85.5%

4.3. *Effect of Ranking Features*

Usually ranking features is used in preprocessing data as a feature subdivision. The concept for use is to count the random instance, then calculate their nearest neighbors and set the vector of weighting features, which can distinguish the features from neighbors of various classes.

Two methods chi-square and relief were compared. Table 8 represents the result of the ranking features. The comparative study presents the effectiveness of the relief method that could detect the defect in the high accuracy from features number ten to the end. The chi-square start to recognize the highest efficiency with twelve features. From the results of Table 8 could conclude that the method of relief ranking features was the best with just ten features that was enough to obtain the highest accuracy.

Table 8. Global accuracy. Effect of ranking features with Manhattan distance, $MinPts = n/2$, $\varepsilon = 0.094$.

# Features	4	5	6	7	8	9	10	11	12	13	14	15	16	17
Chi-square%	83	83	83	83	83	83	85.7	93.7	96.7	96.7	96.7	96.7	96.7	96.7
Relief %	85.7	85.7	85.7	85.7	85.7	85.7	96.7	96.7	96.7	96.7	96.7	96.7	96.7	96.7

4.4. Results

In this section, the results were obtained for the following parameters: relief method, distance from Manhattan, $MinPts = n\ /\ 2$ and $\varepsilon = 0.094$. A 3D visualization was chosen (three principal components). In fact, the 3D results gave a detection accuracy of 96.7% and the 2D results covered an accuracy of less than 93.7%. The results of the BPFO (ball pass frequency outer) simulation showed the fault detected from signal A_{11}, for noise levels 0.1*b (t)*, 0.3 *b(t)* and A_{12} for 0.5*b (t)* (Figures 4–6).

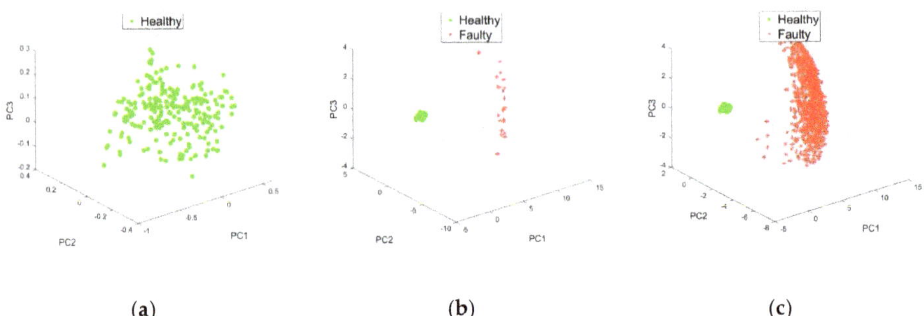

(**a**) (**b**) (**c**)

Figure 4. Effect of amplitude 0.1*b* (*t*): (**a**) A_{10}, (**b**) A_{11} and (**c**) A_{51}.

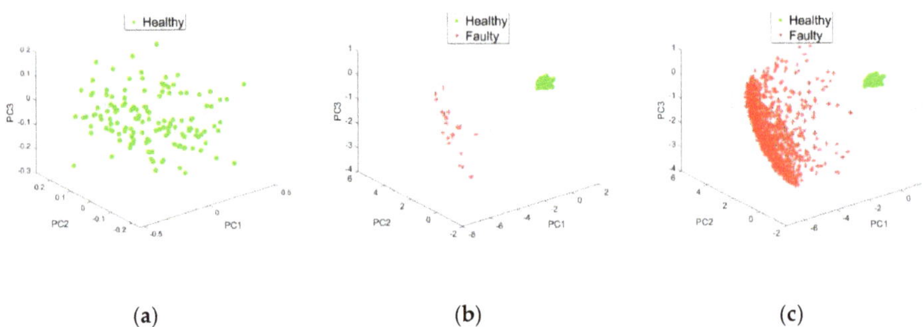

(**a**) (**b**) (**c**)

Figure 5. Effect of amplitude 0.3*b* (*t*): (**a**) A_{10}, (**b**) A_{11} and (**c**) A_{51}.

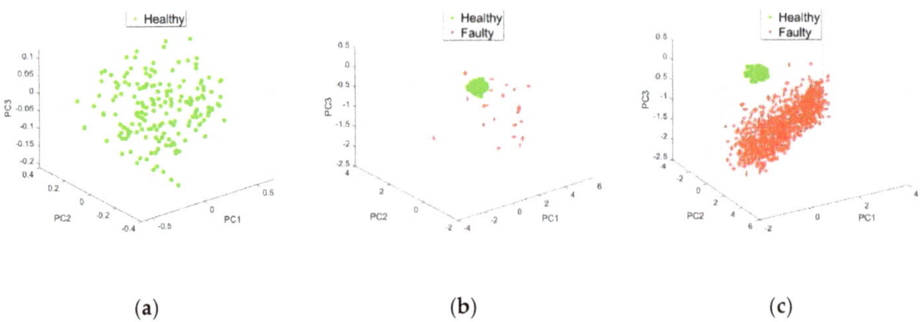

(**a**) (**b**) (**c**)

Figure 6. Effect of amplitude 0.5*b* (*t*): (**a**) A_{11}, (**b**) A_{12} and (**c**) A_{51}.

The follow-up starting after the end of the detection phase. *GV* monitors the growth of the fault with the varied amplitude of signals. The evolution of *GV* was studied for the three noise levels 0.1*b* (*t*), 0.3*b* (*t*) and 0.5*b* (*t*), Figure 7.

Figure 7a represents the Calinski index calculated between the two clusters. The curve values increased with increasing amplitude values. The Calinski index value for the $0.1b\ (t)$ was more significant and the curve was above the others. For a high noise level, the evolution was linear $GV_1 = 0.438k + 7.245\left(R^2 = 0.980\right)$, while for the other two noise levels the evolution was exponential $(R^2 = 0.977\ and\ R^2 = 0.742)$.

Figure 7b represents the Davies–Bouldin index, the curve was the opposite of the Calinski-Harabasz index, which decreased with the increasing amplitude of signals. The results observed here showed a curve of $0.1b\ (t)$, which was above the other curves, and started near to one and ended near-zero. For the three noise levels, the regression was linear. The mathematical model was similar: $GV_2 = -0.0236k + 0.954\left(R^2 = 0.999\right)$, $GV_2 = -0.0239k + 0.997\left(R^2 = 0.997\right)$ and $GV_2 = -0.0252k + 1.064\left(R^2 = 0.994\right)$ respectively for $0.1b\ (t)$, $0.3b\ (t)$ and $0.5b\ (t)$.

Figure 7c represents the density of the defected cluster or the second class. The density decreases over the amplitude of signals until it became constantly equal to zero, contrary to the Davies–Bouldin index decrease, to attend near zero at the end of class. The comparison between the curves showed that the density of $0.1b\ (t)$, bigger than the other noise to signal ratios. The evolution was exponential with the mathematical model: $GV_3 = 1169e^{-193k}\left(R^2 = 0.975\right)$, $GV_3 = 777e^{-O.147k}\left(R^2 = 0.950\right)$ and $GV_3 = 809e^{-0.151k}\left(R^2 = 0.720\right)$, $0.1b\ (t)$, $0.3b\ (t)$ and $0.5b\ (t)$. The correlation was poor for a low noise level.

Figure 7d represents the distance between two clusters, the distance values growing with amplitude. However, the curvy curve had an increasing trajectory form for the three scenarios $0.1b\ (t)$, $0.3b\ (t)$ and $0.5b\ (t)$. Additionally, the distance parameter could observe the trajectory of $0.1b\ (t)$, was above the other curves at the end, but initially, the three curves were conjoined, then started to separate from an amplitude equal to $k = 31$. A linear model mathematic measurement could be done from $k = 31$, $GV_4 = 0.300k - 0.714\left(R^2 = 0.963\right)$, $GV_4 = 0.146k - 0.434\left(R^2 = 0.769\right)$ and $GV_4 = 0.097k - 0.297\left(R^2 = 0.698\right)$.

Figure 7e represents the contour of the second cluster, showing the increase of contour with the amplitude of signals. The comparison of the contour with the Calinski index shows, the Calinski index remained increasing with the number of amplitudes. However, the contour values were similar for noise levels at low amplitudes. The contour was relevant for a certain amplitude level, $k = 31$ for low noise levels and $k = 41$ for higher noise levels. The regression models starting from $k = 31$ were $GV_5 = 0.373e^{-0.154k}\left(R^2 = 0.948\right)$, $GV_5 = 0.059e^{0.240k}\left(R^2 = 0.986\right)$ and $GV_5 = 0.034e^{0.215k}\left(R^2 = 0.924\right)$.

In summary, the Calinski index differentiates noise levels for all amplitudes. However, the mathematical regression model was different. For low noise levels, a linear model was interesting, while for high noise levels, the exponential model was preferred. On the contrary, the Calinski index was little influenced by the noise level, thus the linear regression model was relevant and similar. That could show the importance of the Calinski index, which could separate the curves of different scenarios, the value started with zero and grew directly with the amplitude, while the contour parameter increased slowly with the amplitude. The parameters, density and distance, had values close to 0 either for low amplitudes or high amplitudes. The evolutions were only visible for ranges of amplitudes. According to these simulations the Calinski and Davies–Bouldin indexes were preferred.

This numerical investigation made it possible to fix the internal parameters OPTICS, $\varepsilon = 0.094$ (Equation (4)), $MinPts\ (= n/2)$ and to optimize the methods involved in the AOC-OPTICS process (relief method, t-SNE and Manhattan distance).

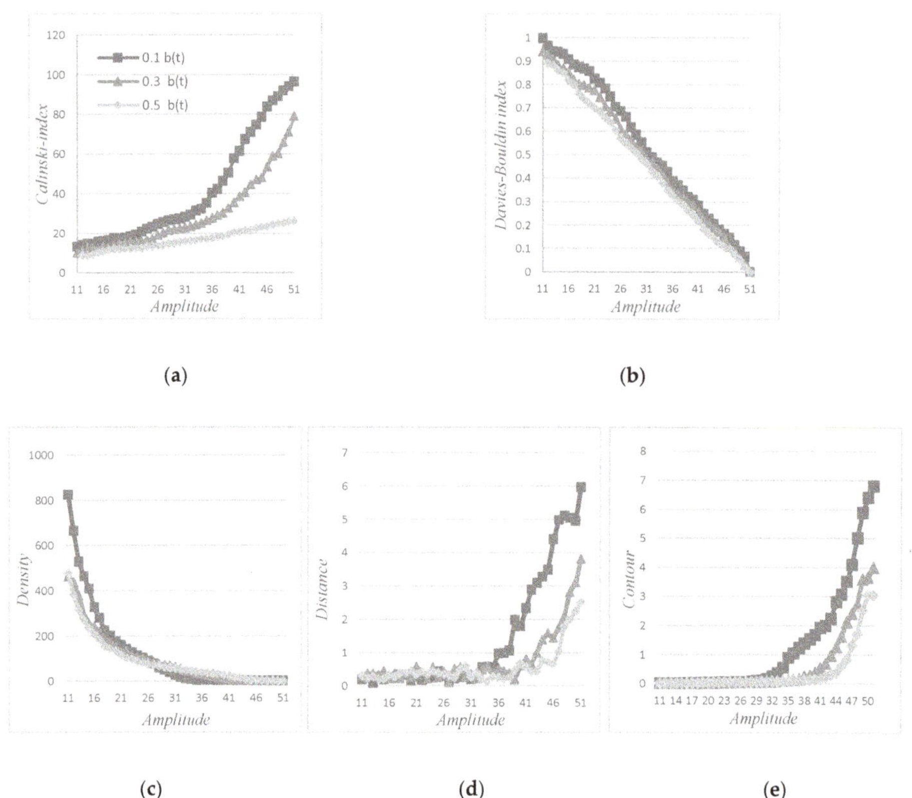

(**a**) (**b**)

(**c**) (**d**) (**e**)

Figure 7. (**a**) Calinski index, (**b**) Davies–Bouldin index, (**c**) density, (**d**) distance and (**e**) contour.

5. Experimental Validation

5.1. Test Bench

The experimental bench consists of a crankcase connected with the electric motor of 10 KW maximum power through a shaft and two rolling bearings: healthy (6206 ball bearing) and degraded (N.206.E. G15 roller bearing), Figure 8. A hydraulic jack via steel cable was used to vary loads on the shaft (Figure 8a). The motor has rotational speed controlled by variable speed drive. The whole device was built to a concrete structure to isolate it from the low frequencies generated by the external environment. A piezoelectric sensor was placed radially on the bearing, considered as the best measuring point. The data were collected with a sampling frequency of 51,200 Hz. Eight defects on the outer ring of the roller bearing were created with an electro pen. The defects were measured using a paste mark "plastiform", Table 9. The resulting profile was characterized as roughness, with a Taylor-Hobson subtronic 3P profilometer (Figure 8b). For the nine states of the defect (one healthy and eight defect sizes), 10 randomly operating conditions were applied among 5 loads ranging from 100 to 220 daN, with a 30 daN step, and 5 rotation speed varies ranging from 1405 to 1560 rpm with a 50 rpm step. The number of combinations was 90 ($k = 90$). For each combination 8 signals were collected ($n = 8$) with 12,800 samples. The total database was made of 720 signals.

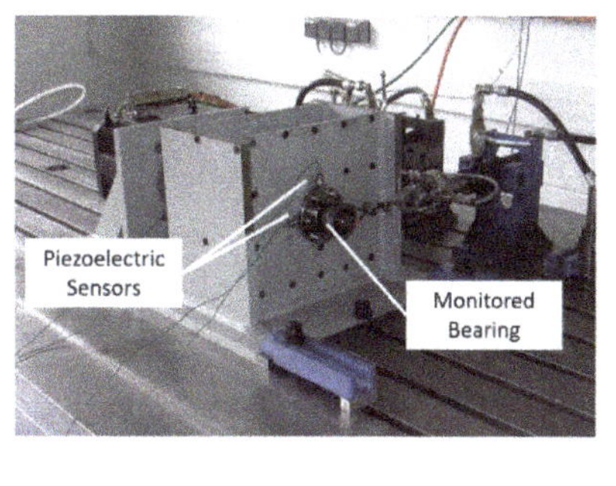

(a) (b)

Figure 8. (**a**) Test bench (**b**) profilometer with Plastiform^{TM} paste.

Table 9. Dimensions of the defects: width (W), arithmetic roughness (Ra) and total roughness (Rt).

Faults#	Rt (μm)	Ra (μm)	W (mm)	Faults #	Rt (μm)	Ra (μm)	W (mm)
0	-	-	-	5	10.55	1.52	1.09
1	2.42	0.33	0.16	6	18.17	1.82	1.78
2	3.00	0.41	0.73	7	18.67	2.36	2.03
3	8.25	0.73	0.45	8	21.42	2.97	2.32
4	10.50	1.32	0.74				

AOC-OPTICS method was applied. The inputs were $\Delta t = 1$ $(default\ value), T_{ini} = 10$ and $n = 8$. Thus, at each iteration k, 8 new signals integrated the algorithm. Eighty signals initiated the monitoring process.

5.2. Results

After the initialization phase ($T_{ini} = 10$), the detection phase operated during the detection of a second class. This detection was made for the iteration $k = 11$. The inputs parameters were $Minpts = n/2 = 4$ and $\epsilon = 0.12$. Results of AOC-OPTICS method are represented in Figure 9. The cluster number 2 appeared at iteration 11 and was confirmed by the following iterations. Despite the variation of loads and speed the accuracy was 100%. The results of our methodology could detect a tiny variation in the state of the bearing. All these results could demonstrate the robustness of the used methodology.

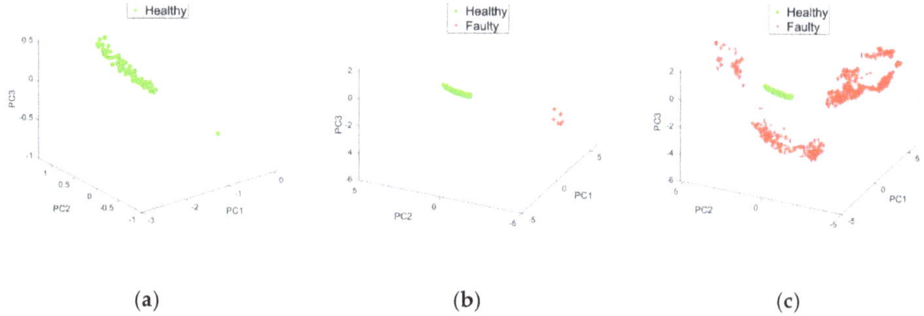

(a) (b) (c)

Figure 9. Experimental validation (**a**) Iteration 10, (**b**) Iteration 11 and (**c**) Iteration 90.

5.3. Follow-Up

The follow-up starts at iteration 11 to iteration 90 for the 5 GV, Calinski and Davies–Bouldin index, density, contour and distance, Figure 10. The behavior of these variables was different and remained similar to the behaviors established during the simulation. The Calinski index increased with the size of defects. At iteration 65, the index had an exponential evolution in the mathematical form $GV_1 = 161.13e^{0.084k} \left(R^2 = 0.982\right)$, Figure 10a. The Davies–Bouldin index decreased proportionally with the fault, Figure 10b. In this case, a linear regression $GV_2 = -0.0107k + 0.947 \left(R^2 = 0.994\right)$ was proposed. The density decreased with the increasing amplitude values to attend around zero from signal number sixty to ninety, the mathematical model was $GV_3 = 197.15.\exp(-0.095k) \left(R^2 = 0.940\right)$. The distance curve was increasing with the increasing of the amplitude values. The evolution was exponential $GV_4 = 0.225e^{0.039k} \left(R^2 = 0.828\right)$. However the monotony was not relevant. There was a lot of variability around the average trend, Figure 10d. The contour shows two trends, Figure 10e. The contour evolved proportionally for the first 60 iterations with a low slope, $GV_5 = 0.014k - 0.0697 \left(R^2 = 0.934\right)$. From the 60th iteration onwards, the evolution remained linear but increased sharply, $GV_5 = 0.4791k - 3.518 \left(R^2 = 0.904\right)$.

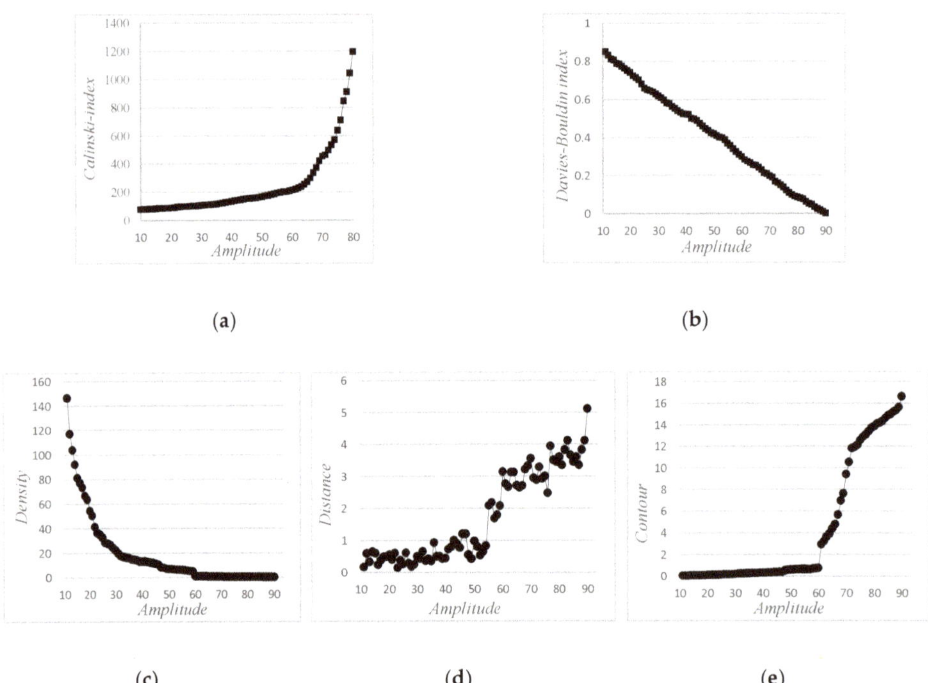

Figure 10. Follow-up of the detected cluster: evolution of (**a**) Calinski index, (**b**) Davies–Bouldin index. (**c**) Density, (**d**) distance and (**e**) contour.

By comparing the evolution of these indicators, the Calinski index and the contour showed some singularities in the evolution at iteration 60 corresponding to defect 5. These parameters indicate the severity degradation stage in the rolling bearing. The Davies-Bouldin index was the index most correlated to the number of iteration ($R^2 = 0.994$). In general, these indicators allowed us to make a prognosis on the evolution of these parameters with the iterations.

6. Conclusions

This paper proposed an automatic online methodology for monitoring ball bearings by optimizing the internal parameters of the OPTICS method and the dimension reduction step. The dynamic monitoring AOC-OPTICS was divided into three phases: the initialization, the detection and following the defect. The methodology was confronted with a simulated fault evolution and then with experimental data. The detection reached an accuracy of 100%. The follow-up was assured by geometrical values whose trend followed linear or exponential mathematical models with correlation coefficients up to 0.994. This methodology brings many improvements: (I) This automated methodology used the best parameters for the detection and following the defects with high accuracy. (II) The variation of speed and load cannot lead to discovering the fault in the rolling bearing. Only the amplitude leads to detecting the faulty state. (III) The relief method is efficient compared to chi-square, which is used to delete unnecessary features, which can make the iteration to be calculated speedily. (IV) The characteristics parameters related to the defect facilitate monitoring of the evolution with the times. (V) The density and Calinski and Davies–Bouldin index represent efficacy more than the other parameters, for monitoring the defect growth trajectory. The major perspective is to add the diagnostic part in the methodology to increase the prognosis. This part must be based on previous knowledge provided by a digital twin or an expert.

Author Contributions: H.H. performed the experiments, conceived the algorithm and analyzed the data. X.C. and L.R. planned the experiments and supervised the research work. H.H. and X.C. wrote the paper and L.R. revised the entire paper. All authors have read and agreed to the published version of the manuscript.

Funding: This research received no external funding.

Conflicts of Interest: The authors declare no conflict of interest.

References

1. Xenakis, A.; Karageorgos, A.; Efthimios, L.; Chis, A.E.; González-Vélez, H. Towards Distributed IoT/Cloud based Fault Detection and Maintenance in Industrial Automation. *Procedia Comput. Sci.* **2019**, *151*, 683–690. [CrossRef]
2. Yan, J.; Meng, Y.; Lu, L.; Li, L. Industrial Big Data in an Industry 4.0 Environment: Challenges, Schemes, and Applications for Predictive Maintenance. *IEEE Access* **2017**, *5*, 23484–23491. [CrossRef]
3. Short, M.; Twiddle, J. An Industrial Digitalization Platform for Condition Monitoring and Predictive Maintenance of Pumping Equipment. *Sensors (Switzerland)* **2019**, *19*, 3781. [CrossRef] [PubMed]
4. Harris, T.A.; Kotzalas, M.N. *Rolling Bearing Analysis*, 5th ed.; Taylor & Francis: Boca Raton, FL, USA, 2006.
5. Sri, J.; Senanayaka, L.; Kandukuri, S.T.; Van Khang, H.; Robbersmyr, K.G. Early Detection and Classification of Bearing Faults Using Support Vector Machine Algorithm. In Proceedings of the 2017 IEEE Workshop on Electrical Machines Design, Control and Diagnosis (WEMDCD), Nottingham, UK, 20–21 April 2017. [CrossRef]
6. Li, Z.; Zhu, J.; Shen, X.; Zhang, C.; Guo, J. Fault diagnosis of motor bearing based on the Bayesian network. *Procedia Eng.* **2011**, *16*, 18–26. [CrossRef]
7. Tian, J.; Azarian, M.H.; Pecht, M. Rolling Element Bearing Fault Detection Using Density-Based Clustering. In Proceedings of the 2014 International Conference on Prognostics and Health Management, Cheney, WA, USA, 22–25 June 2014. [CrossRef]
8. El-thalji, I.; Jantunen, E. A Summary of Fault Modelling and Predictive Health Monitoring of Rolling Element Bearings. *Mech. Syst. Signal Process.* **2015**, *60–61*, 252–272. [CrossRef]
9. Cerrada, M.; Sánchez, R.V.; Li, C.; Pacheco, F.; Cabrera, D.; Valente de Oliveira, J.; Vásquez, R.E. A Review on Data-Driven Fault Severity Assessment in Rolling Bearings. *Mech. Syst. Signal Process.* **2018**, *99*, 169–196. [CrossRef]
10. Brijesh, S.; Dushyanth, N.D.; Soumendu, J.; Sarvajith, M. A Novel Approach for Bearing Fault Detection and Classification Using Acoustic Emission Technique. In Proceedings of the First International Conference on Advances in Computer, Electronics and Electrical Engineering—CEEE 2012, Mumbai, India, 25–27 March 2012.

11. Stein, B.; Busch, M. Density-Based Cluster Algorithms in Low-Dimensional and High-Dimensional Applications. In Proceedings of the Second International Workshop on Text-Based Information Retrieval, Koblenz, Germany, 11 September 2005; pp. 45–55.
12. Wei, Z.; Wang, Y.; He, S.; Bao, J. A Novel Intelligent Method for Bearing Fault Diagnosis Based on Affinity Propagation Clustering and Adaptive Feature Selection. *Knowledge-Based Syst.* **2017**, *116*, 1–12. [CrossRef]
13. Ankerst, M.; Breunig, M.M.; Kriegel, H.P.; Sander, J. OPTICS: Ordering Points to Identify the Clustering Structure. *ACM SIGMOD Record* **1999**, *28*, 49–60. [CrossRef]
14. Shah, G.H.; Bhensdadia, C.K.; Ganatra, A.P. An Empirical Evaluation of Density-Based Clustering Techniques. *Int. J. Soft Comput. Eng. IJSCE* **2012**, *1*, 216–223.
15. Benmahdi, D.; Rasolofondraibe, L.; Chiementin, X.; Murer, S.; Felkaoui, A. RT-OPTICS: Real-Time Classification Based on OPTICS Method to Monitor Bearings Faults. *J. Intell. Manuf.* **2019**, *30*, 2157–2170. [CrossRef]
16. Urbanowicz, R.J.; Meeker, M.; Cava, W.L.; Olson, R.S.; Moore, J.H. Relief-Based Feature Selection: Introduction and Review. *J. Biomed. Inform.* **2018**, *85*, 189–203. [CrossRef] [PubMed]
17. Vakharia, V.; Gupta, V.K.; Kankar, P.K. Bearing Fault Diagnosis Using Feature Ranking Methods and Fault Identification Algorithms. *Procedia Eng.* **2016**, *144*, 343–350. [CrossRef]
18. Xie, Y. A Fault Diagnosis Approach Using SVM with Data Dimension Reduction by PCA and LDA Method. In Proceedings of the 2015 Chinese Automation Congress (CAC), Wuhan, China, 27–29 November 2015. [CrossRef]
19. Deng, F.; Ren, B. Fault Diagnosis of Rolling Bearing Using the Hermitian Wavelet Analysis, KPCA and SVM. In Proceedings of the 2017 International Conference on Sensing, Diagnostics, Prognostics, and Control (SDPC), Shanghai, China, 16–18 August 2017. [CrossRef]
20. Tu, D.; Zheng, J.; Jiang, Z.; Pan, H. Multiscale Distribution Entropy and T-Distributed Stochastic Neighbor Embedding-Based Fault Diagnosis of Rolling Bearings. *Entropy* **2018**, *20*, 360. [CrossRef]
21. Ozkok, F.O.; Celik, M. A New Approach to Determine Eps Parameter of DBSCAN Algorithm. *Intell. Syst. Appl. Eng.* **2017**, *5*, 247–251. [CrossRef]
22. Loohach, R.; Garg, K. Effect of Distance Functions on Simple K-Means Clustering Algorithm. *Int. J. Comput. Appl.* **2012**, *49*, 7–9. [CrossRef]
23. Shukla, S. Analysis of Statistical Features for Fault Detection. In Proceedings of the IEEE International Conference on Computational Intelligence and Computing Research (ICCIC), Madurai, India, 10–12 December 2015.
24. Cai, J.; Xiao, Y. Time-Frequency Analysis Method of Bearing Fault Diagnosis Based on the Generalized S Transformation. *J. Vibroeng.* **2017**, *19*, 4221–4230. [CrossRef]
25. Qian, Y.; Yan, R.; Gao, R.X. A Multi-Time Scale Approach to Remaining Useful Life Prediction in Rolling Bearing. *Mech. Syst. Signal Process.* **2017**, *83*, 549–567. [CrossRef]
26. Vakharia, V.; Gupta, V.K.; Kankar, P.K. A comparison of feature ranking techniques for fault diagnosis of ball bearing. *Soft Comput.* **2016**, *20*, 1601–1619. [CrossRef]
27. Chiementin, X. Localization and quantification of vibratory sources for a predictive maintenance in order to increase the diagnosis and the follow-up of the damage of the rotating mechanical components: Application to rolling bearings. Ph.D. Thesis, University of Reims Champagne-Ardenne, Reims, France, October 2007.
28. Jain, A.K.; Murty, M.N.; Flynn, P.J. Data clustering: A review. *ACM Comput. Surv.* **1999**, *31*, 264–323. [CrossRef]
29. Pandit, S.; Gupta, S. A Comparative Study on Distance Measuring Approaches for Clustering. Int. *J. Res. Comput. Sci.* **2011**, *2*, 29–31.
30. Durak, B. A Classification Algorithm Using Mahalanobis Distances Clustering of Data with Applications on Biomedical Data Set. Master's Thesis, Middle East Technical University, Bahadır, Turkey, January 2011.
31. Krishnan, S.; Kerkhoff, H.G. Exploiting Multiple Mahalanobis Distance Metrics to Screen Outliers from Analog Product Manufacturing Test Responses. *IEEE Des. Test* **2013**, *30*, 18–24. [CrossRef]
32. Egan, W.J.; Morgan, S.L. Outlier Detection in Multivariate Analytical Chemical Data. *Anal. Chem.* **1998**, *70*, 2372–2379. [CrossRef] [PubMed]
33. Taylor, P.; Hadi, A.S.; Simonoff, J.S.; Hadi, A.S.; Simonoff, J.S. Procedures for the Identification of Multiple Outliers in Linear Models Procedures for the Identification of Multiple Outliers in Linear Models. *J. Am. Stat. Assoc.* **2012**, *88*, 37–41. [CrossRef]

34. Soler, J.; Tencé, F.; Gaubert, L.; Buche, C. Data Clustering and Similarity. In Proceedings of the Twenty-Sixth International Florida Artificial Intelligence Research Society Conference, Pete Beach, FL, USA, 22–24 May 2013; pp. 492–495.
35. Xu, R.; Member, S.; Ii, D.W. Survey of Clustering Algorithms. *IEEE Trans. Neural Netw.* **2005**, *16*, 645–678. [CrossRef] [PubMed]
36. Potolea, R.; Cacoveanu, S.; Lemnaru, C. Meta-learning Framework for Prediction Strategy Evaluation. In Proceedings of the International Conference on Enterprise Information Systems, Funchal-Madeira, Portugal, 8–12 June 2010; pp. 280–295.
37. Bellet, A.; Habrard, A.; Sebban, M. *A Survey on Metric Learning for Feature Vectors and Structured Data*; Technical Report; Cornell University: Ithaca, NY, USA, 2013.

Review

A Review of Data Mining Applications in Semiconductor Manufacturing

Pedro Espadinha-Cruz [1,*], Radu Godina [1,*] and Eduardo M. G. Rodrigues [2,*]

[1] UNIDEMI-Research and Development Unit in Mechanical and Industrial Engineering, Faculty of Science and Technology (FCT), Universidade NOVA de Lisboa, 2829-516 Almada, Portugal
[2] Management and Production Technologies of Northern Aveiro—ESAN, Estrada do Cercal 449, Santiago de Riba-Ul, 3720-509 Oliveira de Azeméis, Portugal
* Correspondence: p.espadinha@fct.unl.pt (P.E.-C.); r.godina@fct.unl.pt (R.G.); emgrodrigues@ua.pt (E.M.G.R.)

Abstract: For decades, industrial companies have been collecting and storing high amounts of data with the aim of better controlling and managing their processes. However, this vast amount of information and hidden knowledge implicit in all of this data could be utilized more efficiently. With the help of data mining techniques unknown relationships can be systematically discovered. The production of semiconductors is a highly complex process, which entails several subprocesses that employ a diverse array of equipment. The size of the semiconductors signifies a high number of units can be produced, which require huge amounts of data in order to be able to control and improve the semiconductor manufacturing process. Therefore, in this paper a structured review is made through a sample of 137 papers of the published articles in the scientific community regarding data mining applications in semiconductor manufacturing. A detailed bibliometric analysis is also made. All data mining applications are classified in function of the application area. The results are then analyzed and conclusions are drawn.

Keywords: data mining; semiconductor manufacturing; quality control; yield improvement; fault detection; process control

Citation: Espadinha-Cruz, P.; Godina, R.; Rodrigues, E.M.G. A Review of Data Mining Applications in Semiconductor Manufacturing. *Processes* **2021**, *9*, 305. https://doi.org/10.3390/pr9020305

Academic Editors: Marco S. Reis and Furong Gao
Received: 31 December 2020
Accepted: 3 February 2021
Published: 6 February 2021

1. Introduction

The last few decades have seen the birth of a great diversity of products and services associated with electrical and electronic equipment, and witnessed the presence of electronic and electrical equipment in a large number of products and services, which are subject to constant change [1]. During the last few years, since semiconductor manufacturing processes have gradually diminished in size, the number of transistors that can be fabricated on a sole silicon wafer can amount to a billion units [2]. In order to account for the dynamic evolution of production and distribution and the changes caused by technological advances and inventions, companies that operate in this field need to be flexible and to be able to adapt quickly to a constantly changing environment [3].

Semiconductor production is the process that creates integrated circuits, such as transistors, LEDs, or diodes that can be found in electrical devices and consumer electronics. During the front-end process, the crystalline silicon ingot is produced and the wafers are cut, the electrical circuits are created by photolithography and other chemical processes and, finally, they are electronically tested. In the back-end process, the chunks are cut from the wafer, wired (glued), encapsulated, and tested [4]. The semiconductor manufacturing industrial units (known also as fabs) are one of the highest capital-intensive and entirely automated production systems, in which agnate processes and equipment are utilized to manufacture integrated circuits through a wide range of extensive and complex processes with firmly controlled manufacturing processes, reentering process flows, advanced and complex equipment, and demanding deadlines for complying with constantly unpredictable demands of a constantly increasing product mix [5].

The concept Industry 4.0 involves employing artificial intelligence technologies, data mining techniques, big data and deep learning analysis to the current industrial infrastructure for the purpose of developing innovations that are disruptive [6]. The objective is to strive to put into practice this concept, which will allow flexible decision-making and smart manufacturing systems, as anticipated by the Industry 4.0 concept. Therefore, by turning Industry 4.0 a reality, the role of the Internet of Things (IoT) and additional emergent technologies will have a central role [7]. So far, the tendency to have unmanned operations and increasing automation in semiconductor production systems, as in other production technologies, is constantly growing [8].

Conventionally, semiconductor production systems are known for having a highly complex and lengthy manufacturing process. Typically, semiconductor wafers require a number of process steps that could easily surmount half of a thousand to be produced [9,10]. The level of complexity of every step is frequently equated to that of a medium-sized industrial unit, particularly in such areas such as logistics, planning, control, and data volume, among other steps. Consequently, growing requirements and pressure to perform with a high plant productivity pose a difficult challenge for companies operating in semiconductor manufacturing [1].

The ever-growing demand for integrated circuits that are able to deliver higher performances at lower costs is something semiconductor companies are well familiar with. Therefore, wafer metrology tools are employed for designing and producing semiconductors, cautiously monitoring line widths, film properties, and possible defects in order to improve the production process. Data mining techniques together with metrology tools and wafer verification abilities guarantee a close desired result of the electrical and physical properties of produced semiconductors. Data mining with wafer metrology can accurately and quickly recognize surface pattern defects, particles, and additional conditions that are capable of causing adverse effects on semiconductor performance [11].

Data mining is one of the areas of the knowledge data discovery process and is capable of providing innovative avenues for interpreting data. Data mining comprises the extraction of significant and implicit, previously unidentified, and possibly valuable information from data. Data mining offers the ability to detect patterns that are hidden amid a set of data. Data mining is the process of sorting and classifying data, then finding anomalies, patterns, and correlations in large data sets to predict outcomes. Employing a wide variety of techniques, companies can use this information for problem detection, quality control, increase revenue, cut costs, improve customer relationships, and reduce risk, among others [12]. Since modern semiconductor manufacturing processes suffer from a great degree of complexity, and the amount of data is overwhelming, it is still challenging to reach fast yield improvement by discovering manually useful patterns in raw data [11].

Throughout wafer manufacturing, equipment data, process data, and the historic data will be semiautomatically or automatically collected and grouped in a database in order to be able to diagnose faults, to monitor the process, and to effectively manage the production process. Nevertheless, in such advanced manufacturing units such as semiconductor production, numerous aspects and details are interconnected and have an effect on the yield of the produced wafers [13]. Therefore, data mining techniques are a solution for a significant amount of challenges that the semiconductor manufacturing faces, such as yield improvement [5,11], quality control [14], fault detection [15], predictive maintenance [16], virtual metrology [17], scheduling [18], business improvement [19], and market forecasting [20], among others.

Despite the existence of a high number of studies regarding data mining applications in semiconductor manufacturing, a gap was identified in the literature, in which the necessity to compile and analyze in a more comprehensive way through the compilation in a single paper every published study arose, and expressly perform it without restrictions on location or characteristics. With the intention of filling the identified gap in the research, the aim of this paper is to compile all the existing publications on this topic on Scopus and WoS and to classify and compare them. Therefore, one of the goals of this study is to understand the state of the art regarding data mining solution to existing challenges in semiconductor manufacturing. A bibliometric study is presented, in which are analyzed the number of publications over time, the co-occurrence network, the most cited authors, the distribution of keywords by observed frequency, among other bibliometric metrics. This analysis, besides analyzing bibliometric indicators and making a comparison between distinct features, it also has the purpose to frame these indicators in distinct categories and highlighting every case, not only to seek and detect future research pathways, but also to have a better comprehension of data mining applications in semiconductor industry and to endorse it in order to disseminate its use.

This paper is organized as follows. In Section 2, a brief overview of the semiconductor manufacturing process is given. In Section 3, a structured bibliometric analysis is made. In Section 4, a qualitative organization and analysis data mining application studies in semiconductor manufacturing can be found. In Section 5, a brief result analysis and discussion is made. Finally, in Section 6, overall conclusions are given.

2. Bibliometric Analysis

According to the literature, a systematic literature review neutralizes the perceived weaknesses of a narrative review [21]. A systematic literature review usually has distinct stages of preparation, direction-finding and publishing, and diffusion. Every stage might comprise numerous steps of the review process by being part of a method or system that is created to precisely and objectively focus on the overall question the review is bound to answer. In this study, the research design applied in [21–24] was followed, as seen in Figure 1, by comprising five steps: problem conception; literature search; research evaluation; research analysis; and finally result summarizing.

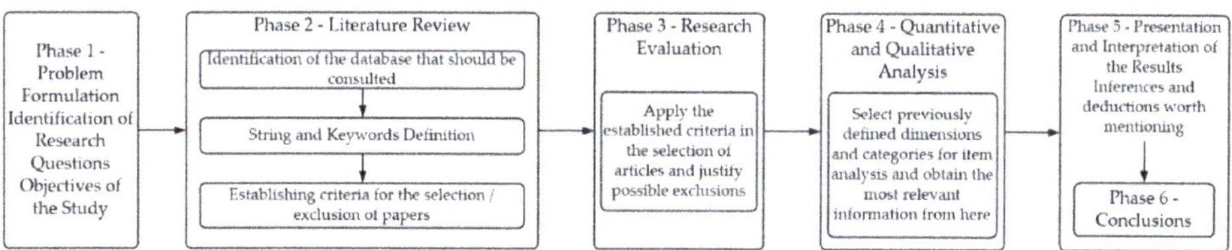

Figure 1. Literature review approach.

The objective of this bibliometric analysis is to know the state-of-the-art of data mining application in the semiconductor manufacturing. In a scenario where companies store large amounts of data, data mining approaches are used to extract useful information and knowledge automatically [25]. To achieve that, data mining approaches use a combination of algorithms and concepts from artificial intelligence, statistics, machine learning, and data management [26]. Accordingly, in this bibliometric analysis we look for data mining applications in semiconductors where authors attempt to extract information and knowledge in semiconductor manufacturing from large datasets.

After the topic of data mining data mining applications in semiconductor manufacturing was selected as an object of intensive study in this literature review, an extensive bibliographic research was carried out on the subject and its surroundings. The purpose of this analysis is to identify and evaluate the adopted methodologies of data mining applications in semiconductor manufacturing, by taking into account all the scientific studies found.

The research methodology was carefully developed in order to allow the identification of relevant patterns and areas for the study under analysis. The literature research process comprises such characteristics as the collected qualitative and quantitative information being well defined and delimited, a detailed analysis being made based on the evidence and characteristics recognized in the subject of the study, the analyzed papers are organized by application areas, all contents are analyzed in a qualitative manner, which favors the identification of important subthemes and the successful interpretation of results. We considered papers that address the application of data mining to exploit data stored during semiconductor manufacturing processes. So, in the first step, the usefulness of each article was verified by reading its summary and introduction, so that those who seemed to be out of the review due to imprecision and a lack of details were excluded. Additionally, despite that some of the data mining algorithms and techniques may be applied by semiconductor manufacturing authors, we excluded any papers that do not approach its use for information and knowledge extraction. After defining the aforementioned delimitations, a more detailed analysis was made on the articles that effectively added value in their incorporation in the review article. The purpose of data mining application has been carefully revised. This more detailed analysis includes: a selective reading and choice of material that suits the objectives and proposed theme; an analytical reading of the texts grouping them by application areas; and concludes with the interpretative reading and writing of the literature review body.

After the main elements of the research process have been well established, it becomes essential to adopt some essential assumptions for the accomplishment of this analysis. First, following the guidelines from [27], only indexed and peer-reviewed articles were taken into account, and the indexing databases considered were Scopus and Web of Science (WoS). The keywords utilized were "Data Mining" and "Semiconductor Manufacturing", which garnered the highest number of results. However, also, all the possible variants, such as "Semiconductor Fabrication", "Semiconductor Production", and "Semiconductor Packaging" were utilized in order to cover all the possible published papers through this combination. Table 1 shows the results from different combinations of keywords in the database.

Table 1. Results from different combinations of keywords in the database.

Search Stream	Results	
	Scopus	**WoS**
"Data Mining" AND "Semiconductor Manufacturing"	142	87
"Data Mining" AND "Semiconductor Fabrication"	11	9
"Data Mining" AND "Semiconductor Production"	8	5
"Data Mining" AND "Semiconductor Packaging"	2	2

The publications considered for this study were publications in English and the type of articles were journal research articles, journal review articles, conference articles, book chapters, and editorials. A few papers were found in Chinese and Polish, but were excluded from this study. In Figure 2 the flowchart of the paper selection process can be observed. In the end, a final sample of 137 papers was used for the article analysis. This sample comprises almost all papers found with the keywords used.

All the selected studies were classified by year and the result can be seen in Figure 3. Three waves can be seen, the first wave that comprises paper from 2004 to 2007 peaked in 2006 with 10 publications and then the interest waned. The second wave peaked in 2014 and comprises the years 2011 until 2015. Finally, the last wave of interest in this topic can be seen, peaking in 2019, with 12 publications. This wave is still ongoing. However, if divided by decades, one can notice that the decade 2010–2020 comprises 64% of all publications, while the previous decade comprises only 33.5%. This interest reveals the growing scientific interest in this topic. This increase coincides with the overall interest in data mining applications for other industries [28,29].

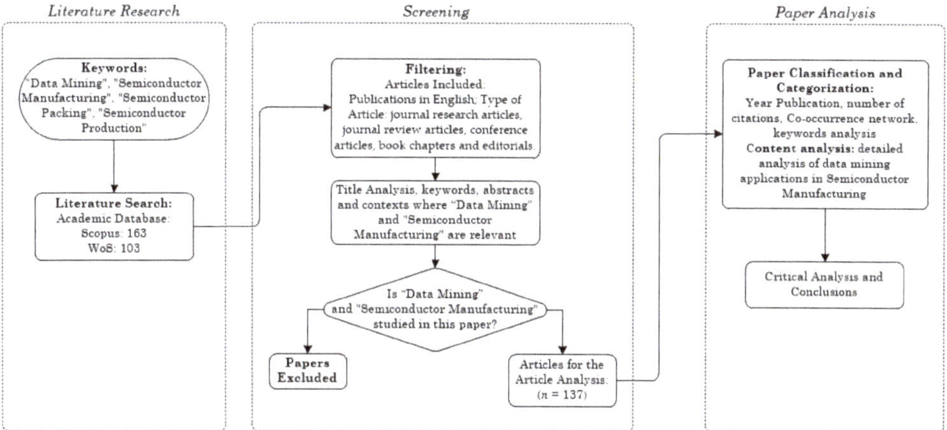

Figure 2. Flowchart of the paper selection process.

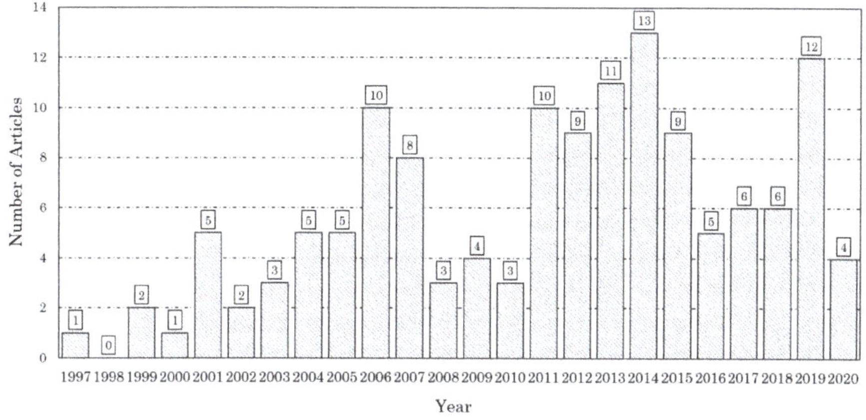

Figure 3. Publications by year of data mining applications in semiconductor manufacturing.

A particular importance has to be given to the papers that garner the highest interest in the community, which is measured by the number of citations that a study has. Figure 4 shows the most cited studies of data mining applications in semiconductor manufacturing, according to Scopus. It can be observed that the first four articles are much more cited than the remaining ones. The most cited paper is proposed by [30] and deals with maintenance. It addresses a multiple classifier machine learning technique for predictive maintenance in the ion implantation process, and, at the time of the writing of this study, it is only 5 years old. The second most cited article is an overview data preprocessing with two examples, with one in semiconductor manufacturing [31]. This study has more than two decades and it is one of the main reasons why it has 185 citations. The third most cited study deals with quality issues and proposes a framework that combines traditional statistical methods and data mining techniques for fault diagnosis and low yield product for the process of wafer acceptance testing and probing [13]. Finally, the fourth most cited study, with 168 citations, addresses a rule-structuring algorithm based on rough set theory to make predictions for the semiconductor industry [32]. This study is focused on decision support systems and has almost two decades. Still, these four studies, which address data mining applications in different contexts and areas of semiconductor manufacturing and distinct subprocesses, are an example of how vast the applications of data mining techniques in this process are. The interest that these studies attracted is a staple in their respective subcategories of semiconductor manufacturing. Lotka's Law states that the large number of small paper producers bring together about as much as the small number of large paper producers [33]. The frequency distribution of scientific productivity according to Lotka's law is shown in Figure 5, Chen-Fu Chien being the most productive author. This can also be observed in Figure 4, in which Chen-Fu Chien is the author of nine of the most cited papers, since Chen-Fu Chien is also a coauthor of the fifth [34] and last [5] most cited papers from this figure.

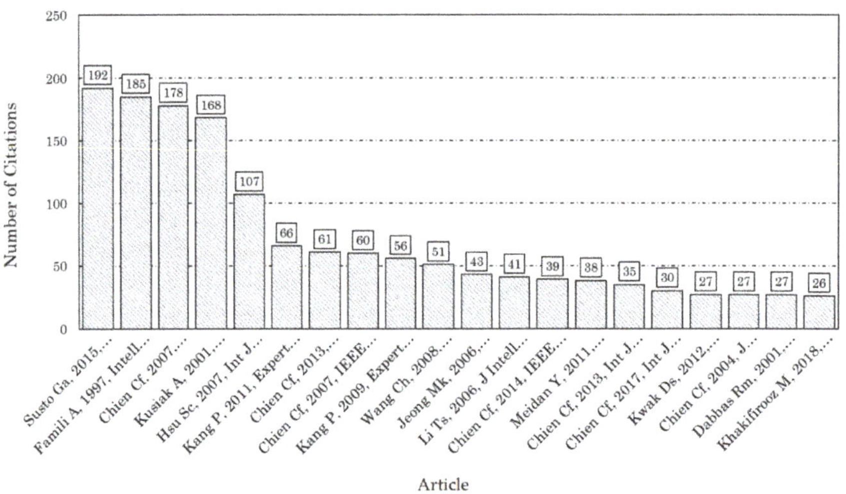

Figure 4. The most cited studies of data mining applications in semiconductor manufacturing.

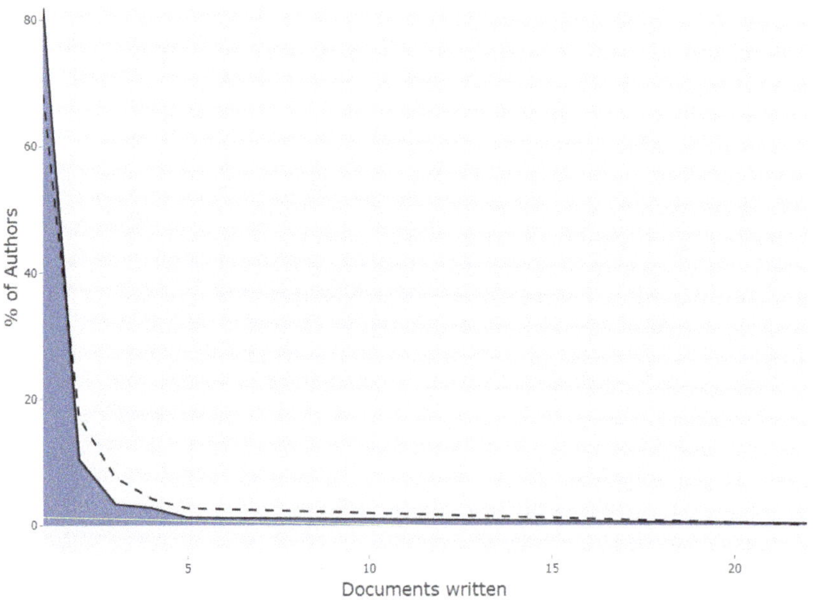

Figure 5. The frequency distribution of scientific productivity according to Lotka's law.

Keyword Analysis

A bibliometric keyword analysis was performed. This analysis was made with the help of VOSViewer software [35] and biblioshiny, which is a web application for Bibliometrix, and R Package [36]. Both have similar but distinct applications. First, the intention was to identify which were the most employed keywords. Therefore, a keyword analysis with VOSViewer software was performed with the main goal to evaluate the specifics of the discussion on how data mining applications in semiconductor manufacturing.

For the goal of this paper, the Keywords Plus function has been employed with the purpose of harmonizing the keywords that other authors have employed in the Abstract and Keyword section of their respective publications. This analysis shows that 2845 keywords were employed in the selected studies. However, only 51 of these terms appear at least 12 times. The six keywords with the highest occurrences are "data" (which appears 264 times), process (which appears 134 times), system (appearing 117 times), approach (appearing 109 times), and, finally, terms "model" and "semiconductor manufacturing" (both appearing 94 times). The network of co-occurrence links between these keywords is also shown in this paper with the intention of complementing the analysis of keywords co-occurrence. The generated keywords co-occurrence network map can be observed in Figure 6. Three different clusters can be observed.

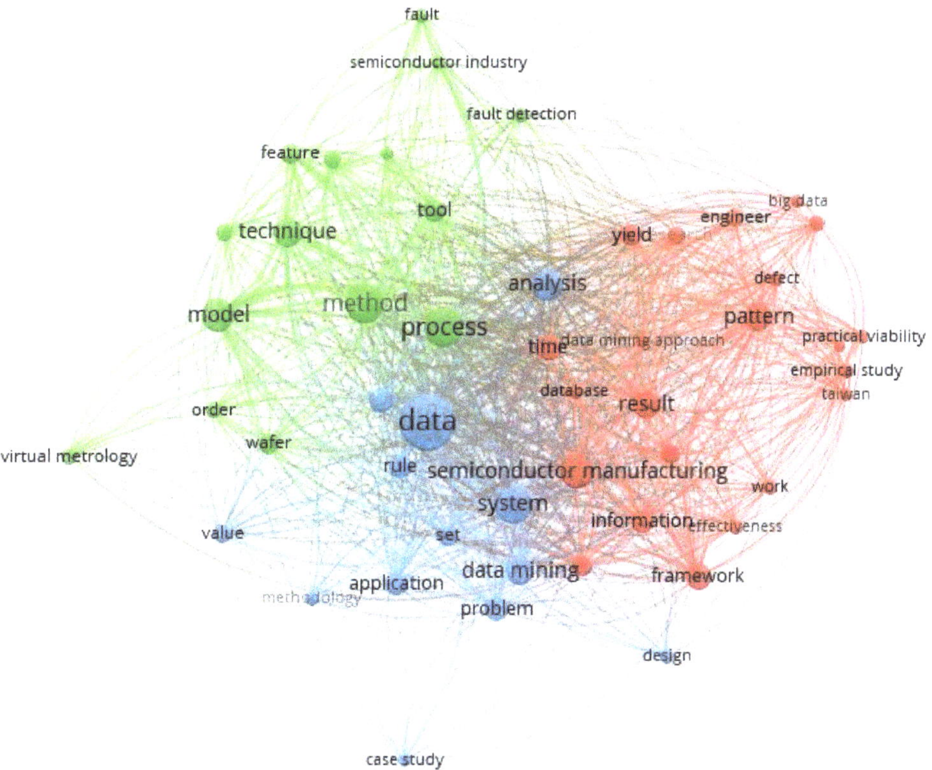

Figure 6. The generated keywords co-occurrence network map by VOSViewer software.

However, another analysis was made with biblioshiny of the Bibliometrix, from the R Package. With this application it is possible to go more in-depth regarding keyword analysis. Here, only keywords inserted by the authors of their respective papers were considered. The top five keywords that are inserted more often are "data mining", "semiconductor manufacturing", "machine learning", "feature selection", and "yield enhancement". However, by making just this simplified analysis not enough can be deduced. In Figure 7 the obtained frequency chart with biblioshiny can be observed with the distribution of the 47 most often found keywords in the selected sample of papers. A total of 349 keywords were found through the simplified technique employed in [37] to represent Zipf's law. This law stated that certain terms occur much more frequently than others and the distribution is similar to a hyperbole 1/n. As the authors from [37], however, the occurrence of the keywords is stratified in decreasing order of frequency and categorized into three areas of analysis. First, the most important zone represents the basic or trivial information area, which shows the most essential terms on the subject. The second zone comprises the terms considered "interesting information". This zone can comprise potentially innovative information and fringe themes. Finally, the last area is the noise zone. This area could represent concepts not yet emerging or even simply, noise.

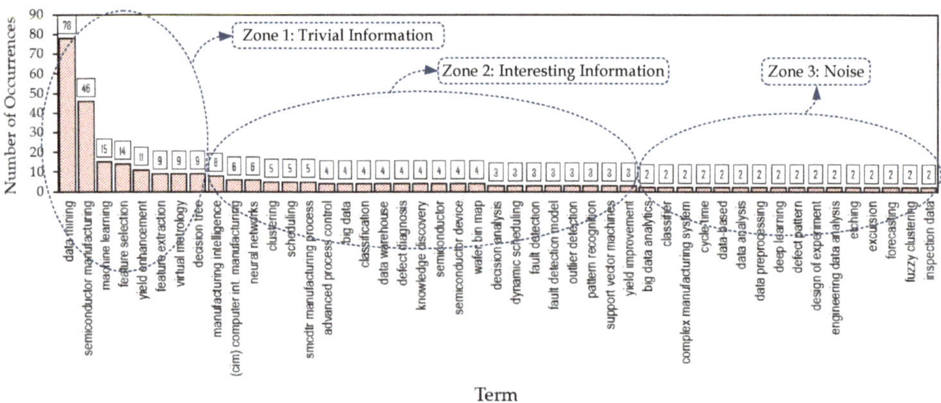

Figure 7. Distribution of keywords by observed frequency.

3. Semiconductor Manufacturing Process

The term "semiconductor" refers to a critical component in millions of electronic devices employed in current daily lives in education, research, communications, healthcare, transportation, energy, and other industries. Smartphones, mobile, wearable devices rely on semiconductors for both core operations and advanced functions and are driving global demand for semiconductors and printed circuit boards (PCBs).

The line width of semiconductors has undergone a drastic reduction, passing from the micrometer to the nanometer scale, while, in parallel, the process power and memory have been increased. Integrated circuits, made of a semiconductor material (such as silicon), are an important part of modern electronic devices in both commercial and consumer industries. These circuits must have the ability to act as an electrically controlled on/off switch (transistor) in order to perform basic arithmetic operations in a computer. To achieve this almost instantaneous switching capability, the circuits must be made of a semiconductor material, a substance with electrical resistance that lies between a conductor and an insulator.

The manufacturing process for semiconductor devices requires several steps that take place in highly specialized facilities. Semiconductor production is a considerably complex process with long lead times that are necessary to deliver the capabilities expected from everyday use of our devices. The semiconductor production times vary depending on the complexity; however, on average, it can take three to five years from initial research to final product.

Highly pure silicon is the most important raw material for the production of microelectronic components such as ICs, microprocessors, and memory chips. Figure 8 shows a summarized version of the manufacturing process. The first step in manufacturing a semiconductor device is to obtain semiconductor materials, such as germanium, gallium arsenide, and silicon, of the desired level of impurities [38,39]. Impurity levels of less than one part in a billion are required for most semiconductor manufacturing [40,41]. Due to the microscopic size of semiconductors, even the slightest hint of contamination can compromise their performance. The partly aggressive liquids required in the further manufacturing process of the microchips for metallizing, developing, etching, and cleaning should be safely conveyed, circulated, and processed [42].

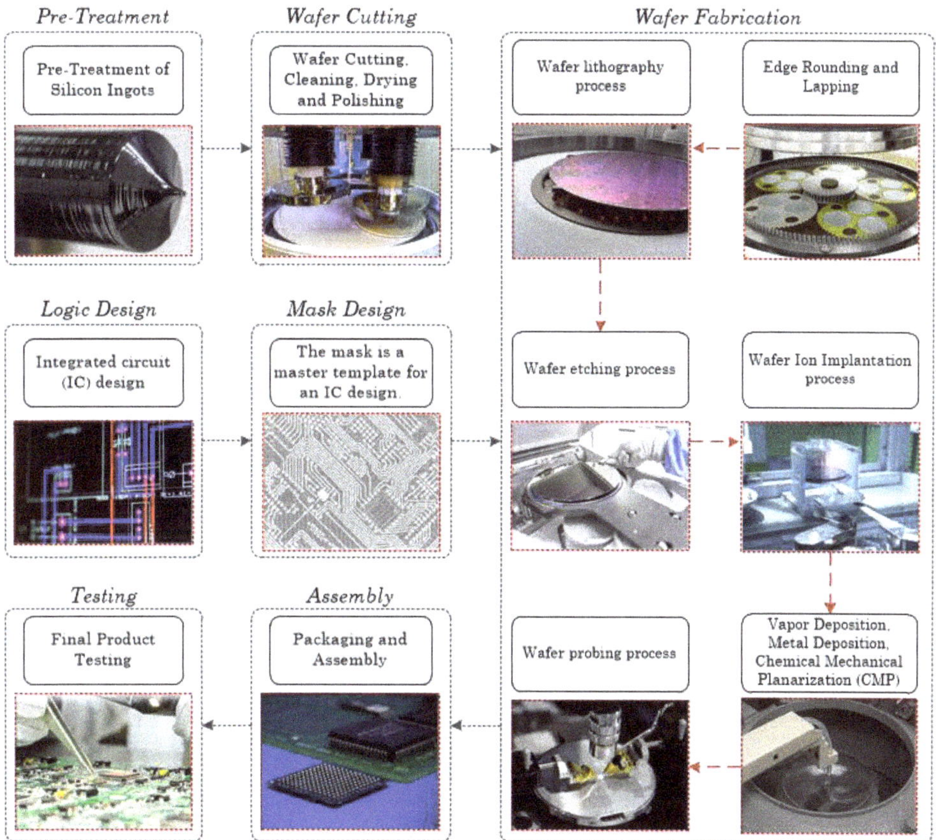

Figure 8. A simplified representation of the semiconductor manufacturing process.

The second main step is the crystal growth of monocrystalline silicon and growth of multicrystalline ingots [43]. Then, from these ingots, wafers are cut, and then shaped, polished, and cleaned with the purpose of being ready for further processing or for device manufacturing [44]. To achieve a functional device with predetermined specifications as a final result, it is necessary to carry out a prior design process for each of the manufacturing steps and a mask design, especially, for the masks used in the photolithographic processes that makes semiconductor manufacturing possible. The mask comprises the master copy of the pattern that will be printed on the wafer [45].

The next important step consists of chemical mechanical planarization or chemical mechanical polishing (CMP) is a process in which topographical irregularities can be removed from wafers with a combination of chemical and mechanical (or abrasive) polishing in order to obtain the smoothest surface possible [46,47]. The process is usually used to planarize oxide, polysilicon, or metal layers in order to prepare them for the subsequent lithographic step [48,49]. During ion implantation, high-energy ions are shot onto the substrate to be doped by the doping agent. The distribution of the implanted atoms in the semiconductor can be specifically influenced by the energy, the entry angle, and the use of masks. With multiple implants carried out one after the other, even complex doping profiles can be produced with good accuracy and replicability [50,51].

As seen in Figure 8, one of the most important steps in semiconductor manufacturing is extreme ultraviolet (EUV) lithography a process that allows carving more electrical circuits in semiconductor silicon wafers. In a lithographic system, images are transferred to silicon with light [52,53]. EUV lithography is considered to be essential to semiconductor manufacturing since it is able to produce a shorter wavelength that allows a greater quantity of electrical circuits to enter a chip [54]. Then, an important step is etching, which is utilized in microfabrication to chemically eradicate layers of a material from the surface of a wafer in order to create a pattern of that material on the substrate [55].

The following step is wafer probing, which is the procedure of electrically verifying each die on a wafer. This is accomplished by utilizing an automatic wafer probing system, which is actively searching for functional defects through by employing special test patterns [56–58]. The next step, semiconductor packaging and assembly process, involves enclosing ICs and encompasses from die-attach adhesives to liquid and film-shaped encapsulation compounds, sealing, lead forming/trimming, deflash, wirebonding, lead finish to heat-conducting materials, and conductive and non-conductive adhesives for sensors, among others. The encapsulation technology protects the sensitive layers from external influences and maintains their efficiency [59,60]. Finally, the final component is carefully tested in order to verify if it meets the requirements of standard specifications. The testing process is employed to test semiconductors in the context of design verification, specialized production, and quality assurance [61].

4. Data Mining Applications in Semiconductor Manufacturing

Data mining techniques can have a vast array of applications in the semiconductor industry. The obtained articles were classified accordingly to areas of application. Five major areas for data mining applications in semiconductor manufacturing emerged: quality control, maintenance, production, decision support systems, and finally, categorized as a whole, measurement, metrology, and instrumentation. However, other applications also exist, such as for human resources and talent recruitment and retainment [62], patent analysis [63], supply chain and inventory management [64], and stock market analysis [20], proving that data mining techniques can truly be employed for a wide range of applications.

Figure 9 shows the schematic representation of these applications. In some cases, only one article exists, and as such the direct reference is provided. In other cases, the identified five major areas are divided by subsections, in which a more detailed analysis is made. Additionally, this section is also useful for practicing engineers, since they can quickly find the semiconductor process step or data mining model they are looking for. They can also find the study that has been implemented and validated in industrial setting and through corresponding references, access to it.

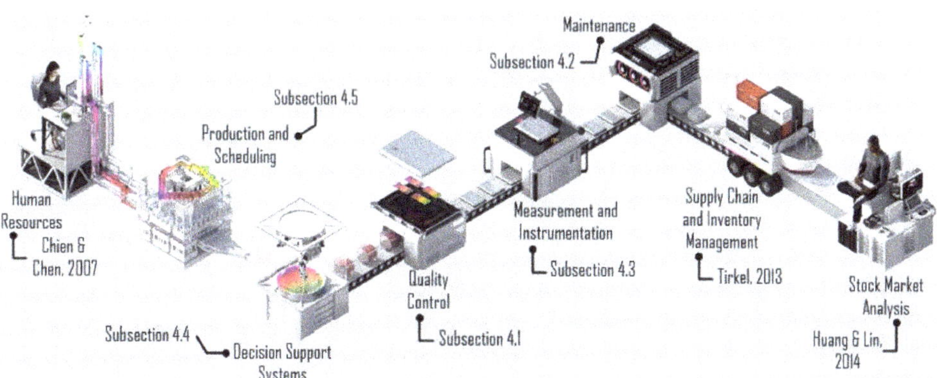

Figure 9. Schematic representation of several data mining applications in semiconductor manufacturing and localization according to categorized areas of application.

4.1. Data Mining Applications for Quality Control

Misaligned image processing can cause thousands of auxiliary operations and damaged wafers during a machine's life during the photolithography process, wafer scrutiny and inspection, or wafer mounting and cutting [65]. Inefficient image processing systems cost semiconductor companies market share and contribute significantly to their overall costs [66]. Data mining techniques are able to provide robust, precise, and fast wafer and chip pattern location for wafer inspection, probing, assembly, cutting, and test equipment to avoid such types of problems. These techniques allow manufacturers to control the quality of wafers and chips with high precision and accuracy, ensuring reliable equipment performance during the semiconductor manufacturing process.

The main purpose of quality prediction tools is to forecast the behavior of the product and then to be able to also forecast the trends of values of its critical parameters, typically accomplished by employ learning functions that have the capacity to stem knowledge from the preceding information. Forecasting quality with the help of data mining techniques normally starts by creating a model based on previous data, for instance labeling samples, and then assess and verify the unidentified samples, or to evaluate, from a given sample, the attributes' value ranges [67].

Table 2 shows the categorized papers by data mining applications for quality control in distinct steps of semiconductor manufacturing. These steps are identified, when possible, and can be found in the summary proposal. The table is subdivided into eight major columns and in a few can be observed the year of publication, reference, and the overall summarized description of the study. One of the remaining columns describes the proposed and/or used data mining algorithm, which can be helpful by quickly identifying a specific algorithm. The next column shows which DM technique is used. The remaining columns show if the sample data is collected from a real production site or if it was simulated, and if it is real, it is identified, when possible, by company and country of origin. Additionally, if experimental validation studies were performed on site, it is also highlighted.

Table 2. Data mining applications for quality control in distinct steps of semiconductor manufacturing.

Year	Overall Proposal	Proposed/Used Algorithm	DM Techniques	Real World Dataset	Real World Validation	Location of Dataset or Company	Refs.
2020	A review of data mining applications for quality control of semiconductor manufacturing	Several	Several	No	No	-	[67]
2020	Correctly identifying actual defective patterns in Wafer Bin Maps (WBM) to support the improvement of production yield	Hybrid clustering algorithm that integrates cluster analysis and spatial statistics	Clustering	Yes	Yes	-	[68]
2020	A new approach of measuring similarity of wafer bin maps in order to improve defect diagnosis and fault detection	Mountain clustering algorithm Weighted Modified Hausdorff Distance (WMHD)	Clustering	Yes	Yes	Taiwan	[10]
2020	An Expected Margin–based Pattern Selection model, that is able to select patterns based on an estimated margin for Support Vector Machines (SVMs) classifiers for wafer quality classification in the photolithography process	Expected Margin-based Pattern Selection (EMPS) Support Vector Machines (SVMs)	Classification	Yes	Yes	South Korea	[69]
2019	Fault detection and diagnosis model directly taken from the variable-length status variables identification (SVID) in the etch process	Convolutional neural networks (CNNs)	Classification	Yes	Yes	South Korea	[70]
2019	Clustering-based defect pattern detection and classification framework for WBMs	Density-based spatial clustering of applications with noise (DBSCAN)	Clustering	Yes	No	-	[71]
2019	An yield prediction model based on the selected critical process steps by taking into account difficulties such as imbalanced data, random sampling, and missing values	Expectation maximization (EM), MeanDiff technique, Synthetic minority over-sampling technique (SMOTE), decision tree, logistic regression, *k*-nearest neighbors (k-NN), and SVM	Classification Regression	Yes	No	-	[9]
2018	A framework based on Bayesian inference and Gibbs sampling to investigate the intricate semiconductor manufacturing data for fault detection	Bayesian inference, Gibbs sampling, high dimensional linear regression, multivariate adaptive regression spline (MARS), Cohen's kappa statistics	Classification	Yes	No	-	[5]
2018	Process errors detection and practical process improvement	Decision tree-based classification C4.5 in KNIME	Association rules	Yes	Yes	France	[19]
2018	A robust incremental on-line feature extraction method by ensuring the accuracy of data analysis and by meeting real-time demands of semiconductor manufacturing process for product quality supervision	PCA (Principal Component Analysis)RIPCA (Robust Incremental Principal Component Analysis) CCIPCA (Covariance-Free Incremental PCA)	(+)Feature selection/Dimensionality reduction	Yes	No	-	[72]
2018	Data mining applications semiconductor manufacturing process quality control	Fisher criterion algorithm, Support Vector Machines (SVMs) and Random Forest	Classification	Yes	No	Northern Ireland	[73]
2018	A mutually-exclusive-and-collectively-exhaustive feature selection framework applied to two cases of datasets, one being from a real manufacturing process	Mutually-exclusive-and-collectively-exhaustive (MECE) Two-phase clustering selection (TPS), stepwise selection (SS) Chi-Square Automatic Interaction Detector (CHAID)	(+)Feature selection/Dimensionality reduction	Yes	No	-	[74]

Table 2. *Cont.*

Year	Overall Proposal	Proposed/Used Algorithm	DM Techniques	Real World Dataset	Real World Validation	Location of Dataset or Company	Refs.
2017	Yield analysis operation performed by engineers with the aim of identifying the causes of failure from wafer failure map patterns and manufacturing historic records. An integrated automated monitoring system with deep learning and data mining techniques is proposed.	Convolutional Neural Networks (CNNs), Support Vector Machine (SVM), Clustering and pattern mining methods of K-Means++ and FPGrowth	Classification Clustering	Yes	No	-	[11]
2017	A data-driven approach for analyzing semiconductor manufacturing big data for low yield diagnosis purposes for detecting process root causes for yield improvement	Random Forest	Regression	Yes	Yes	Taiwan	[75]
2017	Comparison between Angle Based Outlier Detection (ABOD), Local Outlier Factor (LOF), onlinePCA (online Principal Component Analysis) and osPCA (os Principal Component Analysis) for semiconductor Manufacturing Etching process	Angle Based Outlier Detection (ABOD), Local Outlier Factor (LOF), onlinePCA, osPCA	(+) Outlier detection	Yes	No	-	[76]
2015	A statistical comparison of fault detection models for six datasets which were obtained by simulating of a plasma etching machine for a semiconductor manufacturing etching process	Support vector machine recursive feature elimination (SVM-RFE), principal component analysis (PCA), (k-nearest neighbors (kNN), SVMs, neural network (NN), logistic regression, partial least-squares discriminant analysis (PLS-DA), decision tree, squared prediction error, multi-way principal component analysis (MPCA)	Classification (+)Feature selection	No	No	-	[77]
2016	A simulator that carefully mimics data from a real etching process in a wafer production for the identification and prediction of unspecified situations by adopting data mining techniques to derive predictive patterns in order to detect flows and failures	Decision Tree, Naïve Bayes, Support Vector Machines with k-Means and hierarchical clustering	Regression Classification	No	No	-	[78]
2016	A wafer fault detection and essential step identification for semiconductor manufacturing by employing principal component analysis (PCA), AdaBoost and decision trees	Adaptive Boosting algorithm, decision trees, principal component analysis (PCA), SVMs	Classification	Yes	No	-	[79]
2016	Predictive analytics methods and its application in improving semiconductor manufacturing processes by considering several situations in semiconductor fabrication	Artificial neural networks (ANN), Clustering Method- K- Nearest Neighbor, robust regression	Classification	Yes	No	-	[80]
2015	A framework based on a linear model in order to obtain the weight tensor in a hierarchical manner for wafer quality prediction in semiconductor manufacturing	Hierarchical Modeling with Tensor inputs (H-MOTE algorithm), ridge regression, potential support vector machine (PSVM), tensor least squares (TLS)	Regression	Yes	No	-	[81]

Table 2. *Cont.*

Year	Overall Proposal	Proposed/Used Algorithm	DM Techniques	Real World Dataset	Real World Validation	Location of Dataset or Company	Refs.
2015	A data driven framework for degraded pogo pin detection in semiconductor manufacturing integrated circuit product testing process	Linear regression and classification algorithms (unspecified)	Regression Classification	Yes	No	USA	[82]
2016	A multi-feature sparse stacking-based approach for detecting defects and classification in produced semiconductor units	A proposed multi-feature sparse-based classification model Other models for comparison	Classification	Yes	No	Intel (USA)	[83]
2015	A combination of distinct data sources with the intention of identifying yield loss causes. The test is on a production step, comprising an implantation manufacturing step and its quality control step, a test done during the wafer sorting/probing (or wafer test).	K-means algorithm, "a priori" association rules mining algorithm, decision trees	Clustering Association rules	Yes	Yes	France	[84]
2014	A design-of-experiment (DOE) data mining for yield-loss diagnosis for semiconductor manufacturing (lithography, etching, among others) by detecting high-order interactions and show how the interconnected factors respond to a wide range of values	Regression analysis, Kruskal–Wallis test, Dunn's test, Holm–Bonferroni method, closed test procedure	Regression	Yes	Yes	Taiwan	[85]
2014	A yield analysis method employing basic yield and in-line defect information to statistically determine significant root-causes of yield loss in semiconductor manufacturing	Proposed yield accounting system, other unspecified	Classification	Yes	Yes	USA	[86]
2014	A morphology-based support vector machine for similarity search of binary wafer bin maps defect patterns during the probing test for yield enhancement	Support Vector Machines (SVM), morphology-based SVM (MSVM), Receiver Operating Characteristic (ROC), mountain method clustering	Classification	Yes	Yes	Taiwan	[87]
2014	Sequence mining and decision tree induction, to discover frequently occurred patterns of the low performance wafer lots in the semiconductor manufacturing industries	Decision Trees, Sequence Mining	Classification Association rules	No	No	-	[88]
2014	A united outlier detection framework that uses data complexity reduction by employing entropy and abrupt change detection using cumulative sum (CUSUM) method. Over an 8-month use period, the developed method was applied to reactive ion etching (RIE) and photolithography tools and recipes.	Algorithm I—Data Complexity Reduction Using Entropy Algorithm II—Abrupt Change Detection Using CUSUM	(+)Outlier detection	Yes	Yes	IBM (USA)	[89]
2014	A framework for root cause detection of sub-batch processing system in wafer testing and probing process	Random forest (RF), Sub-batch processing model (SBPM)	Regression	Yes	Yes	Taiwan	[90]

Table 2. *Cont.*

Year	Overall Proposal	Proposed/Used Algorithm	DM Techniques	Real World Dataset	Real World Validation	Location of Dataset or Company	Refs.
2013	An online detection and classification system of wafer bin map defect patterns during circuit probing tests	ART1 Neural Network Adaptive Resonance Theory algorithm	Classification	Yes	Yes	Taiwan	[91]
2013	Employment of k-means clustering algorithm by enhancing Support Vector Machines (SVM). Experiments with the real data of a semiconductor test process is given	K-means, Support Vector Machines (SVM), Synthetic Minority Over-sampling Technique (SMOTE)	Clustering	Yes	No	-	[92]
2013	A framework for semiconductor fault detection and classification (FDC) to monitor and analyze wafer fabrication profile data for the CVD Ti/TiN vapor deposition process	Principal component analysis (PCA), Multi-way PCA (MPCA), self-organizing map (SOM) neural network	Classification	Yes	Yes	Taiwan	[93]
2012	An optimization framework for hierarchical multi-task learning, which partitions all the input features into two sets based on their characteristics applied in the process of depositing dielectric materials as capping film on wafers	HEAR algorithm (MTL with Hierarchical task Relatedness) based on block coordinate descent	Classification	Yes	No	-	[14]
2012	A main branch decision tree (MBDT) algorithm that diagnoses the root causes and provides quick responses to irregular equipment operation in the wafer acceptance testing and probing processes with imbalanced classes	Main branch decision tree (MBDT) algorithm	Classification	Yes	Yes	-	[94]
2012	A two-phase morphology-based similarity search for wafer bin maps in semiconductor manufacturing for wafer acceptance testing	Support Vector Machines (SVM)	Classification	Yes	No	-	[95]
2011	A technique based on the data mining technology to automatically generate an accurate model to predict faults during the wafer fabrication process of the semiconductor industries	Principal component analysis (PCA), cluster technique MeanDiff, decision tree, naïve Bayes, logistic regression, and k-nearest neighbor	Regression Classification	Yes	No	-	[96]
2019	An altered AdaBoost tree-based method for defective products identification in wafer testing process	AdaBoost Tree-based method Synthetic Minority Oversampling Technique (SMOTE) + Edited Nearest Neighbor (ENN)—SMOTE-ENN algorithm	Classification	Yes	No	-	[97]
2006	Wavelet-based data reduction techniques for fault detection in rapid thermal chemical vapor deposition processes (RTCVD)	Discrete wavelet transforms, classification and regression tree (CART)	Classification Regression	Yes	No	-	[15]
1999	Effectiveness of association rules and decision trees data mining techniques in determining the causes of failures of a wafer manufacturing process	Association rules and decision trees	Association rules Classification	Yes	No	-	[98]

Table 2. *Cont.*

Year	Overall Proposal	Proposed/Used Algorithm	DM Techniques	Real World Dataset	Real World Validation	Location of Dataset or Company	Refs.
2008	A spatial defect diagnosis system at the probing test which estimates number of clusters in advance and separates both convex and non-convex defect clusters at the same time	Decision trees, a method merging entropy fuzzy c means (EFCM) with Kernel based spectral clustering	Classification	Yes	Yes	Taiwan	[99,100]
2007	A framework that combines traditional statistical methods and data mining techniques for fault diagnosis and low yield product for wafer acceptance testing and probing	Kruskal–Wallis test, K-means clustering, and the variance reduction splitting criterion, decision trees	Clustering Classification	Yes	Yes	Taiwan	[13]
2007	A hybrid data mining method that integrates spatial statistics and adaptive resonance theory neural networks to extract patterns from WBMs	Adaptive resonance theory (ART), Decision trees, Classification and regression tree (CART)	Classification	Yes	Yes	Taiwan	[34]
2007	A Bayesian networks to extract knowledge from data ant the purpose is to implement a data mining task for computer integrated manufacturing (CIM). The end goal is to encounter the cause factors in various parameters which have an effect during the wafer cleaning process	Bayesian networks, directed acyclic graph, decision trees	Classification	Yes	Yes	-	[101]
2007	Data mining technique by utilizing Gradient Boosting Trees for predicting class test yield performance at high volume semiconductor manufacturing after assembly and final testing	Gradient boosting trees (GBT) ensemble algorithm	Regression	Yes	Yes	Intel (Malaysia)	[102,103]
2006	An on-line diagnosis system that relies on denoising and clustering methods for identifying spatial defect patterns in semiconductor manufacturing processes	Integrated clustering scheme combining fuzzy C means (FCM) with hierarchical linkage, decision trees	Clustering	Yes	Yes	Taiwan	[104]
2006	A data mining technique to predict and classify the product yields in semiconductor manufacturing processes in wafer acceptance testing and probing	Genetic programming, Decision trees	Classification	Yes	Yes	Taiwan	[105]
2000	A combination of self-organizing neural networks and rule induction employed in the identification of poor yield factors from collected wafer probing manufacturing data	Self-organizing neural networks and rule induction	Classification Association Rules	Yes	Yes	USA	[106]

This topic is the most popular one, with 47 publications. By observing Table 2, it can be seen that several applications are made in distinct subprocesses such as wafer probing and testing process, etching process, and photolithography, among others. A high and varied number of algorithms are employed. The majority of articles address challenges of correctly identifying defective patterns in order to improve production yield [68]. Yield is a quantitative measure of the quality of a semiconductor process. It is measured as the number of functioning dies or chips on a wafer and can also be seen as the fraction of dies on the yielding wafers that are not rejected during the production process [107]. However, other applications in quality control can also be found, such as a study addressing a design-of-experiment (DOE) data mining for yield-loss diagnosis for semiconductor manufacturing by detecting high-order interactions, for subprocesses such as lithography and etching, among others [85]. These data mining technique are also used with statistical process control. Cumulative sum control charts, known as CUSUM, are a special type of statistical process control tool that is used in [89] as part of and unified outlier detection framework, which takes advantages of data complexity reduction by employing entropy and sudden change detection through the use of CUSUM charts.

4.2. Data Mining Applications for Maintenance

Only a few articles were published addressing maintenance management and prediction, but are important nonetheless. Only five papers were classified and can be observed in Table 3. This table is organized as Table 2. As it can be noticed, these studies are sparse and the majority were published in the last 8 years. However, the most cited article is a study in this area of application. In this study a multiple classifier machine learning methodology for predictive maintenance in the ion implantation subprocess is proposed [30] and a similar study is proposed in [16]. In another study, hidden Markov model-based predictive maintenance for semiconductor wafer production equipment and documented over one year was proposed in [108]. A data mining technique that is able to deliver early warning by identifying tool excursion in real time for advanced equipment control in order to diminish atypical yield loss is proposed in [109] and was validated by practical applications in the field. Finally, the last study addresses spatial pattern recognition in order to improve the resolution and identification of defective and malfunctioning tools in semiconductor manufacturing developed and implemented at Advanced Micro Devices, Inc. (AMD) [110].

Table 3. Data mining applications for maintenance prediction and management in semiconductor manufacturing.

Year	Study Proposal	Proposed/Used Algorithm	DM Techniques	Real World Dataset	Real World Validation	Location of Dataset or Company	Ref.
2017	Hidden Markov model-based predictive maintenance for semiconductor wafer production equipment, recorded over one year	Preliminary fitting of a hidden Markov model (HMM) Genetic, genetic algorithm		Yes	No	-	[108]
2016	Predictive Maintenance with time-series data based on Machine Learning tools in Ion implantation	Supervised Aggregative Feature Extraction (SAFE)		Yes	No	-	[16]
2015	A multiple classifier machine learning technique used for predictive maintenance in Ion implantation process	Support Vector Machines k-Nearest Neighbors	Classification Clustering	Yes	No	-	[30]
2012	Data mining technique that is able to deliver early warning by identifying tool excursion in real time for advanced equipment control in order to diminish abnormal yield loss	Decision trees, Chi-Squared Automatic Interaction Detector, Rough set theory	Classification	Yes	Yes	Taiwan	[109]
2008	Spatial pattern recognition to improve the identification and resolution of rogue and possibly malfunctioning tools in semiconductor manufacturing	Spatial pattern recognition	(+)Feature selection	Yes	Yes	AMD (USA)	[110]

4.3. Data Mining Applications for Metrology, Measurement, and Instrumentation

The high necessity for always striving to make progress regarding the yield of current semiconductor production processes and decrease the time-to-market for more advanced, innovative, and gradually elaborate designs and processes demands for process tools and wafers to be examined and verified with up-to-date measurement systems and equipment. Several papers, namely 19, are categorized in this topic, as depicted in Table 4. This table is organized as Table 2. The topics addressed in this section range from models comprising a precise semiconductor photolithography process control method through virtual metrology by employing significant correlations between focus measurement data encountered by data mining and tool data [111].

In fact, virtual metrology is a recurring topic, and is defined as a set of methods that allow predicting the properties of a wafer through sensor data and machine parameters in the manufacturing equipment, thus avoiding the highly expensive physical measurement of the wafer properties [112–114]. Since machine data is typically sampled much more often when compared to metrology data, and since machine data becomes immediately available when compared to the delays that frequently occur with metrology tools, an accurate virtual metrology is capable of meaningfully developing the process control and monitoring performance through a constantly supply of real-time forecasted metrology data. A few feature extraction methods for virtual metrology with multisensor data are proposed in [17,115,116].

However, other measurement and instrumentation were also proposed and classified. For instance, in [117] a real-time data mining solution with the segmentation, detection, and cluster-extraction (SDC) algorithm that can automatically and accurately extract defect clusters from raw wafer probe test production data is proposed. Additionally, a data mining that employs machine learning methods with the purpose of modeling unknown functional interrelations and to predict the thickness of dielectric layers deposited onto a metallization layer of the manufactured wafers is proposed in [118]. Finally, at IBM, a data mining technique with the purpose of automatically identifying and exploring correlations between inline measurements and final test outcomes in analog and/or radio frequency (RF) devices and by integrating domain expert feedback into the algorithm in order to identify and remove bogus autocorrelations [119]. Practical application and validation of this technique is made.

Table 4. Measurement, metrology, and instrumentation data mining applications.

Year	Study Proposal	Proposed/Used Algorithm	DM Techniques	Real World Dataset	Real World Validation	Location of Dataset or Company	Ref.
2019	Automatic method for extraction of signatures from the raw data generated by non-rotating equipment	Virtual metrology Genetic Algorithms	(+)Feature selection	Yes	No	-	[120]
2019	A Deep Learning method for Virtual Metrology that employs semi-supervised feature extraction reliant on Convolutional Autoencoders for a 2-dimensional Optical Emission Spectrometry data	Convolutional Neural Networks Deep Learning Virtual metrology	(+)Feature selection	Yes	No	-	[115]
2019	A feature extraction technique for virtual metrology with multisensor data in semiconductor manufacturing that relies on deep autoencoder which also offers a clipping fusion regularization on the signals reconstructed by deep autoencoder in the case of an etching process for wafer fabrication	Principal component analysis (PCA) Virtual metrology, unsupervised deep autoencoder (AE)	(+)Feature selection	Yes	No	-	[17]
2016	A Euclidean distance- and standard deviation-based characteristic selection and over-sampling used in a fault detection prediction model and applied to measure performance	Principal component analysis (PCA), SVM (Support Vector Machine), C5.0 (Decision Tree), KNN (K-nearest neighbor), Artificial neural network (ANN)	(+)Feature selection Classification	Yes	No	-	[121]
2017	OpenMV—a low-power smart camera with wireless sensor networks and machine vision applications, it is scripted in Python 3 and comes with an extensive machine vision library	Support vector machine-like (SVM-like) algorithm	Classification	No	No	-	[122]
2014	A precise semiconductor photolithography process control method using virtual metrology using significant correlations between focus measurement data found by data mining and tool data	Virtual metrology Correlation coefficient mining algorithm	(+)Feature selection	Yes	Yes	-	[111]
2014	A Feature Selection wrapper method aiming to find the most important process parameters for smart virtual metrology for High Density Plasma (HDP) Chemical Vapor Deposition	Virtual metrology, Evolutionary Recursive Backward Elimination (ERBE) algorithm, Genetic Algorithms, Support Vector Regression (SVR)	Regression	Yes	Yes	-	[116]
2014	A framework in which the structural information from etching is interpreted as a set of constraints on the cluster membership, an auxiliary probability distribution is then introduced, and the design of an iterative algorithm is prosed for assigning each time series to a certain cluster on every dimension	K-Means algorithm, C-Struts framework, complex-valued linear dynamical systems (CLDS)	Clustering	Yes	No	-	[123]
2013	Data Mining utilizing machine learning techniques for modeling unknown functional interrelations in the high-density plasma chemical vapor deposition process. It predicts the layer thickness through Support Vector Regression	Support Vector Machine (SVM), Support Vector Regression (SVR)	Classification	Yes	No	-	[124]

Table 4. *Cont.*

Year	Study Proposal	Proposed/Used Algorithm	DM Techniques	Real World Dataset	Real World Validation	Location of Dataset or Company	Ref.
2013	Data Mining using Machine learning methods to model to model unknown functional interrelations and to predict the thickness of dielectric layers deposited onto a metallization layer of the manufactured wafers.	Decision Trees (DT) Neural Networks (NN) Support Vector Regression (SVR)	Classification Regression	Yes	No	-	[118]
2011	A qualitative clustering method is given, and a comparison is made between a Virtual Metrology (VM) system running on groups of data with the same targets and one obtained by considering the three chambers of the Chemical Vapor Deposition equipment as separated machines	Back Propagation Neural Networks (BPNN) Partial Least Square (PLS) Regression	Clustering Classification	Yes	No	-	[125]
2011	A real-time data mining model by using a Segmentation, Detection, and Cluster-Extraction algorithm that is able to accurately and automatically extract defect clusters from raw wafer probe test production data	Segmentation, Detection, and Cluster-Extraction (SDC) algorithm	Clustering	Yes	Yes	Malaysia	[117]
2011	A multivariate feature selection able of handling mixed and complex typed data sets as an initial step in yield analysis to reduce the number of variables	Ensemble-Based Feature Selection algorithm, gradient boosted tree (GBT)	Regression	Yes	No	-	[126]
2011	Development of virtual metrology (VM) prediction models using several data mining technique and a VM embedded R2R control system by employing exponentially weighted moving average (EWMA) based on data from a photolithography production equipment	Decision trees, GA with linear regression, GA with support vector regression (SVR), Principal component analysis (PCA), and kernel PCA, multi-layer perceptron (MLP), k-nearest neighbor regression (k-NN)	Regression	Yes	Yes	South Korea	[127]
2011	A data mining method for automatically identifying and exploring correlations between inline measurements and final test outcomes in analog/RF devices and incorporate domain expert feedback into the algorithm for identifying and removing spurious autocorrelations	Multi-objective genetic algorithm (NSGA-II), Genetic algorithms (GA), Multivariate Adaptive Regression Splines (MARS)	Regression	Yes	Yes	IBM (USA)	[119]
2009	A virtual metrology (VM) system for an etching process in semiconductor manufacturing based on various data mining techniques	Genetic algorithm with support vector regression (GASVR), Principal component analysis (PCA), and kernel PCA, Stepwise linear regression	Regression	Yes	Yes	South Korea	[128]
2006	A 2nd Generation Data Mining system in cooperation with Advanced Process Control (APC) system and that aim to stabilize machine fluctuation in Photolithography Process	Regression tree analysis, proposed 2nd Generation Data Mining algorithm	Regression	Yes	Yes	Fujitsu (Japan)	[129]

Table 4. *Cont.*

Year	Study Proposal	Proposed/Used Algorithm	DM Techniques	Real World Dataset	Real World Validation	Location of Dataset or Company	Ref.
2006	A pre-processing procedure used for numerous sets of complex functional data for reducing data size for the support of appropriate decision analysis. This vertical-energy-thresholding (VET) procedure balances the reconstruction error with data-reduction efficiency	Vertical-energy-thresholding (VET), wavelet-based procedure	(+)Dimensionality reduction	Yes	Yes	Nortel (USA)	[130]
2005	An automatic classification of the electrical wafer test maps in order for identifying the classes of failure present in the production lots, especially due to a lithographic process	Commonality analysis (CA), Kohonen's self-organizing feature maps algorithm	Classification	Yes	Yes	STMicroelectronics(Italy)	[131]

4.4. Decision Support Systems

Another trend in semiconductor manufacturing is the use of decision support systems (DSS). A DSS is a system designed to support in solving unstructured and semistructured managerial problems, throughout all the decision process' stages [132]. The DSS use in this area is not novel. Earliest publications in this area date to the 1990s (e.g., [133,134]). DSSs are used to support decision-making in activities like production scheduling, simulation, prediction, material selection, fault detection, quality, etc. DSSs may, sometimes, have a knowledge base, which requires artificial intelligence to provide knowledge to support the decision process. However, the earliest uses of DSS required knowledge modeling by knowledge engineers from documented and expert knowledge. Knowledge extraction from unprocessed data allowed one to discover hidden knowledge in large amounts of data. The use of data mining techniques to uncover knowledge to be modeled in DSS is a trend also present in semiconductor literature. Researchers apply data mining techniques to find patterns and hidden relations that may help in semiconductor decision making. Usually, the goal is to determine links between control parameters and product quality, essentially in the form of decision rules [135].

In Table 5 the literature where data mining is used to support the decision-making process in semiconductors' manufacturing is presented. Analyzing this table, one can see that most contributions address yield management and failure detection issues (see [135–145]). The authors from [146] aim at the same problem, but focus on the development of a computer integrated manufacturing (CIM) system to improve product yield. Other articles provide isolated contributions. In [147], the authors propose the application of data mining techniques to support decision-making in HR management of high-tech companies. In [148], the authors suggest the integration of data mining in semiconductor manufacturing execution systems (MES). Last, in [32] provides a multi-purpose data mining application for predictions in semiconductor manufacturing.

Table 5. Data mining applications for decision support systems.

Year	Study Proposal	Proposed/Used Algorithm	DM Techniques	Real World Dataset	Real World Validation	Location of Dataset or Company	Ref.
2019	The results for yield improvement of our silicon carbide technology using advanced data analytics by outlining how the data was collected, preprocessed and managed in order to turn it much more appropriate for further analysis	Unspecified	(+)Generic	Yes	Yes	Northrop Grumman (USA)	[149]
2018	A new balanced production method for holistic optimization of operation strategies applied to semiconductor manufacturing	DBSCAN clustering algorithm Genetic optimization algorithm	Clustering	Yes	Yes	-	[150]
2015	Development an analytic framework of design for semiconductor manufacturing and validated through a case study in semiconductor manufacturing concerning the layout design of chip size	Model tree (M5), Regression tree (CART) Neural Network (BPNN)	Regression Classification	Yes	Yes	-	[151]
2013	A framework in which the packaging yield is classified using the parametric test data of the previous step of the packaging test in the post-fabrication process for semiconductor manufacturing	Random forests algorithm, support vector machine (SVM)	Classification	Yes	Yes	SK Hynix Semiconductor (South Korea)	[152]
2012	A procedure for the optimization processes named: values-Patient Rule Induction Method (m-PRIM) by addressing the missing-values systematically	Missing Values Patient Rule Induction Method (PRIM)	Association rules	Yes	No	South Korea	[153]
2001	An integrated relational database method for modeling and collecting semiconductor manufacturing data from multiple database systems and transforming it into useful reports	Integrated Relational Manufacturing Database		Yes	Yes	Motorola (USA)	[154]
2012	Knowledge discovery in databases model that relies on decision correlation rules and contingency vectors to enhance semiconductors manufacturing yield	Association and correlation rules, LHS-CHI2 algorithm	Association rules	Yes	Yes	STMicroelectronics, ATMEL	[135]
2011	Rare class prediction for fault case detection in the wafer fabrication process of semiconductor industries	Decision tree induction, naïve Bayes, logistic regression, k-nearest neighbors	Association rules Classification Clustering	Yes	No	SECOM	[136]
2011	Application of rough set theory, support vector machines and decision trees for improving the quality of decisions of class prediction and rule generation encompassed in human resource management.	Rough sets theory, support vector machines, decision trees	Classification	Yes	Yes	UCI data bank	[147]
2011	Development of a rare case prediction for fault case detection in the wafer fabrication process	Decision tree induction, naïve Bayes, logistic regression, k-nearest neighbors	Association rules Classification Clustering	Yes	No	SECOM	[137]

Table 5. *Cont.*

Year	Study Proposal	Proposed/Used Algorithm	DM Techniques	Real World Dataset	Real World Validation	Location of Dataset or Company	Ref.
2010	Propose a system do improve yield, power consumption and speed characteristics using regression rule learning to analyze data collected during wafer production	Regression rule learning, association rules	Association rules	Yes	No	-	[138]
2008	A system to evaluate measurements from a semiconductor production process using feature selection to identify rules	Neural networks, feature selection, simplified fuzzy ARTMAP	Classification	Yes	No	-	[139]
2007	Proposes ensemble classifiers to support decision-making to enhance yield in semiconductor production	Ensemble classification	Regression	Yes	No	.	[140]
2006	Integration of Data Mining techniques in a MES for semiconductor manufacturing	Decision tree	Classification	Yes	No	-	[148]
2006	Combines forward regression and regression tree methods to discover yield loss causes during the yield ramp-up stage	Decision trees, multiple linear regression	Regression	No	No	-	[141]
2005	Uses data mining techniques to design intelligent CIM applied to improve product yield of semiconductor packaging factories.	Decision tree	Classification	No	No	-	[146]
2005	Proposes a model based on decision trees to recognize and classify failure pattern using a fail bit map	Decision tree	Classification	No	No	-	[142]
2004	Proposes a fault detection scheme using a hierarchical fuzzy ruled based classifier to identify defects in wafers	Hierarchical fuzzy rule-based classifier	Classification	Yes	Yes	-	[143]
2003	Proposes a conceptual e-Commerce decision support system that integrates intelligent agents and data mining to help in the sampling process of semiconductor quality	None	(+)Generic	No	No	-	[144]
2001	Proposes the use of neural networks to design in-line measurement sampling methods to monitor and control semiconductor manufacturing	Neural networks	Classification	Yes	No	-	[145]
2001	Proposes a rule-structuring algorithm based on rough set theory to make predictions for semiconductor industry	Rough set theory	Association rules	No	No	-	[32]

4.5. Data Mining Applications for Production and Production Scheduling

Traditional methods for production planning often require complex calculations and do not always allow a prompt reaction to changes or short-term adjustments that may arise. Given the size of the semiconductor production lines in a factory, sensors within production equipment are capable of delivering enormous amounts of data. This data can be, in turn, used not only for machine control, but also for production analysis purposes, especially real-time production planning. This has the potential to bring great advantages, especially in those industrial units in which the production is affected by frequent dynamic changes in the orders to be processed or technical specifications. Additionally, machine learning processes are able to recognize patterns and automatically learn and operationalize practical forecast models from a wide variety of data sources and large amounts of data. Therefore, in the context of semiconductor manufacturing with its complex and numerous subprocesses, numerous data mining applications are proposed for the production and production planning environment.

Table 6 depicts the articles addressing data mining applications for production in semiconductor manufacturing. A total of 16 papers were found in this category. This table is structured as Table 2. It can be noticed that from 2009 until 2015 is when the bulk of these studies were published, then a four-year hiatus was observed. From 2019 can be noticed some interest in the topic.

Many of the studies concerning production planning are focused on reducing cycle time. In [155], a new approach that is capable of integrating data mining that intends to forecast arrival rates and determining the allocation of interchangeable tool sets in order to reduce the work in process (WIP) bubbles for cycle time reduction is proposed. While in another study [64], a cycle time forecasting model is developed by employing knowledge discovery in databases by following cross industry standards for data mining. A data-mining approach for estimating the interval cycle time of each job in a semiconductor manufacturing system is proposed in [156] and a data mining methodology, which identifies key factors of the cycle time in a semiconductor manufacturing plant, which intends to predict its value is addressed in [157].

Scheduling is another concern in semiconductor manufacturing due to its vast number of steps and jobs [158–160], confirmed by the majority of the identified studies in Table 6. Efficient order scheduling structures are required for balancing the production load and capacity throughout all the production stages [161]. A data mining dynamic scheduling strategy selection model that is able to respond to a constantly altering system status for a semiconductor manufacturing system is proposed in [18]. In [162] a data-driven scheduling knowledge life-cycle management for an intelligent shop floor is proposed and validated through a simulation model of the semiconductor production line. As early as in 2004 scheduling challenges were a concern, evidenced by a study proposing an hierarchical clustering method in [163] that is able to discriminate groups according to the similarity of the objects and used to schedule semiconductor manufacturing processes. In [164] a dynamic scheduling model, which is able to optimize the production features subset is proposed, and this model is capable of creating a SVM-based dynamic scheduling strategy classification model for semiconductor manufacturing. A data-based scheduling framework and adaptive dispatching rule for semiconductor manufacturing is addressed in [165] by employing backward propagation neuronetworks (BPNNs). Finally, a shop floor control system in semiconductor production by self-organizing map-based smart multicontroller is given in [166]. This study, as all the scheduling studies, showed a better system performance than the typical fixed decision scheduling rules.

Table 6. Data mining applications for production in semiconductor manufacturing.

Year	Study Proposal	Proposed/Used Algorithm	DM Techniques	Real World Dataset	Real World Validation	Location of Dataset or Company	Refs.
2004	A decision tree algorithm and classification model are proposed. Intelligent computer integrated manufacturing (CIM) system is applied to semiconductor packaging factories. The manufacturing cycle time, the product yield, and the frequency of holding lot were improved	Decision trees	Classification	Yes	Yes	-	[167]
2020	A new approach that is able to integrate data mining that intends to forecast arrival rates and determine the allocation of interchangeable tool sets in order to decrease the work in process (WIP) bubbles for cycle time reduction	Back-propagation neural network (BPNN)	Classification	Yes	Yes	Taiwan	[155]
2019	A data-driven scheduling knowledge life-cycle management for an intelligent shop floor and validated through a simulated model of the semiconductor production line	Extreme learning machine (ELM), Online sequential extreme learning machine (OS-ELM)	Classification	No	No	-	[162]
2015	A data mining based dynamic scheduling strategy selection model which is able to respond to altering system status in semiconductor manufacturing processes	genetic algorithm K-nearest neighbor algorithm	Clustering	Yes	Yes	-	[18]
2015	A variation reduction of Turn Around Time (TAT) in a semiconductor manufacturing through a data mining-based technique for identifying the root cause of TAT variation	Partial Least Squares Regression (PLSR)	Regression	No	No	-	[168]
2014	A data mining framework that is capable of integrating fault detection and classification and manufacturing execution system data for improving the overall usage effectiveness (OUE) for cost reduction in a Chemical Mechanical Planarization (CMP) process	CHAID (Chi-Squared Automatic Interaction Detection) Decision Trees	Classification	Yes	Yes	Taiwan	[169]
2014	A dynamic scheduling model which optimizes production features subset, and creates an SVM-based dynamic scheduling strategy classification model for semiconductor manufacturing	Particle swarm optimization algorithm (BPSO), support vector machine (SVM)	Classification	Yes	Yes	China	[164]
2013	A noted cycle time forecasting model is developed by employing knowledge discovery in databases by following cross industry standards for data mining	Decision trees, Neural networks	Classification	Yes	No	-	[64]
2013	A Data-based scheduling framework and adaptive dispatching rule for semiconductor manufacturing	Backward propagation neuro-network (BPNN), adaptive dispatching rule (ADR)	Classification	Yes	No	-	[165]

Table 6. *Cont.*

Year	Study Proposal	Proposed/Used Algorithm	DM Techniques	Real World Dataset	Real World Validation	Location of Dataset or Company	Refs.
2011	A cycle-time key factor identification and prediction in semiconductor manufacturing by employing data mining and machine learning	Selective naive Bayesian classifier (SNBC) Conditional mutual information maximization (CMIM)	Classification	No	No	-	[170]
2012	A shop floor control system in semiconductor production by self-organizing map-based smart multi-controller showing an improved system performance than fixed decision scheduling rules	Self-organizing map (SOM) neural network	Classification	No	No	-	[166]
2010	Gaussian Processes used for decentralized scheduling with dispatching rule selection in production scheduling for semiconductor manufacturing	Gaussian processes, neural networks	Classification	No	No	-	[171]
2010	A machine learning algorithm capable of implementing an adaptive sequential (A-S) process and accuracy guard band model for improved recipe generation process development in the assembly semiconductor manufacturing processes	Polynomial-based RSM Response Surface Methodology (RSM), Adaptive-sequential (A-S) algorithm	Regression	Yes	Yes	Intel (Malaysia)	[172]
2009	A data-mining approach for estimating the interval cycle time of each job in a semiconductor manufacturing system	Look-ahead self-organization map fuzzy-back-propagation network (SOM-FBPN)	Classification	No	No	-	[156,173]
2009	A data mining methodology which identifies key factors of the cycle time in a semiconductor manufacturing plant which intends to predict its value	Naïve Bayesian classifier (NBC), CRISP-DM (Cross-Industry Standard Process for Data Mining)	Classification	No	No	-	[157]
2004	A hierarchical clustering method that is able to discriminate groups according to the similarity of the objects and used to schedule semiconductor manufacturing processes	Agglomerative hierarchical cluster algorithm	Clustering	No	No	-	[163]

5. Discussion

After analyzing all the studies collected in the sample, a few trends begin to be noticed. First, that studies regarding data mining applications in subprocesses such as ICs and mask design are very scarce. The same occurs with studies addressing wafer cutting, cleaning drying, and polishing, while edge rounding and lapping subprocess has no dedicated study. This is better illustrated by Figure 10 in which a representation of several studies depicting data mining applications in several subprocesses of semiconductor manufacturing can be seen. It is noticeable that the majority of studies are concentrated in 5–6 major steps. A few studies do not specify in which subprocess data mining techniques are applied, and these are not represented in Figure 10.

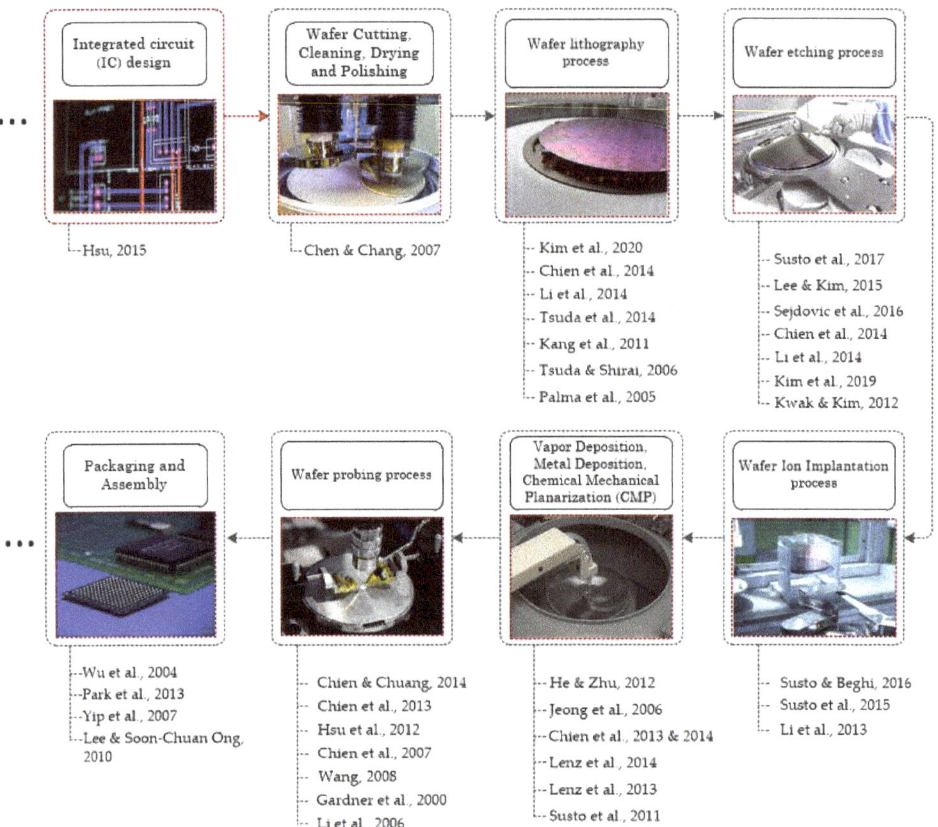

Figure 10. Representation of several studies depicting data mining applications in several subprocesses of semiconductor manufacturing.

Another trend visible in the analyzed literature is the diverse use of data mining techniques. The application of data mining in semiconductor manufacturing has a different focus depending on the subject areas concerning the manufacturing processes. However, most articles address mainly the issues of quality control, maintenance, and production. Predictive techniques, using algorithms as regression or decision trees, are often used in semiconductor literature to estimate wafer quality [81], fault detection [121,136], or cycle-time [170]. Classification techniques in quality control arise as a way to classify defects [83], failures in bin maps [91], or production lots [131]. The exploration of yield loss causes [84] or failure diagnostics [98] is performed using techniques as rule induction, decision trees, and association rules.

Many opportunities and improvements can still be made. For example, the semiconductor companies could employ the internet of things and sensors to empower industrial units with the capability of interpreting data and transmitting analytics, in real time, to an application that could provide insights and alerts to whom it may concern [174]. This will allow these players to gather a high amount of data. However, even though internet of things and data mining applications represent a key opportunity for semiconductor manufacturing companies—one that they should start to pursue as soon as possible, while the use of data mining in the sector is still developing under the current upgrading environment. Nevertheless, the effectiveness and scale of the internet of things implementation, and with it a comprehensive use of data mining techniques, could depend on how fast industry players can overcome some challenges [175]. In order to persevere and being able to accompany the change speed and challenges, semiconductor companies are required to adapt rapidly. Taking into account this dynamic, industrial units should embrace digitalization in an agile manner as well [176].

Limitations and Challenges

Even though employing data mining techniques has been very beneficial for this industry, as shown by all the studies used in this review, several disadvantages of data mining still exist and are as follows:

- Data mining systems can violate privacy. Absence of safety and security can be very detrimental to its users and it can create miscommunication between employees, thus leading to genuine privacy concerns [177].
- Security is an important factor related to every data-oriented technology, and semiconductor manufacturing is not an exception. Data that is very critical might be a target of malicious attacks [178].
- Too much and redundant information collection can be disadvantageous as irrelevant collected information is a challenge [179,180].
- There is a possibility of information misuse through the mining process. Data mining system have to evolve in order to diminish the misuse of the information ratio [181].
- Accuracy of data mining techniques is another limitation [182]. Accuracy is an evaluation system of measurement on how well a data mining model can perform. Many common accuracy and error scores for regression and classification can occur. Therefore, improving accuracy becomes paramount.
- Several challenges of data integration and interoperability in data mining can occur. Data interoperability and data integration affect the performance of an organization. A comprehensive approach has to be made in order to address the challenges in interoperability and integration [183,184].
- Missing and imbalanced data is a challenge in this industry. In cases in which data is imbalanced, the majority of classification algorithms have as a consequence a weak performance. Since wafer yield enhancement is a crucial performance index in semiconductor wafer manufacturing, key process steps must be cautiously selected and managed [9].
- Data processing time is another limitation that has a significant impact on the available time since data preprocessing very often involves more than 50% of time and effort of the entire data analysis process [185].

This evolution of semiconductor manufacturing relies heavily on the big data explosion in order to cope with the abovementioned data limitations and challenges of the semiconductor industry. Especially, supporting greater volumes and lengthier archives of data has allowed many solutions to correctly portray system dynamics, significantly simplify intricate multivariate interactions of parameters, eliminate disturbances, and clean and overcome data quality challenges. Data mining algorithms in such types of solutions must be rewritten in order to benefit from the parallel computation allowed by the high processing capacity and storage power with the purpose of processing data without consuming too much time. However, an enormous amount of data and a wide range of data mining techniques does not mean necessarily more predictive capability and insights [186]. Researchers and practitioners have to adapt data mining techniques in a manner so that these will be customized to specific applications in terms of data quality available data and objective, among others.

Overall, through this review, some light was shed over the possible applications of data mining techniques in semiconductor manufacturing. Yet, given the sheer number of steps that this production process has, and due to its complexity, the number of studies already made is still scarce. Big data and data mining allowed for original and innovative insights through the analysis of large amounts of data and presenting correlations and opportunities that were not previously noticed. However, decision makers must decide and which data should be collected and employed and which questions must be answered [149]. This signifies that the potential to apply these techniques in other subprocesses is enormous and is still left largely unexplored. Finally, by suffering constant and quick evolution, the need to adapt these techniques to the newer processes in semiconductor manufacturing is another opportunity to explore.

6. Conclusions

The production of semiconductors is a highly complex process, which entails several subprocesses that employ a diverse array of equipment. The size of the semiconductors signifies a high number of units can be produced, which require huge amounts of data in order to be able to control and improve the semiconductor manufacturing process. Therefore, in this paper a structured review was made through a sample of 137 papers of the published articles in the scientific community regarding data mining applications in semiconductor manufacturing. A detailed bibliometric analysis was made. All data mining applications were classified in function of the application area. Five distinct areas were identified: quality control, maintenance, production, decision support systems, and finally, categorized as a whole, measurement, metrology, and instrumentation. Results showed that quality was the most popular one, with 47 publications, making 34.3% of all publications. Maintenance was an area in which only a few studies were made, highlighting the gap and the opportunity for more studies to be made in this area.

The work performed in this study concerning data mining applications in semiconductor manufacturing can have theoretical implications. The characterization and categorization of several useful and successful cases can positively contribute to future research efforts of employing such a wide range of techniques with the purpose of increasing the application and diffusion of data mining applications in semiconductor manufacturing. Knowledge of different models and algorithms could have positive implications for the development of theory, for understanding all the possible applications in different areas of semiconductor production, but also for the development of practice, since many of these were implemented and validated on the shop floor. However, as the literature review has shown, many applications can still be made since several studies address only a specific step of semiconductor manufacturing and documentation of real-life application are scarce. Additionally, recent data mining techniques and models have a great opportunity to be used since only a few studies exist. Finally, since the semiconductor manufacturing process is always evolving, the need to adapt these techniques to the newer process is another challenge and opportunity to explore.

Overall, as seen from all the comprised studies from distinct steps of semiconductor production, the scope and functions of data mining techniques can be enhanced and disseminated throughout the entire semiconductor manufacturing process in order to provide, in real time, a proactive adjustment and advanced control decisions for the whole process and the smart facilities. Therefore, more research should be made to employ and facilitate smart production for Industry 4.0 in several industries for digital transformation and for upgrading existing manufacturing units. This will allow for an improving capability for optimizing interrelated decisions and improving decision flexibility.

Author Contributions: Conceptualization and methodology, R.G. and P.E.-C.; software R.G.; validation and investigation, R.G. and P.E.-C.; review and editing, E.M.G.R. All authors have read and agreed to the published version of the manuscript.

Funding: The authors acknowledge Fundação para a Ciência e a Tecnologia (FCT-MCTES) for its financial support via the project UIDB/00667/2020 (UNIDEMI).

Institutional Review Board Statement: Not applicable.

Informed Consent Statement: Not applicable.

Data Availability Statement: Not applicable.

Conflicts of Interest: The authors declare no conflict of interest.

References

1. Biebl, F.; Glawar, R.; Jalali, A.; Ansari, F.; Haslhofer, B.; de Boer, P.; Sihn, W. A Conceptual Model to Enable Prescriptive Maintenance for Etching Equipment in Semiconductor Manufacturing. *Proc. CIRP* **2020**, *88*, 64–69. [CrossRef]
2. Bui, P.-D.; Lee, C. Unified System Network Architecture: Flexible and Area-Efficient NoC Architecture with Multiple Ports and Cores. *Electronics* **2020**, *9*, 1316. [CrossRef]
3. Weber, A. Smart manufacturing in the semiconductor industry: An evolving nexus of business drivers, technologies, and standards. In *Smart Manufacturing*; Soroush, M., Baldea, M., Edgar, T.F., Eds.; Elsevier: Amsterdam, The Netherlands, 2020; Chapter 3; pp. 59–105. ISBN 978-0-12-820028-5.
4. Hurtarte, J.S.; Wolsheimer, E.A.; Tafoya, L.M. Semiconductor Manufacturing Basics. In *Understanding Fabless IC Technology*; Newnes: Burlington, MA, USA, 2007; Chapter 4; pp. 41–45. ISBN 978-0-7506-7944-2.
5. Khakifirooz, M.; Chien, C.F.; Chen, Y.-J. Bayesian Inference for Mining Semiconductor Manufacturing Big Data for Yield Enhancement and Smart Production to Empower Industry 4.0. *Appl. Soft Comput.* **2018**, *68*, 990–999. [CrossRef]
6. Reis, M.S.; Gins, G. Industrial Process Monitoring in the Big Data/Industry 4.0 Era: From Detection, to Diagnosis, to Prognosis. *Processes* **2017**, *5*, 35. [CrossRef]
7. Lin, Y.-C.; Yeh, C.-C.; Chen, W.-H.; Hsu, K.-Y. Implementation Criteria for Intelligent Systems in Motor Production Line Process Management. *Processes* **2020**, *8*, 537. [CrossRef]
8. Chen, T. Strengthening the Competitiveness and Sustainability of a Semiconductor Manufacturer with Cloud Manufacturing. *Sustainability* **2014**, *6*, 251–266. [CrossRef]
9. Lee, D.-H.; Yang, J.-K.; Lee, C.-H.; Kim, K.-J. A Data-Driven Approach to Selection of Critical Process Steps in the Semiconductor Manufacturing Process Considering Missing and Imbalanced Data. *J. Manuf. Syst.* **2019**, *52*, 146–156. [CrossRef]
10. Hsu, C.-Y.; Chen, W.-J.; Chien, J.-C. Similarity Matching of Wafer Bin Maps for Manufacturing Intelligence to Empower Industry 3.5 for Semiconductor Manufacturing. *Comput. Ind. Eng.* **2020**, *142*, 106358. [CrossRef]
11. Nakata, K.; Orihara, R.; Mizuoka, Y.; Takagi, K. A Comprehensive Big-Data-Based Monitoring System for Yield Enhancement in Semiconductor Manufacturing. *IEEE Trans. Semicond. Manuf.* **2017**, *30*, 339–344. [CrossRef]
12. Yang, X.-S. Data mining techniques. In *Introduction to Algorithms for Data Mining and Machine Learning*; Academic Press: London, UK, 2019; Chapter 6; pp. 109–128. ISBN 978-0-12-817216-2.
13. Chien, C.-F.; Wang, W.-C.; Cheng, J.-C. Data Mining for Yield Enhancement in Semiconductor Manufacturing and an Empirical Study. *Expert Syst. Appl.* **2007**, *33*, 192–198. [CrossRef]
14. He, J.; Zhu, Y. Hierarchical Multi-Task Learning with Application to Wafer Quality Prediction. In Proceedings of the 2012 IEEE 12th International Conference on Data Mining, Brussels, Belgium, 10–13 December 2012; pp. 290–298.
15. Jeong, M.K.; Lu, J.-C.; Huo, X.; Vidakovic, B.; Chen, D. Wavelet-Based Data Reduction Techniques for Process Fault Detection. *Technometrics* **2006**, *48*, 26–40. [CrossRef]
16. Susto, G.A.; Beghi, A. Dealing with Time-Series Data in Predictive Maintenance Problems. In Proceedings of the 2016 IEEE 21st International Conference on Emerging Technologies and Factory Automation (ETFA), Berlin, Germany, 6–9 September 2016; pp. 1–4.
17. Choi, J.; Jeong, M.K. Deep Autoencoder With Clipping Fusion Regularization on Multistep Process Signals for Virtual Metrology. *IEEE Sens. Lett.* **2019**, *3*, 1–4. [CrossRef]
18. Wenjing, W.; Yumin, M.; Fei, Q.; Xiang, G. Data Mining Based Dynamic Scheduling Approach for Semiconductor Manufacturing System. In Proceedings of the 2015 34th Chinese Control Conference (CCC), Hangzhou, China, 28–30 July 2015; pp. 2603–2608.
19. Khemiri, A.; Amine Hamri, M.E.; Frydman, C.; Pinaton, J. Improving Business Process in Semiconductor Manufacturing by Discovering Business Rules. In Proceedings of the 2018 Winter Simulation Conference (WSC '18), Gothenburg, Sweden, 9–12 December 2018; pp. 3441–3448.

20. Huang, C.-Y.; Lin, P.K.P. Application of Integrated Data Mining Techniques in Stock Market Forecasting. *Cogent Econ. Financ.* **2014**, *2*, 929505. [CrossRef]
21. Tranfield, D.; Denyer, D.; Smart, P. Towards a Methodology for Developing Evidence-Informed Management Knowledge by Means of Systematic Review. *Br. J. Manag.* **2003**, *14*, 207–222. [CrossRef]
22. Denyer, D.; Tranfield, D. Producing a systematic review. In *The Sage Handbook of Organizational Research Methods*; Buchanan, D., Bryman, A., Eds.; Sage Publications Ltd.: London, UK, 2009; pp. 671–689.
23. Rousseau, D.M.; Manning, J.; Denyer, D. Evidence in Management and Organizational Science: Assembling the Field's Full Weight of Scientific Knowledge Through Syntheses. *Acad. Manag. Ann.* **2008**, *2*, 475–515. [CrossRef]
24. Correia, E.; Carvalho, H.; Azevedo, S.G.; Govindan, K. Maturity Models in Supply Chain Sustainability: A Systematic Literature Review. *Sustainability* **2017**, *9*, 64. [CrossRef]
25. Wang, K. Applying Data Mining to Manufacturing: The Nature and Implications. *J. Intell. Manuf.* **2007**, *18*, 487–495. [CrossRef]
26. Harding, J.A.; Shahbaz, M.; Kusiak, A. Data Mining in Manufacturing: A Review. *J. Manuf. Sci. Eng.* **2006**, *128*, 969–976. [CrossRef]
27. Buchanan, P.D.; Bryman, P.A. *The Sage Handbook of Organizational Research Methods*; Sage Publications Ltd.: London, UK, 2009; ISBN 978-1-4462-4605-4.
28. Yan, H.; Yang, N.; Peng, Y.; Ren, Y. Data Mining in the Construction Industry: Present Status, Opportunities, and Future Trends. *Autom. Constr.* **2020**, *119*, 103331. [CrossRef]
29. Galati, F.; Bigliardi, B. Industry 4.0: Emerging Themes and Future Research Avenues Using a Text Mining Approach. *Comput. Ind.* **2019**, *109*, 100–113. [CrossRef]
30. Susto, G.A.; Schirru, A.; Pampuri, S.; McLoone, S.; Beghi, A. Machine Learning for Predictive Maintenance: A Multiple Classifier Approach. *IEEE Trans. Ind. Inform.* **2015**, *11*, 812–820. [CrossRef]
31. Famili, A.; Shen, W.-M.; Weber, R.; Simoudis, E. Data Preprocessing and Intelligent Data Analysis. *IDA* **1997**, *1*, 3–23. [CrossRef]
32. Kusiak, A. Rough Set Theory: A Data Mining Tool for Semiconductor Manufacturing. *IEEE Trans. Electron. Packag. Manufact.* **2001**, *24*, 44–50. [CrossRef]
33. Kumar, S.; Sharma, P.; Garg, K.C. Lotka's Law and Institutional Productivity. *Inf. Process. Manag.* **1998**, *34*, 775–783. [CrossRef]
34. Hsu, S.-C.; Chien, C.-F. Hybrid Data Mining Approach for Pattern Extraction from Wafer Bin Map to Improve Yield in Semiconductor Manufacturing. *Int. J. Prod. Econ.* **2007**, *107*, 88–103. [CrossRef]
35. Van Eck, N.J.; Waltman, L. Software Survey: VOSviewer, a Computer Program for Bibliometric Mapping. *Scientometrics* **2010**, *84*, 523–538. [CrossRef]
36. Muñoz, J.A.M.; Viedma, E.H.; Espejo, A.L.S.; Cobo, M.J. Software Tools for Conducting Bibliometric Analysis in Science: An up-to-Date Review. *Prof. Inf.* **2020**, *29*, 4.
37. Sordan, J.E.; Oprime, P.C.; Pimenta, M.L.; Chiabert, P.; Lombardi, F. Lean Six Sigma in Manufacturing Process: A Bibliometric Study and Research Agenda. *TQM J.* **2020**, *32*, 381–399. [CrossRef]
38. Wellmann, P.J. Power Electronic Semiconductor Materials for Automotive and Energy Saving Applications—SiC, GaN, Ga_2O_3, and Diamond. *Z. Anorg. Allg. Chem.* **2017**, *643*, 1312–1322. [CrossRef]
39. Garlapati, S.K.; Divya, M.; Breitung, B.; Kruk, R.; Hahn, H.; Dasgupta, S. Printed Electronics Based on Inorganic Semiconductors: From Processes and Materials to Devices. *Adv. Mater.* **2018**, *30*, 1707600. [CrossRef]
40. Satpathy, R.; Pamuru, V. Silicon wafer manufacturing process. In *Solar PV Power*; Satpathy, R., Pamuru, V., Eds.; Academic Press: London, UK, 2021; Chapter 3; pp. 53–70. ISBN 978-0-12-817626-9.
41. Möller, H.J. Wafering of Silicon. In *Semiconductors and Semimetals*; Willeke, G.P., Weber, E.R., Eds.; Elsevier: Amsterdam, The Netherlands, 2015; Volume 92, Chapter 2; pp. 63–109.
42. Geng, N.; Jiang, Z. Capacity Planning for Semiconductor Wafer Fabrication with Uncertain Demand and Capacity. In Proceedings of the 2007 IEEE International Conference on Automation Science and Engineering, Scottsdale, AZ, USA, 22–25 September 2007; pp. 100–105.
43. Satpathy, R.; Pamuru, V. Silicon crystal growth process. In *Solar PV Power*; Satpathy, R., Pamuru, V., Eds.; Academic Press: London, UK, 2021; Chapter 2; pp. 31–52. ISBN 978-0-12-817626-9. Available online: https://doi.org/10.1016/B978-0-12-817626-9.00002-2 (accessed on 2 February 2021).
44. Tilli, M. Silicon wafers preparation and properties. In *Handbook of Silicon Based MEMS Materials and Technologies*, 3rd ed.; Tilli, M., Paulasto-Krockel, M., Petzold, M., Theuss, H., Motooka, T., Lindroos, V., Eds.; Elsevier: Amsterdam, The Netherlands, 2020; Chapter 4; pp. 93–110. ISBN 978-0-12-817786-0.
45. Gallagher, E.; Hibbs, M. Masks for micro- and nanolithography. In *Nanolithography*; Feldman, M., Ed.; Woodhead Publishing: Cambridge, UK, 2014; Chapter 5; pp. 158–178. ISBN 978-0-85709-500-8.
46. Cadien, K.C.; Nolan, L. Chapter 10—Chemical Mechanical Polishing Method and Practice. In *Handbook of Thin Film Deposition*, 4th ed.; Seshan, K., Schepis, D., Eds.; William Andrew Publishing: Norwich, NY, USA, 2018; pp. 317–357. ISBN 978-0-12-812311-9.
47. Bao, H.; Chen, L.; Ren, B. A Study on the Pattern Effects of Chemical Mechanical Planarization with CNN-Based Models. *Electronics* **2020**, *9*, 1158. [CrossRef]
48. Zhang, Y.; Wagner, L.; Golbutsov, P. Importance of Wafer Flatness for CMP and Lithography. In Proceedings of the Metrology, Inspection, and Process Control for Microlithography XI, Santa Clara, CA, USA, 7 July 1997; International Society for Optics and Photonics, 1997; Volume 3050, pp. 266–269. Available online: https://doi.org/10.1117/12.275916 (accessed on 2 February 2021).

49. Ki, M.; Sungmin, K.; Taesung, K. Study on Effect of Back-Surface Treatment of Silicon Wafer in Photo Lithography Process after CMP Process. In Proceedings of the 2015 International Conference on Planarization/CMP Technology (ICPT), Chandler, AZ, USA, 30 September–2 October 2015; pp. 1–3.
50. Jain, A. Ion Implantation for Semiconductor Processing. *Radiat. Eff.* **1982**, *63*, 39–46. [CrossRef]
51. Zolper, J.C. Ion Implantation in Wide Bandgap Semiconductors. In *Processing of Wide Band Gap Semiconductors*; Pearton, S.J., Ed.; William Andrew Publishing: Norwich, NY, USA, 2000; Chapter 7; pp. 300–353. ISBN 978-0-8155-1439-8.
52. Rice, B.J. Extreme ultraviolet (EUV) lithography. In *Nanolithography*; Feldman, M., Ed.; Woodhead Publishing: Cambridge, UK, 2014; Chapter 2; pp. 42–79. ISBN 978-0-85709-500-8.
53. Marconi, M.C.; Wachulak, P.W. Extreme Ultraviolet Lithography with Table Top Lasers. *Prog. Quantum Electron.* **2010**, *34*, 173–190. [CrossRef]
54. Buitrago, E.; Kulmala, T.S.; Fallica, R.; Ekinci, Y. EUV lithography process challenges. In *Frontiers of Nanoscience*; Robinson, A., Lawson, R., Eds.; Materials and Processes for Next Generation Lithography; Elsevier: Amsterdam, The Netherlands, 2016; Chapter 4; Volume 11, pp. 135–176.
55. Kolasinski, K.W. Growth and Etching of Semiconductors. In *Handbook of Surface Science*; Hasselbrink, E., Lundqvist, B.I., Eds.; Dynamics; North-Holland: Amsterdam, The Netherlands, 2008; Chapter 16; Volume 3, pp. 787–870. Available online: https://doi.org/10.1016/S1573-4331(08)00016-4 (accessed on 2 February 2021).
56. Chang, H.-Y.; Pan, W.-F.; Shih, M.-K.; Lai, Y.-S. Geometric Design for Ultra-Long Needle Probe Card for Digital Light Processing Wafer Testing. *Microelectron. Reliab.* **2010**, *50*, 556–563. [CrossRef]
57. Sakamaki, R.; Horibe, M. Realization of Accurate On-Wafer Measurement Using Precision Probing Technique at Millimeter-Wave Frequency. *IEEE Trans. Instrum. Meas.* **2018**, *67*, 1940–1945. [CrossRef]
58. Sakamaki, R.; Horibe, M. Uncertainty Analysis Method Including Influence of Probe Alignment on On-Wafer Calibration Process. *IEEE Trans. Instrum. Meas.* **2019**, *68*, 1748–1755. [CrossRef]
59. Kuo, C.-H.; Hu, A.H.; Hung, L.H.; Yang, K.-T.; Wu, C.-H. Life Cycle Impact Assessment of Semiconductor Packaging Technologies with Emphasis on Ball Grid Array. *J. Clean. Prod.* **2020**, *276*, 124301. [CrossRef]
60. Elshabini, A.A.; Barlow, F.; Wang, P.J. Electronic Packaging: Semiconductor Packages. In *Reference Module in Materials Science and Materials Engineering*; Elsevier: Amsterdam, The Netherlands, 2017; ISBN 978-0-12-803581-8.
61. Sang, H.-Y.; Duan, P.-Y.; Li, J.-Q. An Effective Invasive Weed Optimization Algorithm for Scheduling Semiconductor Final Testing Problem. *Swarm Evol. Comput.* **2018**, *38*, 42–53. [CrossRef]
62. Chien, C.; Chen, L. Using Rough Set Theory to Recruit and Retain High-Potential Talents for Semiconductor Manufacturing. *IEEE Trans. Semicond. Manuf.* **2007**, *20*, 528–541. [CrossRef]
63. Geum, Y.; Jeon, J.; Seol, H. Identifying Technological Opportunities Using the Novelty Detection Technique: A Case of Laser Technology in Semiconductor Manufacturing. *Technol. Anal. Strateg. Manag.* **2013**, *25*, 1–22. [CrossRef]
64. Tirkel, I. Forecasting Flow Time in Semiconductor Manufacturing Using Knowledge Discovery in Databases. *Int. J. Prod. Res.* **2013**, *51*, 5536–5548. [CrossRef]
65. Han, H.; Gao, C.; Zhao, Y.; Liao, S.; Tang, L.; Li, X. Polycrystalline Silicon Wafer Defect Segmentation Based on Deep Convolutional Neural Networks. *Pattern Recognit. Lett.* **2020**, *130*, 234–241. [CrossRef]
66. Hsu, C.-Y.; Chiu, S.-C. A Two-Phase Non-Dominated Sorting Particle Swarm Optimization for Chip Feature Design to Improve Wafer Exposure Effectiveness. *Comput. Ind. Eng.* **2020**, *147*, 106669. [CrossRef]
67. Li, J.; Zhang, H.; Wang, Y.; Cui, H. A Review of the Applications of Data Mining for Semiconductor Quality Control. In *Signal and Information Processing, Networking and Computers*; Wang, Y., Fu, M., Xu, L., Zou, J., Eds.; Lecture Notes in Electrical Engineering; Springer Singapore: Singapore, 2020; Volume 628, pp. 486–492. ISBN 9789811541629.
68. Gallo, C.; Capozzi, V. A Wafer Bin Map "Relaxed" Clustering Algorithm for Improving Semiconductor Production Yield. *Open Comput. Sci.* **2020**, *10*, 231–245. [CrossRef]
69. Kim, D.; Kang, S.; Cho, S. Expected Margin–Based Pattern Selection for Support Vector Machines. *Expert Syst. Appl.* **2020**, *139*, 112865. [CrossRef]
70. Kim, E.; Cho, S.; Lee, B.; Cho, M. Fault Detection and Diagnosis Using Self-Attentive Convolutional Neural Networks for Variable-Length Sensor Data in Semiconductor Manufacturing. *IEEE Trans. Semicond. Manuf.* **2019**, *32*, 302–309. [CrossRef]
71. Jin, C.H.; Na, H.J.; Piao, M.; Pok, G.; Ryu, K.H. A Novel DBSCAN-Based Defect Pattern Detection and Classification Framework for Wafer Bin Map. *IEEE Trans. Semicond. Manuf.* **2019**, *32*, 286–292. [CrossRef]
72. Kong, X.; Chang, J.; Niu, M.; Huang, X.; Wang, J.; Chang, S.I. Research on Real Time Feature Extraction Method for Complex Manufacturing Big Data. *Int. J. Adv. Manuf. Technol.* **2018**, *99*, 1101–1108. [CrossRef]
73. Tong, P.; Lu, J.; Yun, K. Fault Detection for Semiconductor Quality Control Based on Spark Using Data Mining Technology. In Proceedings of the 2018 Chinese Control and Decision Conference (CCDC), Shenyang, China, 9–11 June 2018; pp. 4372–4377.
74. Lee, C.-Y.; Chen, B.-S. Mutually-Exclusive-and-Collectively-Exhaustive Feature Selection Scheme. *Appl. Soft Comput.* **2018**, *68*, 961–971. [CrossRef]
75. Chien, C.-F.; Liu, C.-W.; Chuang, S.-C. Analysing Semiconductor Manufacturing Big Data for Root Cause Detection of Excursion for Yield Enhancement. *Int. J. Prod. Res.* **2017**, *55*, 5095–5107. [CrossRef]
76. Susto, G.A.; Terzi, M.; Beghi, A. Anomaly Detection Approaches for Semiconductor Manufacturing. *Proc. Manuf.* **2017**, *11*, 2018–2024. [CrossRef]

77. Lee, T.; Kim, C.O. Statistical Comparison of Fault Detection Models for Semiconductor Manufacturing Processes. *IEEE Trans. Semicond. Manuf.* **2015**, *28*, 80–91. [CrossRef]
78. Sejdovic, S.; Hegenbarth, Y.; Ristow, G.H.; Schmidt, R. Proactive Disruption Management System: How Not to Be Surprised by Upcoming Situations. In Proceedings of the 10th ACM International Conference on Distributed and Event-based Systems, Irvine, CA, USA, 20–24 June 2016; Association for Computing Machinery: New York, NY, USA, 2016; pp. 281–288.
79. Fan, S.-K.S.; Lin, S.-C.; Tsai, P.-F. Wafer Fault Detection and Key Step Identification for Semiconductor Manufacturing Using Principal Component Analysis, AdaBoost and Decision Tree. *J. Ind. Prod. Eng.* **2016**, *33*, 151–168. [CrossRef]
80. Butte, S.; Patil, S. Big Data and Predictive Analytics Methods for Modeling and Analysis of Semiconductor Manufacturing Processes. In Proceedings of the 2016 IEEE Workshop on Microelectronics and Electron Devices (WMED), Boise, ID, USA, 15 April 2016; pp. 1–5.
81. Zhu, Y.; He, J.; Lawrence, R.D. A General Framework for Predictive Tensor Modeling with Domain Knowledge. *Data Min. Knowl. Disc.* **2015**, *29*, 1709–1732. [CrossRef]
82. Aye, T.T.; Yang, F.; Wang, L.; Lee, G.K.K.; Li, X.; Hu, J.; Nguyen, M.C. Data Driven Framework for Degraded Pogo Pin Detection in Semiconductor Manufacturing. In Proceedings of the 2015 IEEE 10th Conference on Industrial Electronics and Applications (ICIEA), Auckland, New Zealand, 15–17 June 2015; pp. 345–350.
83. Haddad, B.; Karam, L.; Ye, J.; Patel, N.; Braun, M. Multi-Feature Sparse-Based Defect Detection and Classification in Semiconductor Units. In Proceedings of the 2016 IEEE International Conference on Image Processing (ICIP), Phoenix, AZ, USA, 25–28 September 2016; pp. 754–758.
84. Barkia, H.; Boucher, X.; Riche, R.L.; Beaune, P.; Girard, M.A.; Rozier, D. Semiconductor Yield Loss' Causes Identification: A Data Mining Approach. In Proceedings of the 2013 IEEE International Conference on Industrial Engineering and Engineering Management, Bangkok, Thailand, 10–13 December 2013; pp. 843–847.
85. Chien, C.-F.; Chang, K.-H.; Wang, W.-C. An Empirical Study of Design-of-Experiment Data Mining for Yield-Loss Diagnosis for Semiconductor Manufacturing. *J. Intell. Manuf.* **2014**, *25*, 961–972. [CrossRef]
86. Hessinger, U.; Chan, W.K.; Schafman, B.T. Data Mining for Significance in Yield-Defect Correlation Analysis. *IEEE Trans. Semicond. Manuf.* **2014**, *27*, 347–356. [CrossRef]
87. Liao, C.; Hsieh, T.; Huang, Y.; Chien, C. Similarity Searching for Defective Wafer Bin Maps in Semiconductor Manufacturing. *IEEE Trans. Autom. Sci. Eng.* **2014**, *11*, 953–960. [CrossRef]
88. Kerdprasop, K.; Kerdprasop, N. Tool Fault Analysis with Decision Tree Induction and Sequence Mining. *AMM* **2014**, *548–549*, 703–707. [CrossRef]
89. Li, Z.; Baseman, R.J.; Zhu, Y.; Tipu, F.A.; Slonim, N.; Shpigelman, L. A Unified Framework for Outlier Detection in Trace Data Analysis. *IEEE Trans. Semicond. Manuf.* **2014**, *27*, 95–103. [CrossRef]
90. Chien, C.; Chuang, S. A Framework for Root Cause Detection of Sub-Batch Processing System for Semiconductor Manufacturing Big Data Analytics. *IEEE Trans. Semicond. Manuf.* **2014**, *27*, 475–488. [CrossRef]
91. Chien, C.-F.; Hsu, S.-C.; Chen, Y.-J. A System for Online Detection and Classification of Wafer Bin Map Defect Patterns for Manufacturing Intelligence. *Int. J. Prod. Res.* **2013**, *51*, 2324–2338. [CrossRef]
92. Park, E.; Lee, J.-H. Classifying Imbalanced Data Using an Svm Ensemble with K-Means Clustering in Semiconductor Test Process. In *Proceedings of the Sixth International Conference on Machine Vision (ICMV 2013)*; International Society for Optics and Photonics: Bellingham, WA, USA, 2013; Volume 9067, p. 90672D.
93. Chien, C.-F.; Hsu, C.-Y.; Chen, P.-N. Semiconductor Fault Detection and Classification for Yield Enhancement and Manufacturing Intelligence. *Flex. Serv. Manuf. J.* **2013**, *25*, 367–388. [CrossRef]
94. Hsu, C.-Y.; Chien, C.-F.; Lai, Y.-C. Main Branch Decision Tree Algorithm for Yield Enhancement with Class Imbalance. In Proceedings of the Intelligent Decision Technologies; Watada, J., Watanabe, T., Phillips-Wren, G., Howlett, R.J., Jain, L.C., Eds.; Springer: Berlin, Germany, 2012; pp. 235–244. Available online: https://doi.org/10.1007/978-3-642-29977-3_24 (accessed on 5 February 2021).
95. Hsieh, T.; Liao, C.; Huang, Y.; Chien, C. A New Morphology-Based Approach for Similarity Searching on Wafer Bin Maps in Semiconductor Manufacturing. In Proceedings of the 2012 IEEE 16th International Conference on Computer Supported Cooperative Work in Design (CSCWD), Wuhan, China, 23–25 May 2012; pp. 869–874.
96. Kerdprasop, K.; Kerdprasop, N. Feature Selection and Boosting Techniques to Improve Fault Detection Accuracy in the Semiconductor Manufacturing Process. In Proceedings of the IMECS—International Multi Conference Engineering Comput. Scientists, Hong Kong, China, 16–18 March 2011; Volume 1, pp. 398–403.
97. Zuo, L.; Liu, X.; He, J.; Wang, J.; Zheng, P.; Zhang, J. An Improved AdaBoost Tree-Based Method for Defective Products Identification in Wafer Test. In Proceedings of the 2019 IEEE International Conference on Smart Manufacturing, Industrial Logistics Engineering (SMILE), Hangzhou, China, 19–21 April 2019; pp. 64–68.
98. Bertino, E.; Catania, B.; Caglio, E. Applying Data Mining Techniques to Wafer Manufacturing. In Proceedings of the Principles of Data Mining and Knowledge Discovery; Żytkow, J.M., Rauch, J., Eds.; Springer: Berlin, Germany, 1999; pp. 41–50.
99. Wang, C.-H. Recognition of Semiconductor Defect Patterns Using Spatial Filtering and Spectral Clustering. *Expert Syst. Appl.* **2008**, *34*, 1914–1923. [CrossRef]
100. Chih-Hsuan, W. Recognition of Semiconductor Defect Patterns Using Spectral Clustering. In Proceedings of the 2007 IEEE International Conference on Industrial Engineering and Engineering Management, Singapore, 2–5 December 2007; pp. 587–591.

101. Chen, R.S.; Chang, C.C. Using Bayesian Networks to Build Data Mining Applications for a Semiconductor Cleaning Process. *IJMPT* **2007**, *30*, 386. [CrossRef]
102. Yip, W.; Law, K.; Lee, W. Forecasting Final/Class Yield Based on Fabrication Process E-Test and Sort Data. In Proceedings of the 2007 IEEE International Conference on Automation Science and Engineering, Scottsdale, AZ, USA, 22-25 September 2007; pp. 478–483.
103. Yip, W.K.; Lim, C.C.; Lee, W.J. Method for Proposing Sort Screen Thresholds Based on Modeling Etest/Sort-Class in Semiconductor Manufacturing. In Proceedings of the 2008 IEEE International Conference on Automation Science and Engineering, Washington, DC, USA, 23–26 August 2008; pp. 236–241.
104. Wang, C.-H.; Wang, S.-J.; Lee, W.-D. Automatic Identification of Spatial Defect Patterns for Semiconductor Manufacturing. *Int. J. Prod. Res.* **2006**, *44*, 5169–5185. [CrossRef]
105. Li, T.-S.; Huang, C.-L.; Wu, Z.-Y. Data Mining Using Genetic Programming for Construction of a Semiconductor Manufacturing Yield Rate Prediction System. *J. Intell. Manuf.* **2006**, *17*, 355–361. [CrossRef]
106. Gardner, R.M.; Bieker, J.; Elwell, S. Solving Tough Semiconductor Manufacturing Problems Using Data Mining. In Proceedings of the 2000 IEEE/SEMI Advanced Semiconductor Manufacturing Conference and Workshop. ASMC 2000 (Cat. No.00CH37072), Boston, MA, USA, 12–14 September 2000; pp. 46–55.
107. Gruber, H. The Yield Factor and the Learning Curve in Semiconductor Production. *Appl. Econ.* **1994**, *26*, 837–843. [CrossRef]
108. Kinghorst, J.; Geramifard, O.; Luo, M.; Chan, H.-L.; Yong, K.; Folmer, J.; Zou, M.; Vogel-Heuser, B. Hidden Markov Model-Based Predictive Maintenance in Semiconductor Manufacturing: A Genetic Algorithm Approach. In Proceedings of the 2017 13th IEEE Conference on Automation Science and Engineering (CASE), Xi'an, China, 20–23 August 2017; pp. 1260–1267.
109. Hsu, C.-Y.; Chien, C.-F.; Chen, P.-N. Manufacturing Intelligence for Early Warning of Key Equipment Excursion for Advanced Equipment Control in Semiconductor Manufacturing. *J. Chin. Inst. Ind. Eng.* **2012**, *29*, 303–313. [CrossRef]
110. Retersdorf, M.; Anand, A.; Drozda-Freeman, A.; McIntyre, M.; Song, X.; Wang, J. Use of Spatial Pattern Recognition (SPR) for Enhancing the Resolution and Identification of Rogue Tools in Manufacturing. In Proceedings of the 2008 IEEE/SEMI Advanced Semiconductor Manufacturing Conference, Cambridge, MA, USA, 5–7 May 2008; pp. 200–205.
111. Tsuda, H.; Shirai, H.; Kawamura, E. A Precise Photolithography Process Control Method Using Virtual Metrology. *Electron. Commun. Jpn.* **2014**, *97*, 48–55. [CrossRef]
112. Chen, C.-H.; Zhao, W.-D.; Pang, T.; Lin, Y.-Z. Virtual Metrology of Semiconductor PVD Process Based on Combination of Tree-Based Ensemble Model. *ISA Trans.* **2020**, *103*, 192–202. [CrossRef] [PubMed]
113. Cai, H.; Feng, J.; Zhu, F.; Yang, Q.; Li, X.; Lee, J. Adaptive Virtual Metrology Method Based on Just-in-Time Reference and Particle Filter for Semiconductor Manufacturing. *Measurement* **2021**, *168*, 108338. [CrossRef]
114. Park, C.; Kim, Y.; Park, Y.; Kim, S.B. Multitask Learning for Virtual Metrology in Semiconductor Manufacturing Systems. *Comput. Ind. Eng.* **2018**, *123*, 209–219. [CrossRef]
115. Maggipinto, M.; Beghi, A.; McLoone, S.; Susto, G.A. DeepVM: A Deep Learning-Based Approach with Automatic Feature Extraction for 2D Input Data Virtual Metrology. *J. Process. Control.* **2019**, *84*, 24–34. [CrossRef]
116. Lenz, B.; Barak, B.; Leicht, C. Development of Smart Feature Selection for Advanced Virtual Metrology. In Proceedings of the 25th Annual SEMI Advanced Semiconductor Manufacturing Conference (ASMC 2014), Saratoga Springs, NY, USA, 19–21 May 2014; pp. 145–150.
117. Ooi, M.P.; Joo, E.K.J.; Kuang, Y.C.; Demidenko, S.; Kleeman, L.; Chan, C.W.K. Getting More from the Semiconductor Test: Data Mining With Defect-Cluster Extraction. *IEEE Trans. Instrum. Meas.* **2011**, *60*, 3300–3317. [CrossRef]
118. Lenz, B.; Barak, B.; Mührwald, J.; Leicht, C.; Lenz, B. Virtual Metrology in Semiconductor Manufacturing by Means of Predictive Machine Learning Models. In Proceedings of the 2013 12th International Conference on Machine Learning and Applications, Washington, DC, USA, 4–7 December 2013; Volume 2, pp. 174–177.
119. Kupp, N.; Slamani, M.; Makris, Y. Correlating Inline Data with Final Test Outcomes in Analog/RF Devices. In Proceedings of the 2011 Design, Automation Test in Europe, Grenoble, France, 14–18 March 2011; pp. 1–6.
120. Ul Haq, A.A.; Djurdjanovic, D. Dynamics-Inspired Feature Extraction in Semiconductor Manufacturing Processes. *J. Ind. Inf. Integr.* **2019**, *13*, 22–31. [CrossRef]
121. Kim, J.K.; Cho, K.C.; Lee, J.S.; Han, Y.S. Feature Selection Techniques for Improving Rare Class Classification in Semiconductor Manufacturing Process. In Proceedings of the Big Data Technologies and Applications, Gwangju, Korea, 23–24 November 2017; Jung, J.J., Kim, P., Eds.; Springer International Publishing: Cham, Switzerland, 2017; pp. 40–47.
122. Abdelkader, I.; El-Sonbaty, Y.; El-Habrouk, M. Openmv: A Python Powered, Extensible Machine Vision Camera. *arXiv* **2017**, arXiv:1711.10464.
123. Zhu, Y.; He, J. Co-Clustering Structural Temporal Data with Applications to Semiconductor Manufacturing. In Proceedings of the 2014 IEEE International Conference on Data Mining, Shenzhen, China, 14–17 December 2014; pp. 1121–1126.
124. Lenz, B.; Barak, B. Data Mining and Support Vector Regression Machine Learning in Semiconductor Manufacturing to Improve Virtual Metrology. In Proceedings of the 2013 46th Hawaii International Conference on System Sciences, Wailea, HI, USA, 7–10 January 2013; pp. 3447–3456.
125. Susto, G.A.; Beghi, A.; Luca, C.D. A Virtual Metrology System for Predicting CVD Thickness with Equipment Variables and Qualitative Clustering. In Proceedings of the ETFA 2011, Toulouse, France, 5–9 September 2011; pp. 1–4.

126. St. Pierre, E.; Tuv, E. Robust, Non-Redundant Feature Selection for Yield Analysis in Semiconductor Manufacturing. In *Proceedings of the Advances in Data Mining. Applications and Theoretical Aspects*; Perner, P., Ed.; Springer: Berlin, Germany, 2011; pp. 204–217.
127. Kang, P.; Kim, D.; Lee, H.; Doh, S.; Cho, S. Virtual Metrology for Run-to-Run Control in Semiconductor Manufacturing. *Expert Syst. Appl.* **2011**, *38*, 2508–2522. [CrossRef]
128. Kang, P.; Lee, H.; Cho, S.; Kim, D.; Park, J.; Park, C.-K.; Doh, S. A Virtual Metrology System for Semiconductor Manufacturing. *Expert Syst. Appl.* **2009**, *36*, 12554–12561. [CrossRef]
129. Tsuda, H.; Shirai, H. Improvement of Photolithography Process by 2nd Generation Data Mining. In Proceedings of the 2006 IEEE International Symposium on Semiconductor Manufacturing, Tokyo, Japan, 25–27 September 2006; pp. 122–125.
130. Jung, U.; Jeong, M.K.; Lu, J.-A. Vertical-Energy-Thresholding Procedure for Data Reduction with Multiple Complex Curves. *IEEE Trans. Syst. Man Cybern. Part B* **2006**, *36*, 1128–1138. [CrossRef] [PubMed]
131. Palma, F.D.; Nicolao, G.D.; Miraglia, G.; Donzelli, O.M. Process Diagnosis via Electrical-Wafer-Sorting Maps Classification. In Proceedings of the Fifth IEEE International Conference on Data Mining (ICDM'05), Houston, TX, USA, 27–30 November 2005; p. 4.
132. Turban, E.; Aronson, J.; Liang, T.-P. *Decision Support. Systems and Intelligent Systems*, 7th ed. 2007. Available online: https://books.google.pt/books/about/Decision_Support_Systems_and_Intelligent.html?id=m0R5QgAACAAJ&redir_esc=y (accessed on 5 February 2021).
133. Hood, S.J. Detail vs. Simplifying Assumptions for Simulating Semiconductor Manufacturing Lines. In Proceedings of the Ninth IEEE CHMT International Electronics Manufacturing Technology Symposium, Piscataway, NJ, USA, 12–17 February 1989; pp. 103–108.
134. Narayanan, S.; Bodner, D.A.; Sreekanth, U.; Dilley, S.J.; Govindaraj, T.; McGinnis, L.F.; Mitchell, C.M. Object-Oriented Simulation to Support Operator Decision Making in Semiconductor Manufacturing. In Proceedings of the 1992 IEEE International Conference on Systems, Man, and Cybernetics, Chicago, IL, USA, 18–21 October 1992; pp. 1510–1515.
135. Casali, A.; Ernst, C. Discovering Correlated Parameters in Semiconductor Manufacturing Processes: A Data Mining Approach. *IEEE Trans. Semicond. Manufact.* **2012**, *25*, 118–127. [CrossRef]
136. Kerdprasop, K.; Kerdprasop, N. Data Preparation Techniques for Improving Rare Class Prediction. Available online: https://dl.acm.org/doi/10.5555/2039846.2039882 (accessed on 5 February 2021).
137. Kerdprasop, K.; Kerdprasop, N. A Data Mining Approach to Automate Fault Detection Model Development in the Semiconductor Manufacturing Process. *Int. J. Mech.* **2011**, *5*, 10.
138. Weiss, S.M.; Baseman, R.J.; Tipu, F.; Collins, C.N.; Davies, W.A.; Singh, R.; Hopkins, J.W. Rule-Based Data Mining for Yield Improvement in Semiconductor Manufacturing. *Appl. Intell.* **2010**, *33*, 318–329. [CrossRef]
139. Sassenberg, C.; Weber, C.; Fathi, M.; Holland, A.; Montino, R. Feature Selection for Improving the Usability of Classification Results of High-Dimensional Data. *DMIN* **2008**, *2*, 197–201.
140. Braha, D.; Elovici, Y.; Last, M. Theory of Actionable Data Mining with Application to Semiconductor Manufacturing Control. *Int. J. Prod. Res.* **2007**, *45*, 3059–3084. [CrossRef]
141. Chen, A.; Hong, A.; Ho, O.; Liu, C.-W.; Huang, Y.-H. Sample Efficient Regression Trees (SERT) for Yield Loss Analysis. In Proceedings of the 2006 IEEE International Symposium on Semiconductor Manufacturing, Tokyo, Japan, 25–27 September 2006; pp. 29–32.
142. Han, Y.; Kim, J.; Lee, C. Lecture Notes in Computer Science. Automatic Detection of Failure Patterns Using Data Mining. In *Knowledge-Based Intelligent Information and Engineering Systems*; Khosla, R., Howlett, R.J., Jain, L.C., Eds.; Springer: Berlin, Germany, 2005; Volume 3682, pp. 1312–1316. ISBN 978-3-540-28895-4.
143. Lin, S.-Y.; Horng, S.-C.; Tsai, C.-H. Fault Detection of the Ion Implanter Using Classification Approach. In Proceedings of the 2004 5th Asian Control Conference, Melbourne, Australia, 20–23 July 2004; pp. 809–814.
144. Lee, J.H.; Park, S.C. Agent and Data Mining Based Decision Support System and Its Adaptation to a New Customer-Centric Electronic Commerce. *Expert Syst. Appl.* **2003**, *25*, 619–635. [CrossRef]
145. Jang, H.L.; Song, J.Y.; Sang, C.P. Design of Intelligent Data Sampling Methodology Based on Data Mining. *IEEE Trans. Robot. Automat.* **2001**, *17*, 637–649. [CrossRef]
146. Ruey-Shun, C.; Ruey-Chyi, W.; Chang, C.C. Using Data Mining Technology to Design an Intelligent CIM System for IC Manufacturing. In Proceedings of the Sixth International Conference on Software Engineering, Artificial Intelligence, Towson, MD, USA, 23–25 May 2005; pp. 70–75.
147. Chen, L.-F.; Chien, C.-F. Manufacturing Intelligence for Class Prediction and Rule Generation to Support Human Capital Decisions for High-Tech Industries. *Flex. Serv. Manuf. J.* **2011**, *23*, 263–289. [CrossRef]
148. Chen, R.; Tsai, Y.; Chang, C. Design and Implementation of an Intelligent Manufacturing Execution System for Semiconductor Manufacturing Industry. In Proceedings of the 2006 IEEE International Symposium on Industrial Electronics, Montreal, QC, Canada, 9–13 July 2006; pp. 2948–2953.
149. Anaya, A.; Henning, W.; Basantkumar, N.; Oliver, J. Yield Improvement Using Advanced Data Analytics. In Proceedings of the 2019 30th Annual SEMI Advanced Semiconductor Manufacturing Conference (ASMC), Saratoga Springs, NY, USA, 6–9 May 2019; pp. 1–5.
150. Mörzinger, B.; Loschan, C.; Kloibhofer, F.; Bleicher, F. A Modular, Holistic Optimization Approach for Industrial Appliances. *Proc. CIRP* **2019**, *79*, 551–556. [CrossRef]

151. Hsu, C.-Y. An Analytic Framework of Design for Semiconductor Manufacturing. In Proceedings of the Asia Pacific Business Process Management; Bae, J., Suriadi, S., Wen, L., Eds.; Springer International Publishing: Cham, Switzerland, 2015; pp. 128–137.
152. Park, S.H.; Park, C.; Kim, J.S.; Kim, S.; Baek, J.; An, D. Data Mining Approaches for Packaging Yield Prediction in the Post-Fabrication Process. In Proceedings of the 2013 IEEE International Congress on Big Data, Santa Clara, CA, USA, 27 June–2 July 2013; pp. 363–368.
153. Kwak, D.-S.; Kim, K.-J. A Data Mining Approach Considering Missing Values for the Optimization of Semiconductor-Manufacturing Processes. *Expert Syst. Appl.* **2012**, *39*, 2590–2596. [CrossRef]
154. Dabbas, R.M.; Chen, H.-N. Mining Semiconductor Manufacturing Data for Productivity Improvement—An Integrated Relational Database Approach. *Comput. Ind.* **2001**, *45*, 29–44. [CrossRef]
155. Chien, C.-F.; Kuo, C.-J.; Yu, C.-M. Tool Allocation to Smooth Work-in-Process for Cycle Time Reduction and an Empirical Study. *Ann. Oper. Res.* **2020**, *290*, 1009–1033. [CrossRef]
156. Lin, Y.C.; Chen, T.-C. Interval Cycle Time Estimation in a Semiconductor Manufacturing System with a Data-Mining Approach. *Int. Rev. Comput. Softw.* **2009**, *4*, 737–742.
157. Meidan, Y.; Lerner, B.; Hassoun, M.; Rabinowitz, G. Data Mining for Cycle Time Key Factor Identification and Prediction in Semiconductor Manufacturing. *IFAC Proc. Vol.* **2009**, *42*, 217–222. [CrossRef]
158. Pang, J.; Zhou, H.; Tsai, Y.-C.; Chou, F.-D. A Scatter Simulated Annealing Algorithm for the Bi-Objective Scheduling Problem for the Wet Station of Semiconductor Manufacturing. *Comput. Ind. Eng.* **2018**, *123*, 54–66. [CrossRef]
159. Lee, Y.-H.; Chang, C.-T.; Wong, D.S.-H.; Jang, S.-S. Petri-Net Based Scheduling Strategy for Semiconductor Manufacturing Processes. *Chem. Eng. Res. Des.* **2011**, *89*, 291–300. [CrossRef]
160. Chen, T. An Optimized Tailored Nonlinear Fluctuation Smoothing Rule for Scheduling a Semiconductor Manufacturing Factory. *Comput. Ind. Eng.* **2010**, *58*, 317–325. [CrossRef]
161. Wang, P.-S.; Yang, T.; Yu, L.-C. Lean-Pull Strategy for Order Scheduling Problem in a Multi-Site Semiconductor Crystal Ingot-Pulling Manufacturing Company. *Comput. Ind. Eng.* **2018**, *125*, 545–562. [CrossRef]
162. Ma, Y.; Lu, X.; Qiao, F. Data Driven Scheduling Knowledge Management for Smart Shop Floor. In Proceedings of the 2019 IEEE 15th International Conference on Automation Science and Engineering (CASE), Vancouver, BC, Canada, 22–26 August 2019; pp. 109–114.
163. Chun-Hai, H.; Shun-Feng, S. Hierarchical Clustering Methods for Semiconductor Manufacturing Data. In Proceedings of the IEEE International Conference on Networking, Sensing and Control, Taipei, Taiwan, 21–23 March 2004; Volume 2, pp. 1063–1068.
164. Ma, Y.; Chen, X.; Qiao, F.; Tian, K.; Lu, J. The Research and Application of a Dynamic Dispatching Strategy Selection Approach Based on BPSO-SVM for Semiconductor Production Line. In Proceedings of the Proceedings of the 11th IEEE International Conference on Networking, Sensing and Control, Miami, FL, USA, 7–9 April 2014; pp. 74–79.
165. Li, L.; Zijin, S.; Jiacheng, N.; Fei, Q. Data-Based Scheduling Framework and Adaptive Dispatching Rule of Complex Manufacturing Systems. *Int. J. Adv. Manuf. Technol.* **2013**, *66*, 1891–1905. [CrossRef]
166. Shiue, Y.-R.; Guh, R.-S.; Tseng, T.-Y. Study on Shop Floor Control System in Semiconductor Fabrication by Self-Organizing Map-Based Intelligent Multi-Controller. *Comput. Ind. Eng.* **2012**, *62*, 1119–1129. [CrossRef]
167. Wu, R.C.; Chen, R.S.; Fan, C.R. Design an Intelligent CIM System Based on Data Mining Technology for New Manufacturing Processes. *IJMPT* **2004**, *21*, 487. [CrossRef]
168. Chong, I.-G.; Zhu, C.; Wu, Y. Data Mining Analysis of Turnaround Time Variation in a Semiconductor Manufacturing Line. *ICORES* **2015**, *1*, 185–189. [CrossRef]
169. Chien, C.-F.; Diaz, A.C.; Lan, Y.-B. A Data Mining Approach for Analyzing Semiconductor MES and FDC Data to Enhance Overall Usage Effectiveness (OUE). *Int. J. Comput. Intell. Syst.* **2014**, *7*, 52–65. [CrossRef]
170. Meidan, Y.; Lerner, B.; Rabinowitz, G.; Hassoun, M. Cycle-Time Key Factor Identification and Prediction in Semiconductor Manufacturing Using Machine Learning and Data Mining. *IEEE Trans. Semicond. Manuf.* **2011**, *24*, 237–248. [CrossRef]
171. Scholz-Reiter, B.; Heger, J.; Hildebrandt, T. Gaussian Processes for Dispatching Rule Selection in Production Scheduling: Comparison of Learning Techniques. In Proceedings of the 2010 IEEE International Conference on Data Mining Workshops, Sydney, Australia, 13–17 December 2010; pp. 631–638.
172. Lee, W.; Soon-Chuan, O. Learning from Small Data Sets to Improve Assembly Semiconductor Manufacturing Processes. In Proceedings of the 2010 The 2nd International Conference on Computer and Automation Engineering (ICCAE), Singapore, Singapore, 26–28 February 2010; Volume 2, pp. 50–54.
173. Chen, T. A Hybrid Look-Ahead SOM-FBPN and FIR System for Wafer-Lot-Output Time Prediction and Achievability Evaluation. *Int. J. Adv. Manuf. Technol.* **2007**, *35*, 575–586. [CrossRef]
174. Ciacchella, J.; Richard, C.; Zhang, N. IoT Opportunity in the World of Semiconductor Companies. 2018, pp. 1–31. Available online: https://www2.deloitte.com/content/dam/Deloitte/us/Documents/technology/us-semiconductor-internet-of-things.pdf (accessed on 5 February 2021).
175. Bauer, H.; Patel, M.; Veira, J. Internet of Things: Opportunities and Challenges for Semiconductor Companies. 2015. Available online: https://www.mckinsey.com/industries/semiconductors/our-insights/internet-of-things-opportunities-and-challenges-for-semiconductor-companies (accessed on 5 February 2021).
176. Misrudin, F.; Foong, L.C. Digitalization in Semiconductor Manufacturing- Simulation Forecaster Approach in Managing Manufacturing Line Performance. *Proc. Manuf.* **2019**, *38*, 1330–1337. [CrossRef]

177. Javid, T.; Gupta, M.K.; Gupta, A. A Hybrid-Security Model for Privacy-Enhanced Distributed Data Mining. *J. King Saud Univ. Comput. Inf. Sci.* **2020**. [CrossRef]
178. Dogan, A.; Birant, D. Machine Learning and Data Mining in Manufacturing. *Expert Syst. Appl.* **2021**, *166*, 114060. [CrossRef]
179. Hand, D.J.; Adams, N.M. Data Mining. In *Wiley StatsRef: Statistics Reference Online*; American Cancer Society: Atlanta, GA, USA, 2015; pp. 1–7. ISBN 978-1-118-44511-2.
180. García, S.; Luengo, J.; Herrera, F. *Data Preprocessing in Data Mining*; Springer: New York, NY, USA, 2015; ISBN 978-3-319-10246-7.
181. Silva, J.; Cubillos, J.; Villa, J.V.; Romero, L.; Solano, D.; Fernández, C. Preservation of Confidential Information Privacy and Association Rule Hiding for Data Mining: A Bibliometric Review. *Proc. Comput. Sci.* **2019**, *151*, 1219–1224. [CrossRef]
182. Galdi, P.; Tagliaferri, R. Data Mining: Accuracy and Error Measures for Classification and Prediction. In *Encyclopedia of Bioinformatics and Computational Biology*; Ranganathan, S., Gribskov, M., Nakai, K., Schönbach, C., Eds.; Academic Press: Oxford, UK, 2019; pp. 431–436. ISBN 978-0-12-811432-2.
183. Da Silva Serapião Leal, G.; Guédria, W.; Panetto, H. Interoperability Assessment: A Systematic Literature Review. *Comput. Ind.* **2019**, *106*, 111–132. [CrossRef]
184. Kadadi, A.; Agrawal, R.; Nyamful, C.; Atiq, R. Challenges of Data Integration and Interoperability in Big Data. In Proceedings of the 2014 IEEE International Conference on Big Data (Big Data), Washington, DC, USA, 27–30 October 2014; pp. 38–40.
185. Ramírez-Gallego, S.; Krawczyk, B.; García, S.; Woźniak, M.; Herrera, F. A Survey on Data Preprocessing for Data Stream Mining: Current Status and Future Directions. *Neurocomputing* **2017**, *239*, 39–57. [CrossRef]
186. Moyne, J.; Iskandar, J. Big Data Analytics for Smart Manufacturing: Case Studies in Semiconductor Manufacturing. *Processes* **2017**, *5*, 39. [CrossRef]

Article

Enhancing Failure Mode and Effects Analysis Using Auto Machine Learning: A Case Study of the Agricultural Machinery Industry

Sami Sader [1,*], István Husti [2] and Miklós Daróczi [2]

1 Doctoral School of Mechanical Engineering, Szent Istvan University, 2100 Godollo, Hungary
2 Institute of Engineering Management, Szent Istvan University, 2100 Godollo, Hungary; husti.istvan@gek.szie.hu (I.H.); daroczi.miklos@gek.szie.hu (M.D.)
* Correspondence: sami.s.a.sader@phd.uni-szie.hu; Tel.: +36-702235922

Received: 9 January 2020; Accepted: 12 February 2020; Published: 14 February 2020

Abstract: In this paper, multiclass classification is used to develop a novel approach to enhance failure mode and effects analysis and the generation of risk priority number. This is done by developing four machine learning models using auto machine learning. Failure mode and effects analysis is a technique that is used in industry to identify possible failures that may occur and the effects of these failures on the system. Meanwhile, risk priority number is a numeric value that is calculated by multiplying three associated parameters namely severity, occurrence and detectability. The value of risk priority number determines the next actions to be made. A dataset that includes a one-year registry of 1532 failures with their description, severity, occurrence, and detectability is used to develop four models to predict the values of severity, occurrence, and detectability. Meanwhile, the resulted models are evaluated using 10% of the dataset. Evaluation results show that the proposed models have high accuracy whereas the average value of precision, recall, and F1 score are in the range of 86.6–93.2%, 67.9–87.9%, 0.892–0.765% respectively. The proposed work helps in carrying out failure mode and effects analysis in a more efficient way as compared to the conventional techniques.

Keywords: Industry 4.0; auto machine learning; failure mode effects analysis; risk priority number

1. Introduction

Failure modes and effects analysis (FMEA) is a proactive analytical technique for identifying, tracking and mitigating product and process potential failures in a systematic way by determining its potential occurrence, root causes, consequences, and impact [1]. FMEA provides a quantitative score to evaluate failures where every failure is transformed into a numerical value that is called risk priority number (RPN). RPN is the result of multiplying three parameters namely severity, occurrence, and detectability. Severity is the risk or damage that may affect the machine, product, next operator or the end-user. On the other hand, occurrence is the likelihood of this failure that may occur again. Finally, detectability is the degree to which this failure could be detected [2–4]. Higher RPN value represents a higher priority of risk [5]. Appropriate corrective actions are usually determined based on RPN threshold value. If this threshold is reached, a risk mitigation procedure is applied accordingly [6]. Moreover, RPN value is used as a tool for optimal resource allocation by giving focus on risks that have the highest RPN or the most critical issues [3,7].

FMEA was firstly developed by NASA in 1963 to enhance the performance of the devices that are used in the aerospace industry [8]. Later, FMEA was adopted and promoted by Ford Motors in 1977 [3]. Currently, FMEA is being used in the automotive industry to ensure the quality and reliability of production systems [9]. Daimler Chrysler, Ford, and General Motors have developed an international standard called SAE J1739_200006 as general guidance for implementing FMEA techniques to avoid failures and enhance system reliability and safety [10]. FMEA documents are classified into two types namely design FMEA, and process FMEA [11]. Design FMEA is constructed during product design to define product weaknesses, critical components and their respective potential failure modes, root causes, and effects [1]. Meanwhile, process FMEA focuses on potential failures that may occur during the manufacturing process and incurred risks at each process step [3].

FMEA is a robust tool for quality improvement in both manufacturing and services industries. It can be used at the design stage of the product and during its implementation [9]. The aim of this is to avoid the end-user from experiencing unfavorable defects that may affect the reputation of the company negatively [3]. FMEA is also used as a process improvement technique to ensure consistency, reliability and avoid deviations. Moreover, it is also used to define and mitigate risks [12]. On the other hand, FMEA is used to improve maintenance management by analyzing the maintenance requirements of the product and developing the maintenance plans that would be used to ensure that the system is doing what it is meant to do when it was created. Finally, FMEA is used to improve safety by conducting hazards analysis of components that have critical hazards on lives, property, or other losses that are identified and mitigated [7].

However, FMEA is criticized for many conceptual aspects. The most popular disadvantage of this method is the narrative and qualitative nature of its structure. For every product or process, FEMA documents are developed by engineers and experts using linguistic terms that are based on the personal evaluation. The RPN parameters' values are determined by engineers and experts which may include uncertainty and vagueness [12]. Moreover, the parameters that are used in FMEA are represented by (1–10) crisp scale which is an unreliable representation of real-application cases [5,13]. Additionally, Chang, et al. in [3] have criticized the RPN estimation by the inhomogeneous morphologic correlations between the three parameters. This criticism is based on the fact that each of these parameters is obtained and linearly multiplied by the other with an identical scale. This process is done despite the actual impact of every independent parameter and the different qualitative interpretation of the scale. For example, high severity value should result an extremely high RPN value due to the critical hazard on the operator or the machine. In other words, once there is a risk on human, the other parameters shouldn't downgrade the overall value of RPN even if they are low.

Thus, in order to overcome this ambiguity, researchers proposed several approaches to improve the application of FMEA and the development of RPN. Several fuzzy techniques were examined to develop a new risk assessment approach to overcome the weaknesses of FMEA. Haktanır and Kahraman in [13] have summarized several fuzzy techniques and grey theory and proposed interval-valued neutrosophic (IVN) sets-based FMEA to eliminate the inaccuracy of human decisions and evaluations. Ayber and Erginel in [12] have proposed single-valued neutrosophic (SVN) Fuzzy FMEA as a new risk analysis tool to overcome the ambiguity of the linguistic terms. Al-Khafaji, et.al in [14] have proposed a fuzzy multicriteria decision-making model aligned with FMEA principles to obtain an efficient criterion for maintenance management. Liu et al. in [5] have used cloud model theory and hierarchical TOPSIS method to enhance FMEA effectiveness, overcoming bias probability of human judgment, and to facilitate the transformation of qualitative terms to quantitative values. Yang et al. in [2] have utilized a data mining-based method for isolating faults based on FMEA parameters in order to enhance predictive maintenance by using historical big-data to create data-driven models, by which future failure can be predicted efficiently and accordingly avoid failures at a very critical operational item. Keskin and Özkan in [6] have applied a fuzzy adaptive resonance theory (ART) method for FMEA modeling in order to improve the classical methodology of calculating the RPN, which in total minimized cost and efforts needed to respond to corrective actions alerts.

In the aforementioned research, the interpretation of FMEA documents was well addressed and resolved. However, the weakness of FMEA and RPN is not limited to the ambiguity of the FMEA textual description nor its quantitative representation, but it also extends to the importance of being proactive and responsive to failures. The flow of information once a failure is detected until the time it is ranked and resolved is important as well to guarantee minimum impact and limited implications.

Another shortcoming of the conventional FMEA technique comes from the fact that its documents are prepared during the product or process design stages, which makes these documents obsolete after production starts ahead. Therefore, these documents need to be dynamically validated and updated on a continuous basis. Hence, utilizing new technologies is very vital to overcome these weaknesses and keep these documents updated and responsive [2].

In the era of Industry 4.0, connectivity offered instant communication and collaboration among the value chain. Artificial intelligence (AI), the internet of things (IoT), big-data, and cyber-physical systems (CPS) made a great leap in automation and optimization at all levels of manufacturing. Here, automation is not limited to machines and processes, but also to management information systems such as enterprise resources planning (ERP), customer relationship management (CRM) and quality management systems (QMS) [15]. Additionally, the real-time flow of data among the value chain, which is instantly analyzed and transformed to user-friendly information, thanks here to the advanced supercomputing and analyzing power [16], resulted new paradigms of manufacturing systems which are being called nowadays by smart factory, smart machine, smart product and augmented operator [17]. These pillars changed the production systems from being reactive to be proactive and levered the human intervention from doing the work to supervise it while it is being done. Sensors, 3D cameras, radio frequency identifier (RFID), and Wi-Fi made monitoring processes more precise and accurate. Unseen defects or deviation of products or processes can be detected as soon as it is occurring. Defect elimination and processes re-adjustment are made autonomously at the micro and macro levels [15,18,19]. All these technologies, alongside the increased complexity of products and their manufacturing systems, generated a large volume of data, at a high velocity, veracity, and variety. The analysis of such big data requires advanced resources and techniques to classify data and detect patterns that cannot be detected using traditional analytical tools.

Automated machine learning (AutoML) are tools that automate the process of a machine learning workflow, offering the same capabilities of regular machine learning, without explicit knowledge of programming [20]. AutoML aims at reducing the human intervention in data preprocessing, feature selection and algorithm selection so as to make machine learning automated [21]. Google AutoML is a cloud machine learning platform that automates supervised machine learning in a very efficient way. It handles the tasks of data preprocessing, feature extraction, feature engineering, feature selection, algorithm selection, and hyperparameter optimization [22]. Google AutoML automatically develops models based on neural architecture search (NAS). It follows the try and error strategy by developing the model based on a random set of hyperparameters, then evaluates the performance of the model which is resulted by using this set of hyperparameters and finally concludes the most accurate model [22,23].

AutoML is increasingly used in scientific research areas. Faes et al. in [24] have evaluated the performance of AutoML hosted by google cloud platform against other machine learning methods and algorithms. It is claimed that AutoML has higher accuracy in medical image classification and can be used by people who are less experienced in coding and algorithms. Similarly, Hayashi et.al. in [25] utilized Google AutoML to identify pest aphid species and improving crop protection effectiveness. The authors concluded that such a tool provided an accuracy of 0.96 which allowed them to consider the AutoML as a useful and effective tool. Additionally, Li et al. in [26] have used AutoML to automate customer service activities by analyzing different customers' information and respond to their inquiries based on historical frequent inquiries. According to the authors, the solution provided improved responsiveness and minimized the cost of customer service management. Moreover, Galitsky et.al in [27] proposed a novel approach to automate customer complaints processing and classification by training a machine learning algorithm for analyzing dialogues recorded between customers and company-agents.

Based on that, the aim of this paper is to examine a novel optimization approach applied to FMEA and RPN by classifying failures according to updated FMEA documents and generating the RPN automatically without human intervention. A successful FMEA is gained through optimized consistency, responsiveness, and accumulated experience. The suggested approach aims at solving the above-mentioned FMEA weaknesses by two steps: first, reviewing and re-evaluating a dataset containing reported failures manually by experts to ensure accuracy. Secondly, conducting supervised machine learning techniques on the updated data and develop machine learning models that can be deployed to evaluate and classify newly reported failures automatically with minimum processing time and enhanced consistency.

2. Study Background

In this research, CLAAS Hungária Kft (CLH) is adapted as a case study. CLH was established in 1997 in Hungary as a subsidiary company of CLAAS Group. CLAAS group is an international German family-owned business company based in Germany and owns many manufacturing plants worldwide. CLAAS is a world-leading manufacturer of agricultural equipment and machinery such as tractors and combine harvesters. Since establishment, CLH expanded from 350 workers and 8 hectares plant to more than 700 workers working on a 14-hectare plant and became a center of excellence for combine harvester tables and trolley carts production. CLH manufactures supplementary devices such as combine harvester tables, cutting heads, and trolley carts, as shown in Figure 1. These devices are shipped from Hungary either to the mother company that is located in Germany or directly to the end-customers for final assembly with the machine which can be a combine harvester or a tractor. The cost of a single failure is tremendously high, not only due to the machine cost itself but also due to the entailed logistics and the re-work cost.

Figure 1. Sample of devices manufactured at the subsidiary company subject of this study.

CLH's staff has developed "Quality Checklists" for every product, process or manufacturing phase. These quality checklists are developed based on the FMEA documents and are being used at the quality gates in the shop floor in order to ensure that common failure causes are avoided. Moreover, this process aims to make sure that critical device components are installed and configured at the optimal conformance to design. However, as mentioned earlier, FMEA documents are prepared during the product design phase and can be changed once the serial production is initiated. Meanwhile, further failure modes can be detected at the final assembly phase. Therefore, these quality checklists are demanded to be dynamic, updatable and responsive to real quality issues reported during or after production.

This research activity is focusing on a single device that consists of the combine harvester feeder house as shown in Figure 2. Feeder house is a device that is attached to the combine harvester to facilitate the control of the cutting head and the flow of crops from the cutting head to the combine harvester. The device consists of several complex systems such as mechanical, hydraulic, electrical, and electronic systems. This device is wholly manufactured in the subsidiary company in Hungary and dispatched to be assembled to the combine harvester at the mother company in Germany.

Failures or defects which are observed during assembly or reported by end-users are gathered on a daily basis through the global ERP system of the company. After that, this information is extracted and manually and reviewed by an experienced quality management team. This evaluation process aims at analyzing root cause and consequently taking the needed correction actions in order to maintain profitability and high-quality production. The company uses an internally customized FMEA technique to evaluate reported claims by obtaining RPN for every claim according to FMEA documents. The method which is used here aims at generating an RPN value for every claim on a scale from 1 to 300 points, where 300 is the highest priority number.

Figure 2. Feeding house attached to the combine harvester body and ready to be attached with a cutting head.

RPN in this CLH is obtained based on three major factors: (severity, occurrence, and impact). Severity, or gravity as named by the company's internal manuals, represents the risk consequences of the claim from customer and company perspectives. It also includes the cost of resolving this issue and the safety impact on the operator. The weight of this factor ranges between 1 to 10 points, where 1 is the lowest severity and 10 is the highest. In the meanwhile, occurrence represents the number of incidents a specific claim has been witnessed in a specific period. The weighting scale of this factor is also 1–10, where 10 is the highest. Impact is weighted by a scale of 3 points from 1 to 3. Impact represents the repair efforts, time, repetition of the same work, and the overall impact of the claim on the reputation and image of the company. The meaning of every scale value from 1 to 10 is elaborated in detail in [3,9]. The evaluation process is summarized in Figure 3 below.

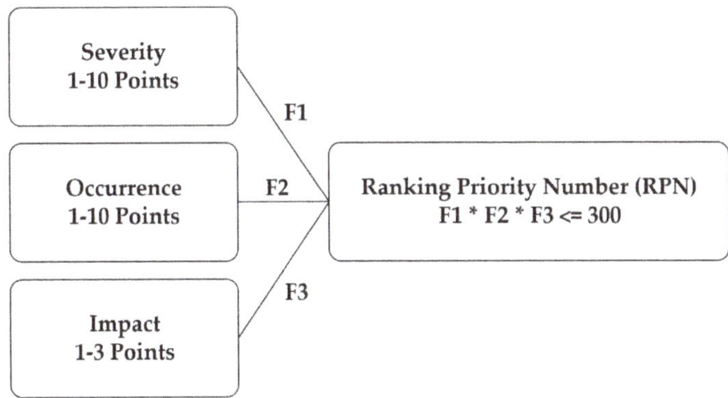

Figure 3. Factors affecting claim ranking and the weight of every factor.

Equation (1) shows the multiplication of the three factors values that results an RPN value between 1 and 300 points. An RPN value above 160 points is classified at a very high priority, while, a value between 100 and 160 points is classified as a high priority. Medium priority is noted if the RPN value is in the range of 35–100, while low priority is noted if the RPN value is less than 35:

$$\text{RPN} = \text{Severity} \times \text{Occurrence} \times \text{Impact} \tag{1}$$

According to the RPN value of every claim, the quality team decides the next handling steps. Further steps could be tracing root cause(s) and ensuring the elimination of such cause(s) and/or updating the quality checklists to ensure further failures will not repeat in the future. Time and experience play a crucial role in this regime. It is important to improve the process of evaluating claims and lever the current experience.

The evaluation and ranking process requires highly experienced people who are fully aware of the FMEA documents and its applications. The volume, velocity, and veracity of claims reported, and their processing time is very critical from a quality management perspective. It is essential in such a high-value industry to resolve issues as soon as they are reported. Early and fast processing of quality issues is translated to a lower quality cost and will positively enhance the general business performance. Moreover, standardization of the evaluation process and consistency of the process is vital to guarantee consistent RPN results every time.

The accumulated experience, time of processing, consistency of the evaluation process can be attained through the proposed solution in this paper; utilizing automated machine learning to classify and analyze claims data. Machine learning capabilities provide the capacity to analyze several input features (columns) at one dimension, aligned with a large volume of data (rows) at the other dimension. This helps in discovering and analyzing unseen factors, considering that the best quality practices focus on the claim root cause analysis. Additionally, utilizing technology whenever possible is very promising in the industry, because of its availability at any time (24/7) under any conditions and its ability to go deeper in analysis beyond human capacity. Delegating such tasks to machines will let human intelligence focus on higher strategic issues and to reach a higher level of efficiency and effectiveness.

3. The Proposed FMEA Analysis Method

In this section, it is suggested to utilize supervised machine learning technology to replace human intervention in processing, evaluating, and categorizing claims. The current flow of claims from involved parties is illustrated in Figure 4. Claims from internal company quality product audit (product audit claims) and issues that were detected during assembly (cross-company claims) are pipelined in the company's ERP system and human intervention is important at one point to evaluate claims manually. Based on the evaluation results, quality management decides how to deal with every single claim to find the root cause of the problem. This is done either by following the eight disciplines of problem-solving (8D) methodology for critical or high-ranking issues or by just updating the shop floor quality checklists in order to ensure the quality of next produced devices. Otherwise, this reported issue is just as it is an accidental incident and occupies a very low RPN value.

Accordingly, a dataset that contains one-year data of claims is extracted from the ERP system of the company. This data is concerning the selected device only (the feeder house shown in Figure 2). Firstly, to ensure the accuracy of the developed models, the data was re-evaluated and validated manually by experienced quality engineers to obtain the three RPN elements (severity, occurrence, and detectability) and to define the root cause and the source manufacturing process (such as cutting, bending, welding, painting, assembly, etc.) of every claim. The evaluation process depends on the experience of the quality team and based on the internal FMEA procedure for every failure mode. The resulted updated dataset is used for training and develop an ML model that is deployed to predict an RPN value for future failures claims and classify its root cause instantly without further human intervention.

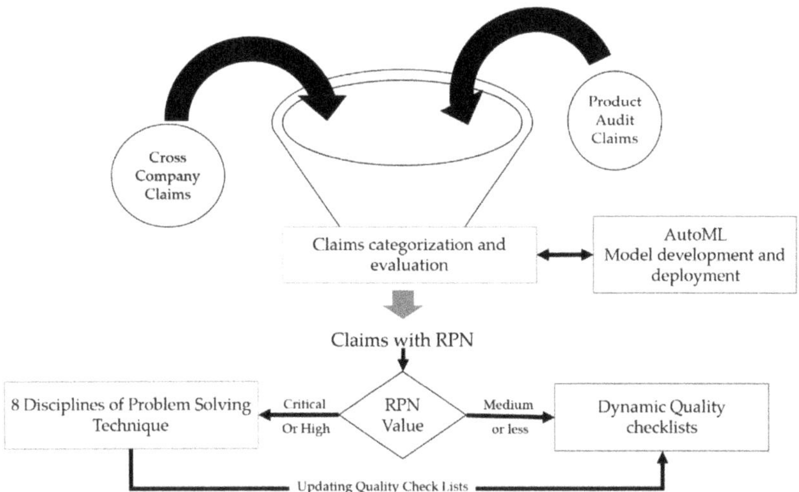

Figure 4. The flow of internal quality audit and cross-company claims to quality management.

In this research, Google AutoML is selected for three reasons: first, its effectiveness, as Google AutoML is free to try and utilizes the latest ML technology developed by Google brain team. Second, its ease of use, which is very important to ensure the sustainability of the project results after the research cooperation ends, keeping in mind that people who are working at the partner company are less experienced with coding and modeling. Therefore, such a friendly system will ensure that the partner company can deal with the work after the end of the project with the least knowledge of coding and data processing. Third, its ability to integrate. It is agreed with the partner company to integrate the developed models in the company's ERP system. Google AutoML offers the ability to deploy and integrate the resulted models by application programming interface (API).

Supervised machine learning is used to improve the failure claims processing process. Claims are reported by engineers at the assembly location in Germany or from service centers to the quality management office through the company's ERP system. Claims are analyzed, categorized and ranked by the quality management team based on FMEA documents. Accordingly, failures are ranked and prioritized based on their importance and critical impact.

The proposed solution aims at developing an automatic claim ranking system to replace human intervention based on developing four machine learning models that can read, analyze, evaluate, and assign relevant ranking values for every processed claim. In order to do so, a dataset of already evaluated claims is used to train the model. Afterward, the model will be deployed to evaluate new claims based on the experience gained by the training data. Figure 5 below elaborates on the process of models' development, its inputs, and outputs. The first step in models' training is to preprocess the input data, feature selection, and data types. The auto machine learning tool results four models that will be able to predict four independent values by which three of them will be multiplied to calculate an RPN value. The fourth decides the source manufacturing process of the same claim.

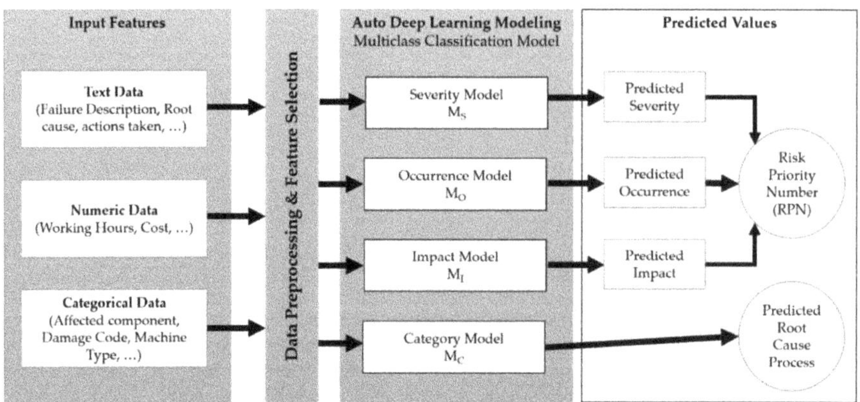

Figure 5. Development of machine learning model.

3.1. Data Pre-Processing

A dataset that contains 1532 rows of failure incidents has been received from the industrial partner of this research project. The data was extracted from the ERP system of the company, it was recorded over one year, and related only to the selected product in Figure 2. Every row in the dataset contains the details of a single incident and described by 23 different input features (columns) that help the quality engineers to recognize the failure mode and therefore, refer to the FMEA documents to assign the proper RPN value that fits this failure mode. For example, a failure is claimed from the assembly line in Germany where the engineers reported an incident of "an insufficiently tightened screws at one component in the device", along with this reported failure, further information are provided such as the serial number of the device, the code number of the component as in the design, further description written in textual format by the labor who solved the issue including his opinion on the issue and its criticality, the damage code as picked from the list of options in the input screen, the expected root cause of the problem is explained in text, the time consumed to fix the problem, and cost involved for rework, and the final conclusion. Table 1 summarizes the input features types and roles in the models. Whereas this dataset is used to develop the machine learning models. The first step is to prepare the data for the AutoML platform. This includes ensuring that all features of the dataset are organized, and data types are well defined. Additionally, the claims are validated manually by the quality management team at the partner company using a specially programmed interface that facilitates the manual validation process. This manual validation of data was made in order to ensure the quality of the input data and therefore ensure the quality of the output models.

Table 1. Dataset input features for the machine learning model.

Data Type	Number of Inputs	Labels	Brief Summary
Textual Text	3	Claim description, Root cause, and remediation action made	• This data is written in natural language by the labors or engineers at the German company, explains the failure, its root cause, and the remediation action made • It will help to recognize the failure mode, its root cause, and its technical solution
Categorial Data	10	Machine code and name, damage name and code, initial criticality assessment, component type, and reporter information	• Contains data about the device affected, the damage category, and its criticality • It will help to identify reoccurrence of similar failure, evaluate its importance, and define the location at which it was detected
Numeric Data	10	Different costs data, number of affected devices	• This data will help to evaluate the consequences of this failure in terms of labor cost, transportation, material cost, and any extra costs

Furthermore, 46 rows are excluded from the training process because of missing critical details such as claim textual description and the root cause input. Moreover, scales (8–10) in severity and (7–10) in occurrence had an insufficient number of claims (lower than 50 rows) for every element, these records are excluded too, as shown in Table 2a. The reason behind that, AutoML platform cannot start training with less than 50 readings per class. Therefore, the dataset is copied three times, and classes with less than 50 readings are eliminated. Finally, 1484, 1425, and 1486 claims are used for models training of severity, occurrence, and impact respectively. The data plot is shown in Figure 6 where the distribution of the data is illustrated.

Table 2. Summary of dataset included in the modeling.

a. FMEA Elements			
Scale	Number of Records		
	Severity	**Occurrence**	**Impact**
1	182	267	866
2	454	199	511
3	167	424	109
4	291	204	
5	218	128	
6	81	203	
7	91	28	
8	0	10	
9	2	0	
10	0	23	
Total	1486	1486	1486
Dataset rows	1484	1425	1486
Training rows	1343	1269	1355
Evaluation rows	141	156	131

b. Category of Claim	
Process Category	**Number of Records**
Category A	89
Category B	104
Category C	464
Category D	123
Category E	206
Category F	86
Category G	303
Category H	68
Others	43
Total	1486
Dataset rows	1443
Training rows	1309
Evaluation rows	134

In addition to RPN evaluation, the research work includes classification of claims according to the respective manufacturing process which is described to be the root cause process of the defect. The names of processes are masked in Table 2b where the process could be any of the known machining processes such as cutting, bending, welding, assembly, etc. Processes with less than 50 records are excluded as well.

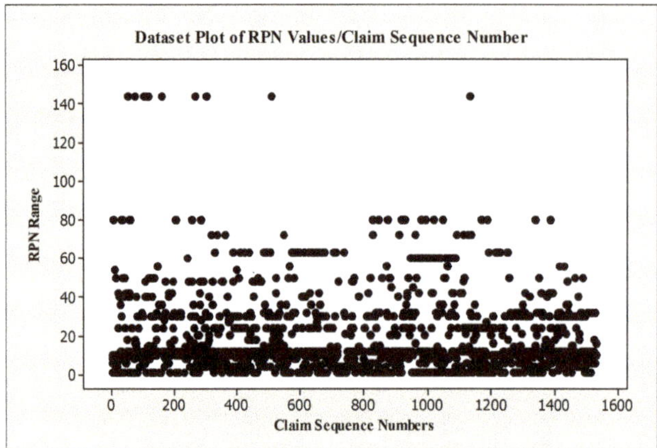

Figure 6. Dataset plot of all claims based on RPN value.

3.2. Data Modeling

The AutoML platform analyzes the dataset and constructs the models automatically. Such analysis includes text processing is made by the Google AutoML in addition to the other categorical and numeric data.

Multiclass classification technique is applied to develop four Machine Learning models, while the suitable algorithm is automatically developed by Google AutoML. Google AutoML is developed to help researchers in handling large data and building high accuracy models with the least coding experience and resource consumption. The datasets are uploaded, the input features are defined, and targeted elements are selected. The prepared datasets are analyzed to obtain three independent models that can evaluate every claim to predict three independent values for severity, occurrence, and impact, from which an RPN can be calculated by applying Equation (1). Additionally, a fourth model (for category) is obtained to identify the manufacturing process which caused this failure to occur. The manufacturing process could be cutting, bending, welding, painting, assembly, packaging, and transportation. The aim of the fourth model could be extended in the future to include more specific processes such as welding machine 1, assembly line 2 and so on. Figure 7 illustrates the four models obtained after training. As the AutoML platform is a cloud system, then the consumption processing can be measured by node hour. The training process consumed 0.944, 1.105, 0.86, and 1.111 node hours for severity, occurrence, impact, and category respectively. Every node hour includes the use of 92 n1-standard-4 equivalent machines in parallel, where a single n1-standard-4 machine operates four virtual CPUs and 16 GB of RAM memory.

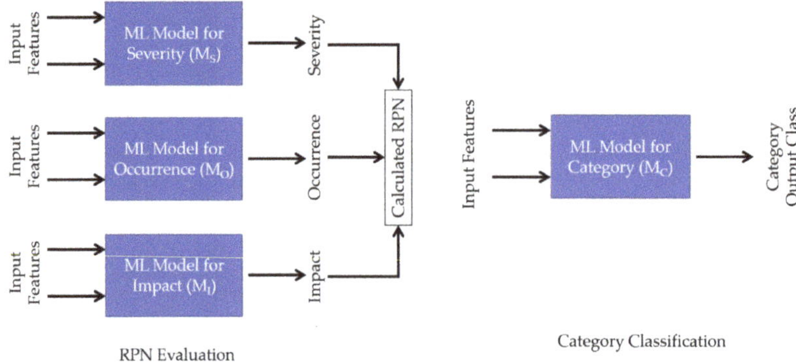

Figure 7. RPN evaluation and Category Classification models.

3.3. Models Evaluation

The models' evaluation is done by three evaluation metrics namely precision, recall, and F1 score. Further evaluation metrics are adopted here such as area under curve (AUC) and the confusion matrices.

Precision is the percentage of true positive predictions to all positive predictions (true positive and false positive). Recall is the percentage of true positive predictions among all actual values (true positive and false negative). While F1 score is a balanced evaluation between precision and recall, and it is used especially when the data in the datasets are not equally distributed over classes.

The area under the precision-recall curve (AUC-PR) and the area under the receiver operating characteristic curve (AUC-ROC) are used to visualize the performance of the models. AUC-PR shows the trade-off between precision and recall for the model. AUC-ROC shows the trade-off between true positive rate and false-positive rate.

The confusion matrices are used here to elaborate on the prediction accuracy and accepted tolerance of every class. For example, higher prediction resulted by any of the RPN elements models (M_S, M_O, and M_I) can be accepted because assigning higher ranking value to an incident increases its priority. The limitation here is the degree of tolerance accepted by the company. To elaborate more, when an incident is evaluated for a severity class of three, it is accepted if the model predicts a value that is higher than the actual value by one step. However, it could be inefficient if the model predicts two or higher steps than the incident deserves.

Finally, the accuracy of the models is affected by the type of every element, the number of rows that are used for training, the accuracy of details provided per row, and the scale of every element (or number of classes per element). It is important here to recall the objective of this work which is to provide a proof-of-concept that machine learning is an effective technique to enhance FMEA and the development of RPN value.

4. Results and Discussion

In this research, four machine learning models are trained and evaluated successfully in this research. Table 3 summarizes the training evaluation results and accuracy metrics for four models of severity (M_S), occurrence (M_O), impact (M_I), and category (M_C). The evaluation sample was automatically split and tested by the AutoML platform.

The performance metrics in Table 3 shows relatively high-quality models, with different levels of precision for each model. The area under the precision-recall curve (AUC-PR) and the area under the receiver operating characteristic curve (AUC-ROC) are close to 1, which indicates high-quality classification models. Moreover, the models' precision rates are 93.2%, 87.6%, 89.9%, and 86.6% for M_S, M_O, M_I, and M_C respectively, which indicates that the models predicted correctly the classes of the validation sample for every model.

Table 3. Models Evaluation.

Dataset Targeted Value	Validation Sample	Score Threshold	Precision	TPR (Recall)	F1 Score	AUC (PR)	AUC (ROC)
Severity (M_S)	141 test rows	0.5	93.2% (96/103)	68.1% (96/141)	0.787	0.895	0.970
Occurrence (M_O)	156 test rows	0.5	87.6% (106/121)	67.9% (106/156)	0.765	0.871	0.955
Impact (M_I)	131 test rows	0.5	89.9% (116/129)	88.5% (116/131)	0.892	0.954	0.973
Category (M_C)	134 test rows	0.5	86.6% (103/119)	76.9% (103/134)	0.814	0.877	0.972

The highest F1 score is recorded for M_I, where the full dataset is used for training, and the classification was only among three classes (1, 2 or 3) while the training dataset for M_I contains 866, 511, 109 readings for every class from 1 to 3, respectively.

The confusion matrices shown in Tables 4–7 below show that the concentration of the true predictions is at the diagonal cells of all models. However, both models M_S and M_O show higher confusion for predicted labels against true labels, in contrast to M_I and M_C models where higher concentration is shown at the diagonal cells. This is highly connected with the data volume and will be improved when a larger volume of data is used for model upgrading.

However, predicting a higher value than the true one (negative true predictions in the confusion matrices) for the three models (M_S, M_O, and M_I) could be accepted, as higher prediction value for severity will increase the RPN value and therefore, the priority to resolve the failure is increased. However, this tolerance is not acceptable for M_C as it deals with a totally different interpretation, it describes the manufacturing process where the root cause of the failure is coming from. The model shouldn't predict a false manufacturing process instead of predicting a true one. In other words, a wrong prediction that a failure is caused by a process (X) is totally rejected if it is actually caused by another different process. However, such a disadvantage can be improved during the transition stage where the process of automatic claims evaluation is running in parallel with the manual traditional one so as to improve the next trained model after a larger dataset size is accumulated.

Table 4. Confusion matrix for the model of severity (M_S).

		Predicted Labels						
	Class	1	2	3	4	5	6	7
True Labels	1	95%	5%	-	-	-	-	-
	2	-	89%	7%	-	4%	-	-
	3	-	7%	60%	13%	13%	-	7%
	4	-	17%	3%	67%	3%	7%	3%
	5	-	20%	13%	-	60%	7%	-
	6	-	-	-	-	12%	88%	-
	7	-	-	22%	-	11%	-	67%

Table 5. Confusion matrix for the model of occurrence (M_O).

		Predicted Labels					
	Class	1	2	3	4	5	6
True Labels	1	81%	6%	3%	9%	-	-
	2	14%	48%	14%	10%	10%	5%
	3	-	7%	87%	4%	2%	-
	4	-	9%	9%	78%	4%	-
	5	-	-	13%	-	73%	13%
	6	-	-	5%	-	5%	90%

Table 6. Confusion matrix for the model of impact (M_I).

		Predicted Labels		
True Labels	**Class**	1	2	3
	1	97%	3%	-
	2	18%	80%	2%
	3	-	33%	67%

Table 7. Confusion matrix for the model of category prediction (M_C).

	Predicted Labels								
	Class	A	B	C	D	E	F	G	H
True Labels	A	95%	-	-	-	-	-	5%	-
	B	14%	86%	-	-	-	-	-	-
	C	-	-	71%	-	-	8%	13%	8%
	D	-	-	-	83%	-	-	17%	-
	E	-	-	-	-	86%	-	14%	-
	F	-	-	25%	-	-	63%	13%	-
	G	5%	10%	-	-	-	2%	83%	-
	H	-	-	-	-	-	-	-	100%

Another approach to evaluate the developed models is to examine the RPN in the original dataset (actual RPN) against the resulted RPN from applying Equation (1) to the three predicted elements, call it (predicted RPN). The frequency histogram in Figure 8 compares the two readings (actual vs. predicted) for the overall dataset. The histogram shows a high overlapping of results between the two RPN values. Applying statistical accuracy measurements between actual and predicted values, resulted in a mean absolute error of 3.86 and a root mean squared error of 12.76 which both represent acceptable accuracy of predicted against actual. Therefore, this is another approach to evaluate the models developed and showing that these models are effective and efficient.

In contrast, this histogram in Figure 8 shows a shortage in predicting higher RPN values when the multiplication result is higher than 80 (the values larger than 140 in the histogram is a clear example). The reason behind this weakness is due to a lack of data at high classes for severity and occurrence in the training dataset. Enhanced accuracy can be reached by enlarging the training dataset and this could be fulfilled when more data is accumulated over time. Given that the dataset used in this activity contained 1532 claims for a single product in one year only. Further improvement can be achieved by reviewing the predictions of the models after a testing period, where an expert engineer can compare both the proposed AutoML approach with the traditional approach and conclude an enhanced and extended dataset that can be used for models retraining. Another approach to improve the models is to minimize the scale of classification for severity and occurrence from 1–10 to become 1–5 scale, such change will improve the model precision and accuracy. Hint, the accuracy for M_I is higher than the others.

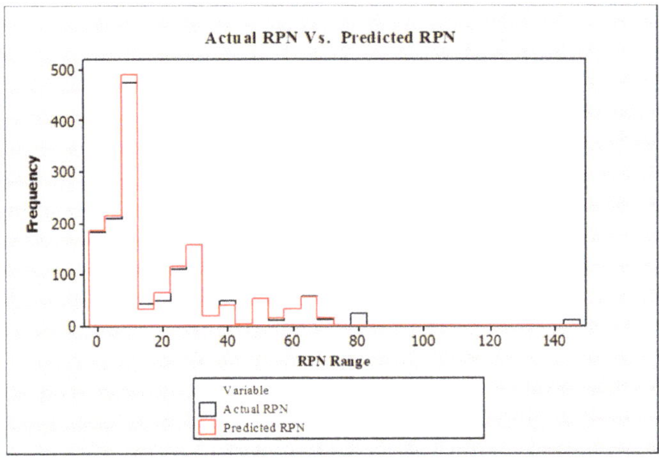

Figure 8. Originally evaluated RPN values frequency Vs Obtained RPN from the predicted Severity, Occurrence and Impact Classes.

Since the results of the proposed method are showing acceptable accuracy, given the dataset volume and used method, the models can be deployed at the partner company. The advantage of the proposed approach as compared to the traditional one is that it replaces the human intervention in the process and automates the decision-making process. In the traditional approach, once a claim is received from the mother company in Germany, a quality engineer in the quality management office in Hungary reviews the claim, decides the failure mode type and then assigns values to the three elements to calculate the RPN. Based on this judgment, further actions are decided. These actions can be by transferring the issue to critical issue resolution by using strategies such as the eight disciplines methodology (8D) if the RPN is above 160 points, or by updating the quality checklists at the production shop floor or could be both. However, this human intervention may imply some implementation error as it depends on the evaluator's experience. For example, assume a claim was evaluated by a quality engineer to be 160 points, while another engineer may underestimate the claim by ranking it to be 140 points based on his experience and memory. In the first situation, the engineer will transferee the claim to a more sophisticated process (8D strategy) which entails using more resources by forming a team to follow up and resolving the issue. On the other hand, the claim is highlighted to the production management (the second case). This is because such a process depends mainly on individual judgment and experience of the staff members who may give inaccurate estimation. Meanwhile, if such a process is done by a machine that makes decision-based on the accumulated leaning process, such uncertainty in decision making can be avoided. Thus, the proposed solution replaces this human intervention with a machine learning algorithm that evaluates claims based on the accumulated and none individualized experience and avoids the uncertainty in the experience of quality engineers. Moreover, the proposed approach can automatically analyze the new claims and construct correlations between incidents and therefore get a better ability for future prediction. Such process saves time, efforts, and improves responsiveness to failures either by alarming the quality management team instantly to serious issues or by automatically updating the quality checklists in the production shop floor by notifying labor and production staff of this issue in a real-time manner. From a business perspective, the proposed solution can be operated at any time and provides higher efficiency and effectiveness.

5. Conclusions

In this research, auto-machine learning was utilized to optimize failure modes handling by automatically identifying the failure mode, obtain its RPN and identify the manufacturing process related to the root cause of the issue. Three multiclass-classification machine learning models were developed to predict values for the RPN three elements namely severity, occurrence and impact. A fourth multiclass-classification model was developed to classify failures to their root cause process. The models' evaluation indicated relatively high accuracy models that can be deployed and integrated to enhance the company's ERP system. One of the features of the selected AutoML platform is its simple integration through the API, which is offered on the cloud. Such technology performs efficiently for large applications at the macro level of the factory. Utilizing such a solution enhanced the capabilities of the quality management team to handle any volume of claims data under high flow velocity. Such a solution will allow the quality team to focus on other strategic issues which will enhance the team's performance and results. The benefits of such technology do not end by this, but also could be furtherly extended to link claims and defects to the relevant manufacturing machine and operator. Once a claim is reported to the quality management it will be processed by the deployed model and instantly will be communicated to the relevant operators or managers and deeper to the shop floor in the factory. One more result for this research is that the manufacturing quality checklists for the selected product can be dynamically updated to include the top ten issues which are updated continuously according to their RPN. Such improvement enhanced the quality of processes and products. The factory in this study uses large screens on the shop floor to display quality checklists at every manufacturing process. These are used to review the quality issues while manufacturing processes are in place. A final check is being made at the quality gate of every process. The operators can watch the screens which are updated every while and learn instantly about recently reported issues and making immediate correction actions. Finally, it is important to highlight the factors that impact the quality and accuracy of the developed models. For example, the accuracy of the model is strongly dependent on the quality of the data originated at the first point where the failure or defect was initially detected. Empty data rows, ambiguous information, or mistyping could forfeit important features and therefore, result in inaccurate prediction and reduce the system credibility. Therefore, a recommendation was suggested to the company to develop the data gathering platform (ERP system) in order to ensure higher quality prediction in the future. Furthermore, it is also essential to keep updating and maintaining the model by conducting periodical review sessions for the predicted RPN values and correct them when needed. Retraining the model using a larger volume of data will accumulate the model experience and improve model accuracy.

Author Contributions: S.S. (Conceptualization, Methodology, Investigation, Writing—Original Draft, Writing—Review & Editing). I.H. (Conceptualization, Supervision, Validation). M.D. (Supervision, Project administration). All authors have read and agreed to the published version of the manuscript.

Funding: This work was supported by the Stipendium Hungaricum Programme and by the Mechanical Engineering Doctoral School, Szent István University, Gödöllő, Hungary.

Acknowledgments: Special thanks to CLAAS Hungaria Kft, especially to Mr. Robert Csombordi, Head of Quality Management, for their endless support in conducting this research work.

Conflicts of Interest: The authors declare no conflict of interest.

References

1. Cicek, K.; Celik, M. Application of failure modes and effects analysis to main engine crankcase explosion failure on-board ship. *Saf. Sci.* **2013**, *51*, 6–10. [CrossRef]
2. Yang, C.; Zou, Y.; Lai, P.; Jiang, N. Data mining-based methods for fault isolation with validated FMEA model ranking. *Appl. Intell.* **2015**, *43*, 913–923. [CrossRef]
3. Chang, C.L.; Wei, C.C.; Lee, Y.H. Failure mode and effects analysis using fuzzy method and grey theory. *Kybernetes* **1999**, *28*, 1072–1080. [CrossRef]
4. Arabian-Hoseynabadi, H.; Oraee, H.; Tavner, P.J. Failure Modes and Effects Analysis (FMEA) for wind turbines. *Int. J. Electr. Power Energy Syst.* **2010**, *32*, 817–824. [CrossRef]
5. Liu, H.-C.; Wang, L.-E.; Li, Z.; Hu, Y.-P. Improving Risk Evaluation in FMEA With Cloud Model and Hierarchical TOPSIS Method. *IEEE Trans. Fuzzy Syst.* **2019**, *27*, 84–95. [CrossRef]
6. Keskin, G.A.; Özkan, C. An alternative evaluation of FMEA: Fuzzy ART algorithm. *Qual. Reliab. Eng. Int.* **2009**, *25*, 647–661. [CrossRef]
7. Pillay, A.; Wang, J. Modified failure mode and effects analysis using approximate reasoning. *Reliab. Eng. Syst. Saf.* **2003**, *79*, 69–85. [CrossRef]
8. Yang, C.; Shen, W.; Chen, Q.; Gunay, B. A practical solution for HVAC prognostics: Failure mode and effects analysis in building maintenance. *J. Build. Eng.* **2018**, *15*, 26–32. [CrossRef]
9. Chin, K.-S.; Chan, A.; Yang, J.-B. Development of a fuzzy FMEA based product design system. *Int. J. Adv. Manuf. Technol.* **2008**, *36*, 633–649. [CrossRef]
10. Xu, K.; Tang, L.C.; Xie, M.; Ho, S.L.; Zhu, M.L. Fuzzy assessment of FMEA for engine systems. *Reliab. Eng. Syst. Saf.* **2002**, *75*, 17–29. [CrossRef]
11. Hassan, A.; Siadat, A.; Dantan, J.Y.; Martin, P. Conceptual process planning an improvement approach using QFD, FMEA, and ABC methods. *Robot. Comput. Integr. Manuf.* **2010**, *26*, 392–401. [CrossRef]
12. Ayber, S.; Erginel, N. *Developing the Neutrosophic Fuzzy FMEA Method as Evaluating Risk Assessment Tool*; Springer: Cham, Switzerland, 2020; pp. 1130–1137. ISBN 9783030237554.
13. Haktanır, E.; Kahraman, C. *Failure Mode and Effect Analysis Using Interval Valued Neutrosophic Sets*; Springer: Cham, Switzerland, 2020; pp. 1085–1093. ISBN 9783030237554.
14. Al-Khafaji, M.S.; Mesheb, K.S.; Jabbar Abrahim, M.A. Fuzzy Multicriteria Decision-Making Model for Maintenance Management of Irrigation Projects. *J. Irrig. Drain. Eng.* **2019**, *145*, 04019026. [CrossRef]
15. Lee, S.M.; Lee, D.; Kim, Y.S. The quality management ecosystem for predictive maintenance in the Industry 4.0 era. *Int. J. Qual. Innov.* **2019**, *5*, 4. [CrossRef]

16. Duan, Y.; Edwards, J.S.; Dwivedi, Y.K. Artificial intelligence for decision making in the era of Big Data – evolution, challenges and research agenda. *Int. J. Inf. Manag.* **2019**, *48*, 63–71. [CrossRef]
17. Keller, M.; Rosenberg, M.; Brettel, M.; Friederichsen, N. How Virtualization, Decentrazliation and Network Building Change the Manufacturing Landscape: An Industry 4.0 Perspective. *Int. J. Mech. Aerosp. Ind. Mechatron. Manuf. Eng.* **2014**, *8*, 37–44.
18. Gilchrist, A. *Industry 4.0: The Industrial Internet of Things*; Apress: Berkeley, CA, USA, 2016; ISBN 978-1-4842-2046-7.
19. MacDougall, W. *Industrie 4.0: Smart Manufacturing for the Future*; Germany Trade & Invest: Berlin, Germany, 2013.
20. Lee, K.M.; Yoo, J.; Kim, S.W.; Lee, J.H.; Hong, J. Autonomic machine learning platform. *Int. J. Inf. Manag.* **2019**, *49*, 491–501. [CrossRef]
21. Liu, B. A Very Brief and Critical Discussion on AutoML. *arXiv* **2018**, arXiv:1811.03822, 1–5.
22. AutoML: In Depth Guide to Automated Machine Learning. 2020. Available online: https://blog.aimultiple.com/auto-ml/ (accessed on 23 January 2020).
23. Gangele, J. How Does AutoML Works?—Jeetendra Gangele—Medium. Available online: https://medium.com/@gangele397/how-does-automl-works-b0f9e45fbb24 (accessed on 24 January 2020).
24. Faes, L.; Wagner, S.K.; Fu, D.J.; Liu, X.; Korot, E.; Ledsam, J.R.; Back, T.; Chopra, R.; Pontikos, N.; Kern, C.; et al. Automated deep learning design for medical image classification by health-care professionals with no coding experience: A feasibility study. *Lancet Digit. Heal.* **2019**, *1*, e232–e242. [CrossRef]
25. Hayashi, M.; Tamai, K.; Owashi, Y.; Miura, K. Automated machine learning for identification of pest aphid species (Hemiptera: Aphididae). *Appl. Entomol. Zool.* **2019**, *54*, 487–490. [CrossRef]
26. Li, Z.; Guo, H.; Wang, W.M.; Guan, Y.; Barenji, A.V.; Huang, G.Q.; McFall, K.S.; Chen, X. A blockchain and automl approach for open and automated customer service. *IEEE Trans. Ind. Inform.* **2019**, *15*, 3642–3651. [CrossRef]
27. Galitsky, B.A.; González, M.P.; Chesñevar, C.I. A novel approach for classifying customer complaints through graphs similarities in argumentative dialogues. *Decis. Support Syst.* **2009**, *46*, 717–729. [CrossRef]

MDPI
St. Alban-Anlage 66
4052 Basel
Switzerland
Tel. +41 61 683 77 34
Fax +41 61 302 89 18
www.mdpi.com

Processes Editorial Office
E-mail: processes@mdpi.com
www.mdpi.com/journal/processes

www.ingramcontent.com/pod-product-compliance
Lightning Source LLC
LaVergne TN
LVHW070200170726
843515LV00007B/1961

* 9 7 8 3 0 3 6 5 2 0 7 3 5 *